Copernicus: Platonist Astronomer-Philosopher

Matjaž Vesel

Copernicus: Platonist Astronomer-Philosopher

Cosmic Order, the Movement of the Earth, and the Scientific Revolution

PETER LANG
EDITION

Bibliographic Information published by the Deutsche Nationalbibliothek
The Deutsche Nationalbibliothek lists this publication in the Deutsche Nationalbibliografie; detailed bibliographic data is available in the internet at http://dnb.d-nb.de.

Library of Congress Cataloging-in-Publication Data

Vesel, Matjaž, 1965-
[Astronom, filozof. English]
Copernicus : Platonist astronomer-philosopher : cosmic order, the movement of the Earth, and the scientific revolution / Matjaž Vesel.
pages cm
Includes bibliographical references and indexes.
ISBN 978-3-631-64242-9
1. Copernicus, Nicolaus, 1473-1543. 2. Astronomers—Poland—Biography. 3. Astronomy, Medieval. I. Title. II. Title: Platonist astronomer-philosopher.
QB36.C8V4113 2014
520.92—c22
[B]

2013049788

Published with the financial support of Research Centre of the Slovenian Academy of Sciences and Arts.

Translation from Slovene: Manca Gašperšič
English editing: Cornelia Lambert
Page layout: Brane Vidmar

Cover image:
Copernicus' seal and his Uppsala notes.
Design by Mateja Goršič.

ISBN 978-3-631-64242-9 (Print)
E-ISBN 978-3-653-03793-7 (E-Book)
DOI 10.3726/978-3-653-03793-7

Peter Lang Edition is an Imprint of Peter Lang GmbH.

Peter Lang – Frankfurt am Main · Bern · Bruxelles · New York · Oxford · Warszawa · Wien

www.peterlang.com

Acknowledgements

My first and greatest acknowledgement goes to other scholars in the field. Without their research and publications, past and present, this book would not be possible. I discussed parts of the book in different stages of its elaboration at different occasions with many scholars. Not all of them can be listed here, but some deserve to be mentioned by name. I would like to thank my colleagues from the Institute of Philosophy of SRC SASA, Ljubljana, Slovenia, Jelica Šumič Riha and Vojislav Likar, for their comments and suggestions on the earlier, Slovenian version of the book. Among more specific debts I accumulated during the work on the English version are those to my American colleagues from the Department of the History of Science, University of Oklahoma, Norman. I would like to express my gratitude to the head of the Department and President of the Mellon Fellowship Committee Steven J. Livesey, for hosting me there as a Mellon Fellow on several occasions. I would also like to thank the staff of the History of the Science Collections, especially JoAnn Palmeri, acting curator, and Kerry Magruder, curator, for providing me with an excellent research environment. During my stays in Norman I had the opportunity to discuss my ideas with students and faculty of the Department, especially Kathleen Crowther, Rienk Vermij and Peter Barker; the latter read and commented on the entire book. Last but not least, I would also like to thank my translator Manca Gašperšič and Cornelia Lambert for English editing and proofreading.

A Note on Primary Texts and Translations

Primary sources are cited by standard references to book, chapter and, if applicable, line. Included in brackets are references to the translator and page number of the English translation (if available) and a reference to editor and page number of the critical edition or original text. Whenever possible I checked existing translations against the (mostly Latin) original. Any modifications of existing translations are intended to give a more literal rendition. These changes of existing translations are – for practical reasons – unmarked.

List of figures

Table of contents

INTRODUCTION

To us inhabitants of the earth it seems as though the earth stands still, while the stars make a full circle around it every twenty-four hours. The stars appearing in the east in the evening are closer to the west in early morning; the sun rising in the east in the morning, moves across the sky towards the mountains and sets behind them in the evening; the moon follows a similar path.

Little wonder then that we cannot shake the habit of talking about the ways of the stars in such a manner; of talking about the sun as though it were a tiny lantern that indeed travelled across the skies. Yet, upon closer reflection, reason prevails once again over perception and teaches us: the stars are immeasurably distant; but how are they to circle the earth in twenty-four hours, given this tremendous distance?

On account of such remoteness, it may flash across everyone's mind that there must be other explanations for the ways of the stars. And it is as follows: The earth, this gigantic orb soaring free in the heavens, makes a full rotation about its axis in twenty-four hours, leading us, its inhabitants, to believe that the stars revolve around us. This appearance is not deceptive in itself, for in the morning the sun is indeed in the east, and in the evening it truly appears where it is seen, in the west. However, the cause of such a change is not the sun but the earth revolving from east to west. No one in their right mind can have any doubt about it today; for aside from the stars' remoteness, the science of the stars has furnished proofs that can in no way be faulted.

Dr. Gregorij Pečjak, *Stoletna pratika dvajsetega stoletja, 1901–2000*, pp. 51–2.

***

Within the horizon of the world we live in – that defined by our senses – the earth is at rest. We cannot see or sense its movement; we feel it firmly fixed beneath our feet. On the other hand, we can observe the sun rise every clear morning and set in the evening, just as we can notice the stars rise in the east every evening and then move towards the west. Within the horizon of man's existence, this appearance is a truth as clear as sunlight. Even from the practical point of view, the earth appears motionless to us. For maritime transport and other purposes of orientation it is completely sufficient to have star charts and other maps based on the assumption that the earth is the stationary centre of the universe.

Yet despite the undetectability of its motion, and very limited practical use of this fact, we know that the earth moves. As the heirs of the Scientific Revolution, whose results have definitively affirmed that the earth is one of the wandering stars, we assume our planet's rotation about its axis and revolution around the sun to be completely self-evident and irrefutable. Not the slightest doubt can topple the evidence furnished by "the science of the stars" as supported by the sound theories of mathematical physics of modern astronomy and cosmology. Moreover, on the basis of verified scientific theories, we also know today that the earth travels around the sun along an elliptical orbit and that its motion is caused by the force of gravity. In an age when science has become man's second nature, the assertion that the earth stands still sounds rather unusual, if not downright insane.

However it was not always so. Until the period known as the Scientific Revolution or, until the sixteenth and particularly seventeenth centuries (and even later), a vast majority condemned as foolish anyone who would claim the opposite: that the earth moves. In all great civilizations, including the Greek, Arab and Christian ones, the earth as the stationary centre of the universe was – with negligible exceptions – the foundation of everyday life, religion and philosophy. The notion of a moving earth was inconceivable and incomprehensible to the inhabitants of Greek *poleis* in the fifth century BC, Arab merchants in the ninth century AD, and Christian monks in the twelfth century AD alike. Its movement could not, and still cannot, be perceived in any *Lebenswelt.* Quite to the contrary: people of all ages, cultures and religions could always clearly see that the earth stands still while celestial bodies travel around it. The deities of all major religions, too, have created a world with the motionless earth at its centre, while philosophers and scholars of every worldview and orientation have been, regardless of their customary and irreconcilable differences, unanimous: The earth is the solid ground beneath our feet.

***

The questions I answer in this book are fairly simple: Why did Nicolaus Copernicus (1473–1543) assert, first in the unpublished *Commentariolus* (ca. 1510) and later in the impressive *De revolutionibus* (1543), that the earth moves while the sun is at rest in the centre of the universe? Why did he think that astronomy of his time was in the need of a reform based on the concept of terrestrial motion? How did he introduce this concept into astronomical discourse, and what argumentative strategy did he use to do so? And finally, what does his assertion mean for the history of human, particularly scientific and philosophical, thought?

Copernicus' name has long been a synonym for the radical revolution in astronomy that took place in the sixteenth and seventeenth centuries; many historians of science have even identified the year of 1543 as the beginning of new scientific thought not only in astronomy but in natural science as a whole. In his work *The Copernican Revolution*, Thomas Kuhn extended the scope of the revolution even further.[1] In his view, Copernicus brought about change not only in mathematical astronomy and science but philosophy in the broadest sense.

Before Copernicus, the earth was the fixed centre around which a host of other celestial bodies revolved. He proposed to improve the accuracy and simplicity of astronomical theories by attributing to the sun a number of astronomical functions that had previously been attributed to the earth, and one century later, the sun, at least in astronomy, replaced the earth as the centre of planetary motions while the earth lost its singular astronomical status and became one of the moving planets. Many subsequent astronomical achievements depend on this shift; for this reason, this radical change in the fundamental concepts of astronomy is known as the Copernican Revolution.

The aforementioned change, however, says Kuhn, is only the first of the revolution's meanings. The publication of *De revolutionibus* was soon followed by other radical changes in human understanding of nature. Many innovations that culminated one hundred and fifty years later in Isaac Newton's conception of the universe were unintentional by-products of Copernicus' astronomy. Although the sole reason that Copernicus proposed the earth's motion was to improve the techniques used in predicting the astronomical positions of the celestial bodies, his suggestion only raised new problems for other strains of science. Until these were solved, his concept of the universe remained incommensurable with those of

---

[1] See T. S. Kuhn, *The Copernican Revolution*, p. 1 et seq.

other scientists. The reconciliation of these sciences with Copernicus' theory was an important cause of the intellectual ferment in the seventeenth century which we know today as the Scientific Revolution. The Scientific Revolution not only reconciled Copernicus' astronomy and physics, but assigned to science the role, which it has since played in the development of Western society and its thought.

Yet, according to Kuhn, even this does not completely exhaust the meanings of the Copernican Revolution. Copernicus lived and worked in an age permeated with rapid changes in political, economic and intellectual life that formed the basis for modern European and American civilization. His planetary theory and associated concept of a heliocentric universe vastly contributed to the transition from medieval to modern Western society, because they appeared to affect human relations to the universe and God. Copernicus' theory, which started as a strictly technical, mathematical revision of classical astronomy, became the focus of momentous discussions in religion, philosophy and social theory that determined the substance of human thought for two centuries after the discovery of America. Those who did not believe that their terrestrial home was no more than a planet blindly orbiting one of a myriad of stars sought to determine their place in the cosmic scheme differently from their predecessors who saw the earth as a singular and pivotal centre of God's creation. "Copernicus' revolution" therefore also contributed to the change in values of the Western European civilization.

Nevertheless, does the fact that it was Copernicus who put the earth in motion, thus making it focal to seventeenth century thought, *eo ipso* mean that he himself was also part of the Copernican Revolution and the Scientific Revolution? Was he himself already subject to a radical conceptual "revolution which implies a radical intellectual 'mutation'"? Did he himself already "destroy one world and [...] replace it by another," or did he himself reshape the "framework of our intellect," "restate and [...] reform its concepts," "evolve a new approach to Being, a new concept of knowledge, a new concept of science"?[2] According to traditional historiography of science, or the "vulgar triumphalist view" as Robert Westman puts it,[3] this is most certainly true. The "triumphalists" clearly deemed Copernicus a revolutionary who decisively and completely broke with medieval and ancient philosophic and scientific principles. He overturned the finite cosmos of Aristotle (384–322 BC) and Ptolemy (ca. 90–ca. 168 AD), and replaced it with

2 This is the most concise and general definition of the Scientific Revolution as understood by A. Koyré. See his "Galileo and Plato," p. 405.

3 See R. S. Westman, "Proof, Poetics, and Patronage: Copernicus's Preface to *De revolutionibus*," p. 169.

the infinite universe of stars. Relying on extensive calculations, he shattered the perfect crystalline spheres which had carried the planets around the earth since ancient times. And finally, he made a bold move of radical simplification and reduced the cumbersome number of epicycles introduced by Ptolemy to predict the planetary positions from eighty (or so) to thirty-four.

However, more careful readings of Copernicus' work undertaken by historians and philosophers of science after World War II revealed an entirely different and considerably less revolutionary image. Westman highlights the four most important conclusions of the research: although Copernicus' universe was appreciably bigger in size than that of Ptolemy, it was still finite; although Copernicus regarded the sun as the motionless centre of the planetary motions, he did not place it exactly at the centre of the universe but slightly off it. Furthermore, Copernicus was not a diligent observer: he made about twenty-seven new observations, but none were important for his theory; he did not eliminate the celestial spheres, although he was ambiguous about their true nature; he indeed eliminated several epicycles, but by abolishing the Ptolemaic mechanism called the equant, he also added new ones.[4] The conclusions made by historical epistemology during the first decades after World War II can be summarized thus: with Copernicus, the All (*to pan*, i. e. the Universe or the Whole) had undergone an effective change – but not in *all* aspects.

In contrast to the image of Copernicus as a hero of the Scientific Revolution, the results of this research reveal the true meaning of his *De revolutionibus* in terms of the impact his work had on the Copernican Revolution and hence on the whole Scientific Revolution of the seventeenth century. By putting the earth in motion, Copernicus also started the wheels of the Scientific Revolution. The motion of the earth, a concept so utterly inconceivable to Aristotelian natural philosophy, came to dominate the thought of the Scientific Revolution, creating ever new challenges and generating ever new solutions that ultimately culminated in the Newtonian concept of the universe. Although the Scientific Revolution is to a great extent a Copernican Revolution – if we disregard the Copernicanism of Galileo Galilei (1564–1642), Johannes Kepler (1571–1630) and Isaac Newton (1642–1727), we basically disregard the Scientific Revolution itself – Copernicus' work alone does not represent a revolutionary breakthrough. As Kuhn explains:

> The principal difficulties of *De revolutionibus* and the ones that we may not evade arise rather from the apparent incompatibility between that text

4 See ibid., p. 170.

> and its role in the development of astronomy. In its consequences the *De revolutionibus* is undoubtedly a revolutionary work. From it derive a fundamentally new approach to planetary astronomy, the first accurate and simple solution of the problem of the planets, and ultimately, with other fibres added to the pattern, a new cosmology. But to any reader aware of this outcome, the *De revolutionibus* itself must be a constant puzzle and paradox, for, measured in terms of its consequences, it is a relatively staid, sober, and unrevolutionary work. Most of the essential elements by which we know the Copernican Revolution – easy and accurate computations of planetary position, the abolition of epicycles and eccentrics, the dissolution of the spheres, the sun a star, the infinite expansion of the universe – these and many others are not to be found anywhere in Copernicus' work. In every respect except the earth's motion, the *De revolutionibus* seems more closely akin to the works of ancient and medieval astronomers and cosmologists than to the writings of the succeeding generations who based their work upon Copernicus' and who made explicit the radical consequences that even its author had not seen in his work.[5]

According to Kuhn, the true significance of *De revolutionibus* therefore lies "less in what it says itself than in what it caused others to say."[6] The book caused a revolution that it itself barely foreshadowed. Thus *De revolutionibus* is "a revolution-making rather than a revolutionary text."[7]

Therefore, the significance of Copernicus' work alone lies not so much in its inherent scientific achievements as it does in its effects on future scholars or, first and foremost, in the destruction of ancient geocentrism and medieval anthropocentrism. Since the time of Copernicus, as Alexandre Koyré states in *The Astronomical Revolution*, "man has ceased to be the centre of the Universe, and the Cosmos ceased to be regulated around him."[8] Pursuant to Koyré, Copernicus' work involved "the destruction of a world that everything – science, philosophy, religion – represented as being centered on man, and created for him; the collapse of the hierarchical order […]."[9] Nevertheless, the old world's response was long overdue:

---

5 T. S. Kuhn, *The Copernican Revolution*, p. 135.

6 Ibid.

7 Ibid.

8 *The Astronomical Revolution*, p. 15.

9 Ibid., p. 16. In the early twentieth century, a similar opinion was voiced by Sigmund Freud (1856–1939). In his view, Copernicus' greatest achievement was that by having

> Only at a much later date, when it became evident that this work of Copernicus was not intended for mathematicians alone; when it became clear that the blow to the geocentric and anthropocentric Universe was deadly; when certain of its metaphysical and religious implications were developed in the writings of Giordano Bruno [(1548–1600)], only then did the old world react.[10]

According to Koyré, the first stage of the astronomical revolution – the destruction of geocentrism and anthropocentrism – was followed by Kepler's step, in which celestial dynamics replaced the kinematics of circles and spheres used by Copernicus and ancient astronomers. Kepler's work enabled a partial transcendence of the obsession with circularity and the triumphant entrance of the astronomy of ellipses into the universe. With Giovanni Alfonso Borelli (1608-1679), the unification of celestial and terrestrial physics was finally completed by the abandonment of the circle in favor of the straight line leading to infinity. The ideas of Kepler and Borelli were then further refined by Newton.[11]

Kuhn's and Koyré's basic argument that Copernicus started the work but never brought it to completion may also be expressed in a different, simpler and more eloquent manner: Copernicus was truly a man of the sixteenth century, not the seventeenth. This is already evident from his approach towards his own project. Copernicus never thought of himself as a revolutionary; quite to the contrary. His aim as a genuine Renaissance astronomer was by no means "to reverse the entire science of astronomy,"[12] but to restore it by relying on the true,

---

banished the earth from the centre of the universe and set it in motion around the sun he delivered a major blow to the vanity of man. According to Freud the next two blows to the vanity of men were Darwinian evolution and his psychoanalysis. See S. Freud, *Introductory Lectures on Psychoanalysis*, p. 326. For a more Freudian perspective on scientific revolution(s), see F. Weinert, *Copernicus, Darwin and Freud*, and J. Laplanche, "The Unfinished Copernican Revolution."

[10] *The Astronomical Revolution*, p. 17. The dates of birth and death of Giordano Bruno are my addition.

[11] See ibid., p. 10.

[12] This was believed to be Martin Luther's (1483–1546) description of Copernicus in of one of the so called *Table-talks* (1539). However, as recent findings have revealed, Luther may not have been referring to Copernicus but to Celio Calciagini (1479–1541). See P. Barker, "The Lutheran Contribution to the Astronomical Revolution: Science and Religion in the Sixteenth Century," p. 34, and n. 16. The thesis that Luther attacked Copernicus was developed in the 1860s by some German Catholic historians for apologetic purposes. On this, see A. Kleinert, "'Eine handgreifliche Geschichtslüge':

authentic foundations, principles and postulates of the ancient philosophy and astronomy, which were violated by the predominant Ptolemaic astronomical tradition. However, in order to achieve this goal, he had to sacrifice one of the fundamental premises of the entire astronomy and cosmology of his time – the stationary earth at the centre of the universe. In its stead he introduced into astronomy a new and, by all contemporary articulations of knowledge, absurd concept of a moving earth. But Copernicus did so by harkening back to and rescuing the time-honored, yet forgotten ancient tradition of terrestrial motion. Copernicus was a man of the Renaissance who, in accordance with the spirit of his day, developed the new by rediscovering and renewing (*renovatio*) the old.[13]

***

For the most part, Kuhn and Koyré view Copernicus from the perspective of the Scientific Revolution; they both inquire how significant his work was for the revolution's achievements. Such a perspective and investigation is, of course, completely legitimate, and we shall deal with them further on in this book. However, before addressing the question of his "revolutionariness," it is, in order to fully grasp Copernicus' significance for the Scientific Revolution, first necessary to view him from the perspective of his self-understanding, through his own eyes, and from the perspective of the understanding of his contemporaries, against the background of the state of astronomy and philosophy in the sixteenth century, which itself was also the culmination of a centuries-long development. To put it differently, the true significance of any project, scientific or not, can only be appreciated when set within a sufficiently long as well as adequately studied historical period. Certain theses, emphases and facts can be endowed with true significance and meaning only within a history of long duration.

Historical and epistemological studies performed in the recent decades and even years have revealed a much clearer picture of astronomy, philosophy, and theology of the sixteenth century that facilitates a much more accurate understanding and evaluation of Copernicus' project then those done by Kuhn and Koyré. For the time being, let me mention only one crucial example. On the basis of Copernicus' critical account on the state of mind in astronomy in his

---

Wie Martin Luther zum Gegner des copernicanischen Weltsystems gemacht wurde." It is nevertheless true, as M.-P. Lerner puts it, that Luther would have, if he had actually been familiar with Copernicus' ideas, reacted with similar enmity. See his "La doctrine copernicienne et sa proscription 1616," p. 21, and "'Der Narr will die gantze kunst Astronomiae umkehren': sur un célèbre *Propos de table* de Luther," pp. 41–65.

13 Georg Joachim Rheticus (1514–1574), Copernicus' one and only immediate pupil, employs in his *Narratio prima* (1540) the term *astronomiae emendatio*.

dedicatory preface *To His Holiness, Pope Paul III. Nicolaus Copernicus' Preface to his Books On the Revolutions* or *Preface* to *De revolutionibus*,[14] Kuhn derives a thesis in *The Copernican Revolution* that the Copernican Revolution occurred because the Ptolemaic astronomic paradigm was in a state of crisis. In *The Structure of Scientific Revolutions*, he then extends his conclusion to the whole of science:

> And Copernicus himself wrote in the Preface to the *De revolutionibus* that the astronomical tradition he inherited had finally created only a monster. By the early sixteenth century an increasing number of Europe's best astronomers were recognizing that the astronomical paradigm was failing in application to its traditional problems. That recognition was prerequisite to Copernicus' rejection of the Ptolemaic paradigm and his search for a new one. His famous preface still provides one of the classic descriptions of a crisis state.[15]

But Kuhn is mistaken. He takes Copernicus both too lightly and too seriously at the same time. Copernicus' depiction of the "crisis state" in the *Preface* is not directed entirely against the "Ptolemaic paradigm." Quite to the contrary: one of the aims of Copernicus' *reform* (rather than "rejection") of the astronomy was to develop one variant of *Ptolemaic* astronomy. On the other hand the *Preface* is no more than a rhetorical radicalization and dramatization of some issues known for centuries. But contrary to Kuhn's belief, no other astronomer before Copernicus recognized that the Ptolemaic astronomical paradigm was "failing." The biggest question for which I shall try to find an adequate answer is: Why did Copernicus, and nobody before him, all of a sudden find these centuries-old issues so troublesome? What happened that motivated Copernicus to venture into so radical a reform of Ptolemaic astronomy? Why and how did he become a Copernican? Or, as Bernard Goldstein puts it in his very important article "Copernicus and the Origin of his Heliocentric System": "[w]hat was the question for which heliocentrism was the answer?"[16]

In very recent years, at least three substantial and very important books (and several important articles), all of them excellent achievements in their own way, addressed more or less explicitly this question. Robert Westman, *The Copernican Question: Prognostication, Skepticism, and Celestial Order* (2011), asks himself

[14] I shall refer to this Copernicus' text as *Preface* or *Dedication to the Pope*, and to his original *proemium* as *Introduction*.

[15] *The Structure of Scientific Revolutions*, p. 71.

[16] "Copernicus and the Origin of his Heliocentric System," p. 219.

basically the same question as Goldstein, and looks for the answer, as reveals the subtitle of the book, in astrology.[17] According to Robert Westman, Copernicus' envisioned the reform of astronomy as a defense of the astrology which came under attack by Pico della Mirandola (1463–1494) in his *Disputationes adversus astrologiam divinatricem* published in 1496. The major issue of Pico's criticism was the inability of astronomers and astrologers to establish certain, fixed order of planetary spheres. Westman in my opinion poses the right question and rightfully underlines the role of *Disputationes* as one of Copernicus' major sources on uncertainty in regard of the arrangement of the cosmos, but I do not believe Copernicus envisioned astronomical reform in order to save astrology. On the other hand, André Goddu, in *Copernicus and the Aristotelian Tradition* (2010), somehow implies that Copernicus' project grew out of his Aristotelian background. I think we should rather look in the opposite direction, that is, towards Plato and Platonism. Goddu himself documents Copernicus' familiarity with works of various Platonists and Plato and even proposes original argument for the influence of Plato's ideal of dialectic from the *Parmenides* on Copernicus' argumentation in *Commentariolus*, but apart from that he makes little use of the Platonic tradition in explaining the origin and the development of Copernicus' heliocentrism. This is not the case in Anna De Pace's excellent *Niccolò Copernico e la fondazione del cosmo eliocentrico con testo, traduzione e commentario del Libro I de Le rivoluzioni celesti* (2009). De Pace provides new sources and evidence to show, quite convincingly, how much Copernicus owes to Plato and Platonism. I agree with her general interpretation of Copernicus as Platonist, however, I would argue, she does not go far enough. She gives too much importance to some particular issues, such as Copernicus' Platonist theory of gravity, which in my opinion is not crucially important for him, and underemphasizes others, such as more metaphysical, Pythagorean-Platonic (read: mathematical) concerns with the cosmic order and ideas about the purpose of that order for humankind.

These latter issues are, as I will show, the main factors which not only stimulated Copernicus' criticism of Ptolemaic astronomy, but even shaped his heliocentric solution. In other words, Copernicus' Platonism explains *all of the fundamental aspects* of his project. His Platonism brings unity and coherence to his work and links into a consistent philosophical stance seemingly unrelated issues, such as the equant problem and the problem of the order of the planetary spheres.

---

[17] See *The Copernican Question*, p. XV: "[W]hat, after all, was the question that Copernicus was trying to answer?"

Copernicus' Platonist conceptions also provide historical and contextual background of his achievement, that is, it explains the nature of what I believe to be a genuine Copernican Revolution. By affirming the earth's motion, and particularly by transposing the argumentation *pro et contra* earth's movement from the sphere of natural philosophy to that of mathematical reasoning, Copernicus produced something that he failed to thoroughly consider. Although that gesture eluded more explicit reflection – Galileo Galilei mentions it in passing – it unambiguously opened up a radically new epistemological horizon: a horizon in which mathematical reasoning takes precedence over our reliance on sensory experience or appearances and establishes itself as decisive criterion of truth. Here, in my opinion, lies the very essence of Copernicus' own Copernican Revolution and his greatest contribution to the Scientific Revolution.

***

Copernicus' major book, *De revolutionibus*, is not easy reading. Rather, it is a comprehensive work divided into six books, and roughly structured on the model of Ptolemy's *Almagest*.[18] The main part, more than ninety percent of the text (including tables), covers technical, mathematical astronomy in the last five books, while a minor fragment, the first book, focuses on the "general structure of the universe (*constitutio vniversi*)." There Copernicus demonstrates that the earth revolves around the sun and rotates around its own axis, and that the inclination of its axis is despite the earth's revolution around the sun, always oriented towards the same point in the starry vault, which means that it also moves in "declination." At the end of the first book, he goes on to explain the basic concepts of trigonometry, which, in his opinion, the reader should understand to follow his demonstration. The second book focuses on the mathematical problems of spherical astronomy; it contains, among others, Copernicus' catalogue of stars. The third book deals with the apparent position of the sun, the fourth investigates the motion of the moon and the theory of eclipses, and the fifth and the sixth are devoted to planetary motion in celestial longitude and latitude.[19]

---

[18] Compare with Rheticus', *Narratio prima*, "Clarissimo viro D. Ioanni Schonero" (Rosen, p. 109; Hugonnard-Roche and Verdet, p. 42): "My teacher has written a work of six books in which, in imitation of Ptolemy, he has embraced the whole of astronomy [...]." I shall quote English translation of *Narratio prima* by E. Rosen, in *Three Copernican Treatises*. When appropriate, Latin text will be given from critical edition *G. I. Rhetici Narratio prima*, established by H. Hugonnard-Roche, J.-P. Verdet (in collaboration with M.-P. Lerner and A. Segonds).

[19] For a comprehensive analysis of the last five books of *De revolutionibus*, see O. Neugebauer and N. Swerdlow, *Mathematical Astronomy in Copernicus' De revolutionibus*.

The *De revolutionibus* was for a long time classified as a "book that nobody read."[20] Given the absence of any spectacular and dramatic reaction in the first decades following its appearance, historians surmised that the book was published in a small edition and available only to a few astronomers. In the 1970s, Owen Gingerich set out on a quest to determine how many copies of the first two editions (Nuremberg, 1543, and Basel, 1566, respectively) have survived to the present day. Upon discovery, he examined the marginalia entered by their owners to gauge the reception of Copernicus' work. His research yielded stunning results. In *An Annotated Census of Copernicus' De revolutionibus* (2002), he states that there are no less than 274 extant copies of the first edition and 327 copies of the second edition.[21] Since the publication of Gingerich's work, four more copies of the first edition have surfaced and another nineteen of the second, including two recently discovered second-edition copies located in Slovenia.[22] Gingerich estimates that 400 to 500 copies were printed for the first edition and 500 to 550 for the second. From this it may be inferred that the work was fairly well known and that the reasons for the absence of a sharp response in the first decades after its publication that seemed rather predictable in light of Copernicus' "ridiculous" and "absurd" thesis should be sought elsewhere.

One of the reasons for this was the anonymous text in the *De revolutionibus*, *Ad lectorem*. It argues that the motions of the earth are presented merely as possible but not necessarily true astronomical hypotheses.[23] But Gingerich's more than three decade long exploration has, in combination with other studies, also revealed something else. Based on the marginalia made by the owners of the copies of the *De revolutionibus* it is obvious that most first readers of Copernicus' work were not so much interested in the first, cosmological book as they were in

[20] This is how Arthur Koestler characterized the work in the popular *Sleepwalkers*, p. 194.

[21] See also O. Gingerich's recent book, *The Book Nobody Read: Chasing the Revolutions of Nicolaus Copernicus*, that tells the story of his research.

[22] According to present knowledge, there are two second edition copies of *De revolutionibus* in Slovenia. One is kept by the National and University Library and the other by the Library of the Franciscan Monastery in Ljubljana.

[23] We know now that the author of the *Ad lectorem* was Andreas Osiander (1498–1552). First to publicly announce that Copernicus was not the author of the address was Johannes Kepler in *Astronomia nova,* which was published in 1509. But first to notice that the author of the *Ad lectorem* was Osiander and not Copernicus was Philipp Apian (1531–1589). This was revealed by Michael Maestlin (1550–1631) in his notes on the *Ad lectorem* (1570). See O. Gingerich, *An Annotated Census of Copernicus' De revolutionibus (Nuremberg, 1543 and Basel, 1566)*, p. XVII.

the technical and mathematical aspects presented in the last five books. The first, cosmological book only became serious reading – and a vehicle of the Scientific Revolution – in the last quarter of the sixteenth century and, particularly, the early seventeenth century. Within the first decades of its publication, *De revolutionibus* was embraced only by a few.[24]

This is another indication that Copernicus' concerns were very specific and that one should seek the answer to the question why and how did he become a Copernican in Book I of *De revolutionibus. De revolutionibus* starts with cardinal Schönberg's letter to Copernicus, continues with (Osiander's) *Ad lectorem* and Copernicus' *Dedication to Pope Paul III*, followed by the first eleven chapters of Book I, in which Copernicus presents the earth's triple motion. In these eleven chapters of first book Copernicus deconstructs the accepted and established astronomical and philosophical truths in a masterly manner. He exposes the weak spots of geocentric cosmology and physics, reveals the conflict between the Ptolemaic arguments and Aristotelian physics, and demonstrates – or at least he presents it that way – that geokinetism and heliocentrism are compatible with some sort of plausible physics, which resonates, at least at first sight, with commonly accepted principles. He maintains that, if we should doubt him, then we have all the more reason to doubt Ptolemy's geocentrism and Aristotelian physics. In the meantime, he quietly and almost undetectably dismantles several fundamental concepts of Aristotelian cosmology and physics.

The first four chapters of the first book contain nothing that would cause concern to traditional Aristotelians and Ptolemaists. There Copernicus demonstrates that the universe and the earth are spherical, and that the celestial bodies move in uniform circular motions. The first seed of doubt is planted in Chapter 5: if the earth, like the universe and the stars, is spherical, and if the stars and the heavens move in circular motions, why should not the earth move as well? For regardless of whether it seems to us as though the earth rotates about its axis and the heavens remain at rest or vice versa, the observable phenomena are the same.

Why then, is the earth held as a fixture in the middle of the universe by the "consensus of many centuries"? In Chapter 6, Copernicus shows that there is no geometric proof that the earth is really fixed in the centre of the heavens.

---

24 For a short overview of first Copernicans, see, for example, K. A. Tredwell, and P. Barker, "Copernicus' First Friends: Physical Copernicanism from 1543 to 1610," and especially R. S. Westman, *The Copernican Question*, pp. 140–503. Despite significant amount of studies dedicated to the reception of Copernicus' ideas there is still a lot of work to be done.

In comparison to the immensity of the universe, it is so negligibly small that the proofs which Ptolemy furnished in the *Almagest* to demonstrate its central position in the universe are invalid. In Chapter 7, Copernicus lists the standard proofs of Aristotelian physics and sensory experience in favor of a motionless earth in the middle of the world; in Chapter 8, he refutes them by demonstrating that they are at least as much, if not even more, incoherent than claiming the earth's rotation. He then proceeds to open Chapter 9 with a question. Convinced he has demonstrated in the previous chapter that it is more probable for the earth to move than the heavens, he now inquires whether the earth can also undergo motions other than rotation about its axis. If it can perform the first motion (rotation), why can it not also perform the second, the motion around the sun (revolution) and thus be thrown off the centre of the world. In Chapter 10, he argues that this is indeed so on the basis of the harmonious arrangement of the planetary spheres, which arises from accepting this motion. The development of the concept of the earth's motion and its introduction into astronomy is brought to its logical conclusion in Chapter 11, where Copernicus devises a "demonstration" of the earth's triple motion.

***

Here I shall follow his logic, and the epistemological implications, of introducing the first two motions of the earth in Book I of the *De revolutionibus*. However I will not analyze all chapters (leaving out Chapters 6 and 11) and will investigate others only partially. But I shall not confine myself only to *De revolutionibus*. Copernicus' *Commentariolus* reveals many secrets and shall be treated accordingly. I believe all of the essential features of Copernicus' project are contained in it.

The first and second parts of the book therefore deal with text analysis of Book I of *De revolutionibus* and the third part places Copernicus' project into a context. The last section moves beyond the confines of contextualized reading with an analysis of Copernicus' gesture "from the outside" to shed light on what he had inadvertently achieved. In my view Copernicus must have believed that the introduction of the concept of the earth's motion into astronomy was justified by purely mathematical reasons, reasons arising from its movement around the sun rather than its rotation. The earth's movement around the sun fulfills a special task: it enables Copernicus to establish a harmonious, well-proportioned universe. This means that Copernicus pushed the earth in circular movement around the sun because mathematics demanded so. This gesture has huge philosophical and epistemological implications.

Part One, "The motion of the earth: a solution to the problems of astronomy," can be subdivided into three sections: the preliminary, rather general presentation and analysis of Copernicus' reasons to abandon geostatic astronomy and introduce the concept of the earth's motion (Chapters 1–3); the presentation of traditional arguments against the earth's motion (Chapter 4); and Copernicus' general strategy to surmount these obstacles (Chapter 5).

In the first three chapters, I analyze in a preliminary manner the reasons cited in the *Preface* to the *De revolutionibus*, as well as *Introduction* and introductory paragraphs in the *Commentariolus*, as the main factors that had led Copernicus to astronomical reform on the basis of the concept of the earth's motion. Thus I will be able to present the basic characteristics of astronomy in the period before Copernicus, defined by three fundamental postulates: the celestial bodies move in uniform circular motions; the celestial bodies are carried by real, three-dimensional physical spheres, which are the causes of their movement; the motionless earth is at the relative or absolute centre of the celestial bodies, depending on the cosmological system advocated by different scholars (eccentric or homocentric astronomy). In Chapter 1, I present the origin of the requirement for the uniform circular motion of celestial bodies. The second postulate, the existence of real spheres, is, according to standard interpretation, crucial for assessing the validity of Copernicus' criticism of the equant, a concept specific to Ptolemaic astronomy that contravenes the ancient principle of uniform circular motion. Since this issue is by Copernicus characterized as the first fundamental factor that has convinced him of the need to reform Ptolemaic astronomy, it will be thoroughly examined in Chapter 2. Yet according to Copernicus, the aforementioned factor, which in itself does not lead to heliocentrism, is not the only critical point of astronomy. Rather, it is the motionless earth at the centre of the universe – the third postulate of pre-Copernican astronomy – that poses the one epistemological barrier that prevents the Ptolemaic astronomy from deducing the "principal consideration," namely, the true, certain structure of the universe or form of the world (*forma mundi*). In Chapter 3, I present Ptolemy's discussion on the order of the celestial spheres from the *Almagest* and give a very brief, preliminary account on why his universe is a monster rather than a harmoniously arranged whole with all parts linked together in a relationship of *symmetria*. Copernicus firmly believed that he was able to solve the *forma mundi* problem with the concept of the earth's motion – the very concept which all articulations of knowledge (theology, philosophy, and *sensus communis*) regarded as absurd and inconceivable. Chapter 4 will therefore focus on the theological, philosophical and sensory objections to the motion of the earth. And finally, in Chapter 5, I analyze Copernicus' *Dedication to Pope Paul III* as an attempt at a general advance neutralization of such and similar objections to the earth's motion. There Copernicus states that mathematics is written for mathematicians (*mathemata mathematicis scribuntur*), and insists that the Scriptures should be explicated in accordance with scientific results, rather than interpreted *ad litteram*.

In Part Two, "Argumentation in favor of the earth's motion," we follow Copernicus' argumentative strategy and his reasons to justify the earth's motion in Book I of *De revolutionibus*. I claim that the entire Book I of *De revolutionibus* is best understood as his dialectical refutation of Ptolemy's and Aristotle's arguments rather than a doctrinal exposition of Copernicus' views.

In Chapter 6, I analyze Copernicus' defense of the earth as a uniform terraqueous globe whose centre of magnitude is coincident with its centre of gravity. Thus Copernicus prepares the earth to launch into motion about its axis and around the sun. In Chapter 7, we examine his defense of the traditional postulate that the celestial bodies move in uniform circular motions. Given that spheres as the carriers of celestial motions move only in uniform circular motions, their apparently "irregular" movements can be nothing other than the outcome of a combination of a larger number of uniform and circular motions, that is a combination of a larger number of celestial spheres that constitute a model for an individual planet. Celestial motions "in themselves" are "regular"; they only appear "irregular" to us. This leads Copernicus to scrutinize the relation of the earth to the heavens, that is, the question whether diurnal motion is attributed to the heavens or to the earth. He then introduces into his discussion the so called "optical argument": we observe the same phenomena, regardless of whether we presume that the earth rotates about its axis and the heavens are at rest or that the earth stands still and the heavens perform a diurnal revolution. In Chapter 8, we follow the logic of the "optical argument" and its role in Copernicus' reasoning. Failing to provide a sufficient basis to determine whether motion can be attributed to the earth or to the heavens, it is given negative rather than a positive role; Copernicus takes it as a universal epistemological principle that dispels any preconceptions on the mobility and immobility in the universe and thus provides the necessary ground for the discussion on the possibility of the earth's motion. In Chapter 9, I analyze Copernicus' formulation of traditional objections to the earth's rotation, his rejections of them, and his positive arguments in favor of the earth's rotation. The dialectical nature of Copernicus' argumentation is most clearly at work in the cluster of Chapters 7 and 8 of the Book I of *De revolutionibus*. Here Copernicus prepares ground for his introduction of the movement of the earth around the sun. As I show in Chapter 10, Copernicus is actually convinced "in advance" that the earth moves (also) about its axis, because its second motion around the sun enables him to establish a harmonious universe in which all planetary spheres are arranged into well-proportioned whole on the basis of one, universal principle. His argumentation in favor of the movement of the earth around the sun is therefore based on purely mathematical reasons.

In Part Three, "Copernicus in context," we rise to the heavens, so to speak, and look upon Copernicus from a bird's eye view to better understand the motives for his astronomical reform and its nature, and finally to better evaluate his role in the Scientific Revolution.

Chapter 11 is dedicated to astronomy before Copernicus. We follow the attempts of astronomers and philosophers from Roger Bacon (ca. 1214–1294) to Georg Peurbach (1423–1461) and Johannes Müller von Königsberg, known as Regiomontanus (1436–1476), to provide philosophically sound astronomy. From the thirteenth century onward the Ptolemaic astronomical tradition faced continuous attacks for violating such basic postulates of Aristotelian physics as that every circular motion ought to take place around the centre of the earth. Regiomontanus, the most expert astronomer of Ptolemaic tradition, even went so far as to work out some homocentric planetary models.

In Chapter 12, divided into four sections, I attempt to establish what philosophical and intellectual changes occurred in Europe after Peurbach and Regiomontanus that motivated and enabled Copernicus' project. Why did he write the *Commentariolus*? I argue that Copernicus' motivation for heliocentric astronomical reform lies in his adoption of Platonism and that his Platonism explains more or less his whole project.

In the first section I examine Copernicus' education which took him from Cracow to Bologna and Padua, and show his space of possibilities, focusing especially on Platonists and Platonism. The second section focuses on the problems of existing astronomy which Copernicus evokes as motives for his astronomical reform: the disagreement regarding the basic principles of astronomy (homocentric astronomy as opposed to eccentric astronomy) and especially the problems of Ptolemaic astronomy, which are basically two: the equant problem and the *forma mundi* problem. Since, as the examination of the sources shows, all the aforementioned issues were familiar to very capable astronomers before him and to his contemporaries, the question that naturally arises is: Why, all of a sudden, were all of these problems so critically important to him? How are they related?

My thesis is: through his Platonism. Copernicus is not just an astronomer; he is, as Galileo will later call him, an astronomer-philosopher. More specifically, he is a Platonist astronomer-philosopher.

In the third section, I analyze several classes of evidence which support this thesis: from Copernicus' education to his personal insignia, to his annotations of Plato's work, finally to the most important one: the philosophical correspondence between Plato' and Copernicus' views. Here are the most important ideas, motives and conceptions that Copernicus took over from Plato (not necessarily always directly from him): the demand for the uniform circular motion of the heavenly bodies; the idea that the orderly motion of the heavens is due to divine management; the idea that men should become godlike by studying the order of the cosmos; the idea that the role of astronomy is to enable humankind to reach the highest Good and, consequently, understanding of astronomy as the consummation of the liberal arts, if not even of the whole knowledge. Copernicus also believes that the real subject of astronomy, that is, the relationships among celestial objects, are accessible to reason and thought alone. And he shares with Plato and Platonists two related ideas on the specifics of the cosmic order: that the heaven is arranged by mathematical relationships holding among its parts and that there is one natural bond that links all parts of universe together. For a Platonist, order, which is based on unity, means a well-proportioned arrangement of parts within a whole. Copernicus expressed these ideas in his unpublished *Introduction* as well as in the key passages of his published *Dedication to the Pope Paul III*, and in Chapter 10 of Book I of *De revolutionibus*.

Finally, in the fourth section, I show how Copernicus' Platonism explains his critical attitude towards specific issues in Ptolemaic astronomy, and how are all those issues related. Copernicus' critical appraisal of the existing state of astronomy is not motivated by astronomical but philosophical concerns. Copernicus' criticism originates in Platonist conception of cosmic order, the role of the cosmic order for humankind's access to the highest Good and the role of astronomy in that pursuit. Thus the question of why Copernicus was bothered with the equant and the order of planetary spheres (the traditional, standard interpretations are examined in Part One) gets a completely new dimension. For Copernicus, the equant is not problematic for mechanical reasons, as the standard interpretation has it, but because it contravenes Plato's axiom of uniform circular motion. And at that time the unsolved question of the structure or form of the world (*forma mundi* problem) was deemed critical because without a certain, harmoniously ordered universe humankind does not have access to the highest

Good. On the other hand Copernicus' adherence to the Platonist conception of cosmos, one arranged by mathematical relationships and governed by unity (in other words, the cosmic *harmonia* and *symmetria*) stimulated him to search for a single principle that could arrange the cosmos according to these standards. He found it in the distance-period relationship. When he realized that the distance-period relationship works for all planetary spheres and arranges them in a well-proportioned whole, if the reference point is no longer the earth but the sun, he had the Platonist solution to the major problem of Ptolemaic astronomy at hand. The only problem with that solution was that it is geokinetic and heliocentric, that it demands earth to move around its axis and around the sun, with the sun the reference point and center of all planetary movements. Since some ancient philosophers believed that the earth moved, Copernicus felt confident enough to write *Commentariolus*, explaining his new system.

Chapter 13, the last chapter of the book, explains the changes to astronomy achieved by Copernicus' project. The first, most obvious change Copernicus brings to the astronomical landscape is his Platonism. While astronomers and philosophers before him tried to establish astronomy that complied with Aristotelian philosophy, Copernicus' philosophical concerns are purely Platonist. The major question that remains to be answered is whether his Platonism opens any new, maybe even revolutionary, epistemological and philosophical horizons. Is Copernicus already a part of the Scientific Revolution?

Copernicus introduced several minor changes to mathematical astronomy. His mathematical planetary models, for example, which are mostly taken over from Islamic astronomers, represent a variation of traditional Ptolemaic ones. His universe is immense, much bigger than the Aristotelian cosmos, and the universe's planetary orbs are separated by vast empty spaces; it is not an Aristotelian plenum any more. He seems to believe that the cause of the circular motion of (planetary or other) spheres is not their inner nature or the substance they are composed of, but their shape. His concept of gravity is clearly Platonist: according to him like is attracted to like. It is important to stress, however, that his minor physical and cosmological changes are for the most part Platonist in their nature, yet that they are not entirely consistent; sometimes they are even contradictory. Copernicus does not have a consistent physics and cosmology to support his heliocentrism. This is, third, a result of his deeper commitment to the Platonist idea of cosmic order. That is all that really matters for him. He wants to establish a well-proportioned universe, governed by unity, regardless of physical and cosmological consequences. One of the major consequences of his mathematical

Platonist commitment is that he subverts the traditional hierarchy of sciences (astronomy, a mathematical art, takes precedence over physics or philosophy of nature, which is in Aristotelianism over and above astronomy and supplies it with the first principles). Closely related to this is most fundamental consequence of his gesture: the criterion for what is true and what is mere appearance is no longer in the domain of appearances or sensory perception, but in pure thought, in pure mathematical reasoning. Copernicus prefers what reason tells him over that which his senses experience; for him, reason precedes what can be perceived and becomes mistress of his belief. He argues not from appearances but by going behind them. Mathematical reasoning which goes beyond mere appearance is the only way to establish the harmony of the world. He must, in order to see the objective truth of the universe, displace himself from the earth and take a view from the sun. Since this is impossible, he must rely on pure mathematical thought. This is the moment, I believe, of Copernicus' true revolution, and his genuine contribution to the Scientific Revolution.

PART ONE

THE MOTION OF THE EARTH: A SOLUTION TO THE PROBLEMS OF ASTRONOMY

Why does Copernicus need the earth's motion? What benefit can he derive from a concept against which the "consensus of many centuries" speaks so clearly and convincingly as he himself states in the *Dedication to the Pope* or *Preface* to *De revolutionibus*? What does he mean by claiming that it is possible "to ascertain whether more firm demonstrations than those [of my predecessors]" – or "the teachers of mathematics in the schools" – "could be found for the revolution of the celestial orbs on the assumption of some motion of the earth"[1]?

In the same – his last – text, written in June 1542,[2] Copernicus recognizes the existing problems of astronomy as the main impulse for seeking "a different explanation of the revolutions of the spheres of the universe," one that would be based on the concept of the earth's motion. His thesis, that the earth moves whereas the sun is at rest in the centre of the universe, is according to his own presentation of the genesis of heliocentric astronomy a response to the troubled state of astronomy that results from the discord among astronomers. The latter manifests itself in two ways. First, the astronomers disagree in their investigations

---

1 *Preface* (Rosen, p. 5; Zekl, p. 74). All quotes from Copernicus' *De revolutionibus* are taken from Edward Rosen's translation in N. Copernicus, *On the Revolution*. References to Latin edition of Book I of *De revolutionibus* are to Hans-Günter Zekl's edition in *Das neue Weltbild*, which includes also Copernicus' original *proemium* to *De revolutionibus* and his translation of *Letter to Lysis*. In a few cases when I refer to other books of *De revolutionibus*, I give Latin text and page numbers of the edition of *De revolutionibus* edited by Ryszard Gansiniec.

2 This is the date on which most scholars agree. See, for example, E. Rosen, *Commentary*, pp. 338 and 343; P. Barker and B. Goldstein, "Patronage and the Production of *De revolutionibus*," and O. Gingerich, *An Annotated Census of Copernicus' De revolutionibus*, p. 108. On the chronology of the composition of different books of *De revolutionibus*, see F. Schmeidler, *Kommentar zu De revolutionibus*, pp. 2–5.

to determine the exact length of the tropical year, and second, they fail to use the same principles, assumptions, and demonstrations:

> [f]or, in the first place, they are so uncertain about the motion of the sun and moon that they cannot demonstrate and observe a constant length even for the tropical year. Secondly, in determining the motions not only of these bodies but also of the other five planets, they do not use the same principles, assumptions, and demonstrations of the apparent revolutions and motions.[3]

This is not the first time that Copernicus had given such a critical appraisal of the state of astronomy. Before his dramatic critique of astronomy in the *Preface*, Copernicus expressed his disappointment with the existing astronomy in two previous texts. The first one was his original prefatory text to *De revolutionibus*, the so called *Introduction* (written sometime between 1516 and 1525 and later replaced by the *Preface*) and the second one was the *Commentariolus*, his first attempt at geokinetic and heliocentric astronomy (made sometime around 1510, possibly much earlier, but certainly before 1514).

In the *Introduction* the state of astronomy is described as intolerable for two reasons. First, astronomers do not agree among themselves on the principles and assumptions of astronomy. According to Copernicus, astronomy, "this divine rather than human science, which investigates the loftiest subjects,"[4] is plagued with perplexities, because "the principles and assumptions, called 'hypotheses' by the Greeks, have been a source of disagreement [...] among most of those who undertook to deal with this subject, and so they did not rely on the same ideas (*rationes*),"[5] and because "the motion of the planets and the revolution of the stars

---

3 *Preface* (Rosen, p. 4; Zekl, p. 70).

4 *Introduction* (Rosen, p. 7; Zekl, p. 82).

5 Ibid. All Latin quotations from *proemium* are taken from N. Copernicus, *Das neue Weltbild*, pp. 80–4. Compare with the edition and French translation of the *proemium* of A.-P. Segonds and M.-P- Lerner, "Éditer et traduire Copernic," pp. 386–9. Copernicus obviously understands *principia*, *assumptiones* and *hypotheses* as denoting one and the same concept. He sometimes uses the term *axioma* and once *primarius lapis*. I believe this rather free use of different terms could be due to his reading of Proclus' *Commentary on Plato's Timaeus*. Proclus uses "axioms" and "hypotheses" arbitrarily to denote all starting points of discussion. On this, see M. Martijn, *Proclus on Nature*, pp. 109–14. Copernicus uses the term *hypotheses* (and *principia, assumptiones, axioma*) to denote two categories of concepts: (1) principles of natural philosophy (his own or of ancient authors), such as motions of the earth, stability of the fixed stars and the sun, circular and uniform motion of celestial objects, etc., that are taken to be true; (2) geometrical models by which appearances can be saved, that is, mathematical models

could not be measured with numerical precision and lead to complete knowledge."[6] Even Ptolemy himself "who far excels the rest by his wonderful skill and industry" fails in this attempt. Ptolemy, Copernicus argues, "brought this entire art almost to perfection with the help of observations extending over a period of more than four hundred years,"[7] but it is clear that "many things [...] do not agree with the conclusions which ought to follow from his system."[8]

Second, Copernicus notes that certain astronomical parameters are not congruent with observations. It is obvious that "certain other motions [of celestial bodies] have been discovered which were not yet known to him."[9] "The motion of the heavenly bodies," Copernicus writes, seemingly quoting Plutarch (ca. 46–120 AD) from memory,[10] "has eluded the skill of the mathematicians."[11] Their inability to determine the exact motion of the celestial bodies is particularly problematic in determining the exact length of the sun's tropical year. How to determine a year has always generated "different opinions [...], so that many have abandoned all hope that the certain [or firm] determination of it (*certam eius rationem*) could be found."[12]

representing movements of celestial objects. The first category is more fundamental than the second one. As shall we see below, such is the principle of circular and uniform motion of heavenly bodies. For a confirmation that Copernicus sees natural philosophical hypotheses that include a moving/resting earth on a par with hypotheses that include circles, see quotation from *Preface*, infra, p. 75. See also A. Goddu, "Hypotheses, Spheres, and Equants in Copernicus' *De revolutionibus*," pp. 71–8, and A.-P. Segonds and M.-P- Lerner, "Éditer et traduire Copernic," p. 398, n. 13. For an illuminating reflection on the relationship between these two kinds of hypotheses, see Galileo, "Considerations on the Copernican Opinion," pp. 75–8.

6 *Introduction* (Rosen, pp. 7–8; Zekl, p. 82).

7 Ibid. (Rosen, p. 8; Zekl, p. 84).

8 Ibid.

9 Ibid.

10 See E. Rosen, *Commentary*, p. 345.

11 *Introduction* (Rosen, p. 8; Zekl, p. 84).

12 Ibid.

CHAPTER ONE

Uniform Circular Motion of Celestial Bodies

In the *Proemium* or *Introduction,* Copernicus is not so much interested in disagreements among astronomers regarding the "principles and assumptions" of astronomy as he is in the incompatibility of astronomical data with actual observed celestial phenomena. These included the exact length of the tropical year and the related issue of determining the sun's motion, which has always been "subject to different opinions."

In fact, Copernicus has very specific reasons for emphasizing the problem of the determination of the exact length of the tropical year, some of which cropped up only after his completion of the *Commentariolus.* He does not mention the issue of the tropical year but only until after around 1514, when Paul of Middelburg (1446–1534) asked him to help with reform of the Julian calendar.[13] Copernicus did not take part in it formally, but at the end of the *Preface,* he

13 On calendar reform as the background of Copernicus' claims both in the *Introduction* and *Preface*, see the article by M. A. Granada, and D. Tessicini, "Copernicus and Fracastoro: The Dedicatory Letters to Pope Paul III, the History of Astronomy, and the Quest for Patronage," pp. 464–70, where Copernicus' allusions to the problems regarding the accurate determination of the duration of the year are appropriately placed in the context of his search for a patron and the approval of geokinetic astronomy. See also D. J. K. O'Connell, "Copernicus and Calendar Reform," and M. di Bono, *Le sfere omocentriche di Giovan Battista Amico nell'astronomia del Cinquecento*, pp. 20–3.

justifies *De revolutionibus* as his own contribution to calendar reform. This serves to further legitimize his introduction of the "absurd concept" of the earth's motion in astronomy. In other words, Copernicus incorporated the issue of the length of the tropic year, and its necessity to the solution of calendrical reform, as part of a strategy to persuade the reader of his *Introduction* (and particularly the *Preface)* of the rationality of introducing the concept of the earth's motion in astronomy. There is, in my opinion, one additional, and perhaps most important reason Copernicus is upset because of calendar problems, but this shall be discussed in due time.

On the other hand, he says nothing about this problem in the *Commentariolus*. Therein his critical presentation of astronomy focuses exclusively on what both his *Preface* and *Introduction* refer to as "principles," "assumptions," and "demonstrations," or "ideas [or explanations]" (*principia, assumptiones, demonstrationes, rationes*). "I understand that our predecessors assumed," Copernicus writes, "a large number of celestial orbs principally in order to save [or account for] the apparent motion of the planets through regular motion, for it seemed highly unreasonable that a heavenly body, which is perfectly spherical, should not always move uniformly."[14] In other words, the ancient astronomers discovered that by "the arrangement and combination of uniform motions in different ways it could be brought about that any body would appear to move to any position."[15]

What is Copernicus talking about?

---

14 *Commentariolus* (Swerdlow, p. 433; Zekl, p. 2). English translation of *Commentariolus* will be taken from N. M. Swerdlow, "The Derivation and First Draft of Copernicus' Planetary Theory: A Translation of the *Commentariolus* with Commentary." I will quote the Latin text of *Commentariolus* from N. Copernicus, *Das neue Weltbild*, edited by H. G. Zekl. Swerdlow actually translates "Valde enim absurdum videbatur coeleste corpus in absolutissima rotunditate non semper aeque moveri." as "[…] for it seemed highly unreasonable that a heavenly body should not always move uniformly in a perfectly circular figure." He reads the text as implying the absolute roundness, perfect circularity of uniform motion. See also H. Hugonnard-Roche and J.-P. Verdet, *Introductions à l'astronomie de Copernic. Le* Commentariolus *de Copernic. La* Narratio prima *de Rheticus*, p. 71: "Il semblait, en effet, tout à fait absurd qu'un corps céleste ne se meuve pas toujours uniformément sur un cercle parfait." Against this interpretation speaks, I think, the link between circular motion and spherical shape of the object moved, which Copernicus establishes, for example, in *De revolutionibus* I, 8 (Rosen, p. 16; Zekl, p. 114): "Why then do we still hesitate to grant it [the earth] the motion appropriate by nature to its [spherical] form (*mobilitatem illi formae suae a natura congruentem concedere*) rather than attribute a movement to the entire universe […]."

15 *Commentariolus* (Swerdlow, p. 433; Zekl, p. 2).

Copernicus is referring to one of the axioms of Greek astronomy and of all subsequent astronomy up to that of Johannes Kepler, whereby the apparent motions of the celestial bodies were explained by drawing on the assumption of their regular or uniform circular motion. There are two extant accounts of such an astronomical program: that of Geminus (ca. 10 BC–ca. 60 AD), and that of Simplicius (ca. 490–ca. 560). In his *Introduction to the Phenomena*, a rudimentary astronomical work from the first century BC, Geminus holds the Pythagoreans to be the first to study the motions of the celestial bodies on the basis of their presumed uniform and circular motions. Simplicius, on the other hand, in his *Commentary on Aristotle's On the Heavens* dating back to the sixth century AD, recognizes Plato (427–347 BC) as the pioneer of such astronomical investigation.[16] Regardless of whose opinion one chooses to believe (or not to believe), both sources clearly reveal the essentially metaphysical motive for the implementation of an astronomical program of this kind: the heavens are the realm of divine and eternal perfection, where the only possible motion is uniform circular motion.[17]

Geminus' statement that "[…] the hypothesis that underlies the whole of astronomy posited for astronomy is that the sun, the moon, and the five planets move circularly and at constant speed in the direction opposite to that of the cosmos"[18] is thus placed in the context of the Pythagorean tradition, which drew from the hypothesis that the movements of the sun, the moon, and the five planets are circular and uniform. With regard to "things divine and eternal," the Pythagoreans did not accept such disorder as moving sometimes more slowly, sometimes more swiftly, sometimes standing still, and then resuming forward movement again. Such irregularities in movement (i.e. slowness and swiftness) are caused by the constraints of life in the sublunary world, whereas for the stars, there is no cause of swiftness or slowness. For this reason, Geminus continues,

16 See Geminos, *Introduction to the Phenomena* I, 19–21 (Evans and Berggren, pp. 117–8). For the Greek text, see G. Aujac's critical edition, Géminos, *Introduction aux phénomènes*, pp. 5–6. Cf. Simplicius, *On Aristotle 'On the Heavens 2.10-14'*. Greek text (with French translation) is more easily available in Autolycus de Pitane, *La sphère en movement, Levers et couchers héliaques, Testimonia*, with the relevant passage on p. 160.

17 Both texts are relatively late and therefore relatively unreliable as sources for reconstructing details of development of the early Greek astronomy. For the discussion on these matters, see, for example, A. Gregory, *Plato's Philosophy of Science*, pp. 97–100, and J. Evans and J. L. Berggren, *Geminos's Introduction to the Phenomena*, pp. 49–58.

18 Geminos, *Introduction to the Phenomena* I, 18 (Evans and Berggren, p. 117).

they raised the question as to how "the phenomena [might] be accounted for" or "saved" (*sozein ta phainomena*) by means of circular and uniform motion.

According to Simplicius' account, Plato used similar reasoning.[19] Quoting from Sosigenes (the end of the second century AD), who invoked Eudemus' *History of Astronomy* (dated the fourth century BC),[20] Simplicius states that it was Plato who proposed that mathematicians investigate which models or hypotheses of uniform and regular circular motion could "save the phenomena," or account for apparent planetary motions. Plato started with the assumption that the celestial bodies must, due to their divine, perfect and absolute nature – and in opposition to what is observed in the heavens – actually move in a uniform and regular circular motion:[21] "And I have also said previously that Plato without hesitation assigned to the heavenly motions circularity, uniformity, and order and put forward to the mathematicians this problem: by making what hypotheses about uniform, circular, and ordered motions will it be possible to preserve the phenomena involving the planets?"[22] Plato's answer to what is presumably his own presentation of the problem was rather lacking. The closest approximation to "astronomy" that would meet the requirements was developed in the *Timaeus*. There Plato explains that the planetary motions result from two uniform revolutions which take place at the same time but in opposite directions and on different planes. From this he concludes that all planets describe so-called "spirals."[23] However, such a combination of uniform circular movements only "saves" the (diurnal and annual) motion of the sun, while providing no adequate explanation for the "wanderings" of the moon, and even less so for the "wanderings" of the five planets. Nonetheless, as Andrew Gregory writes, "Plato's belief in the regular, circular and ordered motion of the heavenly bodies arguably sets the stage for astronomy for the next two millennia [...]."[24]

---

19 On Simplicius and his reconstruction of the history of astronomy from his *Commentary On the Heavens*, see A. C. Bowen, *Simplicius on the Planets and Their Motions.*

20 On Greek historiography of science and especially on Eudemus as historian of science, see L. Zhmud, *The Origin of the History of Science in Classical Antiquity.* History of astronomy is treated in Chapter 7, pp. 228–76.

21 A very suggestive passage is in his *Laws* 998a–b. For analysis of this passage, see A. Gregory, *Plato's Philosophy of Science*, pp. 95–6, and infra, p. 329.

22 Simplicius, *On Aristotle's On the Heavens 2.10–14*, 492, 31–493, 4 (Mueller, p. 33).

23 See *Timaeus* 38c–39a.

24 A. Gregory, *Plato's Philosophy of Science*, p. 101.

Concentric spheres: Eudoxus, Callippus, Aristotle. According to Simplicius' account, Plato's disciple (and perhaps also colleague) Eudoxus of Cnidus (d. 355 BC) was the first to rise to his mentor's scientific and philosophical challenge to explain the apparently irregular motion of the planets and thus "save" the perfection of the heavens, and preserve the incorruptibility of the stars and their regular or uniform circular motion.[25] His model of concentric spheres was, according to Simplicius, improved by Callippus of Cyzicus and Aristotle. With some modifications, it was incorporated into the Aristotelian philosophical system,[26] in which all heavenly bodies must move around the middle of the All (*to pan*), that is, around the centre of the earth.

In Eudoxus' version of the system of concentric spheres,[27] each planet is set in motion by a certain number of revolutions that result from a combined action of spheres (see *Figure 1*). The spheres transmit the motion, which takes place about different axes; planets are attached to the innermost sphere, while the fixed, immovable stars are attached to the outermost sphere with the earth at its centre. The sphere of the fixed stars performs its diurnal revolution from east to west in uniform motion. In an attempt to account for the motion of each "wandering star" (Mercury, Venus, Mars, Jupiter, and Saturn), as well as the moon and the sun, Eudoxus uses a number of concentric spheres nested one within another, but revolving around a central earth but about different axes, in different directions, and at different speeds. Hence, each planet depends on the system of three or four spheres. The axis about which a sphere rotates is contiguous to the encompassing sphere in two points; the latter sphere has its own motion. A planet attached to the equator of the sphere closest to the centre moves in a motion that results from the movement of all spheres in its system.

It seems that Eudoxus regarded his spheres as mathematical models for "saving the phenomena" and did not believe in the actual existence of the spheres.

---

25 W. Knorr, "Plato and Eudoxus on the Planetary Motions," argues that Plato's challenge was not the motivation behind Eudoxus' technical studies. For more information on what these "irregularities" are and whether or not the retrograde motions of the planets were already observed by Eudoxus and Callippus, see also B. R. Goldstein, "The Status of Models in Ancient and Medieval Astronomy," and "Saving the Phenomena: The Background to Ptolemy's Planetary Theory."

26 Aristotle dealt with Eudoxus' and Callippus' systems in the *Metaphysics* XII, 8. However, we have most information on their astronomy from Simplicius' commentary on Aristotle's treatise *On the Heavens*.

27 On Eudoxus' astronomy, see, for example, M.-P. Lerner, *Le monde des sphères*, vol. 1, pp. 24–8; I. Yavetz, "On the Homocentric Spheres of Eudoxus."

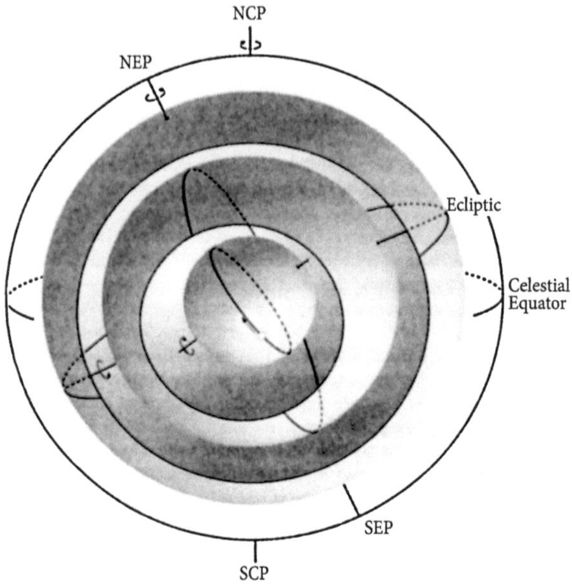

Figure 1.
Homocentric model for the motions of a planet. Reconstruction by M. Lattis, Between Copernicus and Galileo, *p. 52.*

Regardless of its essentially mathematical articulation, however, the Eudoxian system later assumed an accurately defined cosmological framework as it functioned within the Aristotelian geocentric world. Due to its failure to account for certain phenomena, Eudoxus' concentric theory was modified by Callippus. Aristotle adopted the basic characteristics of their geometric constructions, but also introduced reacting or unrolling spheres, so that the number of spheres in the entire system increased to forty-seven or fifty-five.[28] According to Aristotle, the system so constructed was in agreement with the phenomena.

Aristotle's system differed from those of Eudoxus and Callippus in at least two other important aspects. First, where Eudoxus and Callippus tried to account for the observable planetary motions by assigning spheres to each individual planet, Aristotle included all concentric spheres in a single geocentric system.[29] But most importantly, he considered the homocentric theory concerning the circular motion of the first heaven and planet carrying spheres to be real, not just geometric constructions. According to Aristotle, who took the investigation

[28] For a concise overview of the development of homocentric theory from Eudoxus to Aristotle, see J. Evans, *The History and Practice of Ancient Astronomy*, pp. 305–12.

[29] Here I can only give a very rough outline of Aristotle's "astronomy" from the *Metaphysics*. For more on this, see, for instance, M.-P. Lerner, *Le monde des sphères*, vol. 1, pp. 48–54, and J. Evans, *The History and Practice of Ancient Astronomy*, p. 313.

of the heavens to be in the domain of physics and natural philosophy, geometric spheres are turned into material bodies composed of the "fifth" or "first" body, i.e. aether. Eternal circular motion, in his opinion, requires a specific essence, and aether is a body which is "running continually for time everlasting."[30]

Aristotle's cosmos of concentric spheres therefore necessitated a fundamental and radical distinction between the "terrestrial" and "celestial" realms. Whereas simple bodies or elements of the sublunary realm (earth, water, air, and fire) adhere to the principles of motion, generation and corruption, which apply only to them, the superlunary realm consists of the fifth substance, later called *quinta essentia*, which is subject to radically different rules. While the "terrestrial" elements undergo a continuous process of change, the "celestial" element remains unaltered. The only kind of mutability that can be ascribed to the heavens is change in position, movement from one place to another, or locomotion. But even in this respect, the world of the stars differs from the sublunary realm: without support "from the outside," each individual motion in the sublunary realm has a temporal and spacial beginning and end, whereas the celestial motion of the stars must always be conceived of as eternal. Only in this case can the superlunary sphere transcend the processes of "terrestrial" change and corruption.

Assuming that the world is finite – and for various reasons the world of Aristotle can be nothing but finite; one of them is that the infinite cannot move[31] – the "first heavens," the sphere of the fixed stars, can only be assigned circular motion. Rectilinear motion cannot be infinite and continuous. Arriving at this conclusion, Aristotle demonstrates the primacy of circular or rotational motion.[32] Circular motion (rotation, or rotary locomotion; *periphora, revolutio*) is considered primary because: (1) it is simple and more complete; (2) it is eternal or everlasting; it is one and continuous since any one point is the starting-point, middle-point, and finishing-point alike; (3) circular motion is the standard for time-measurement of all other motions, and must as such be primary, since everything is measured by what is primary; and (4) only circular motion is uniform.

Circular motions are therefore the only motion which the ancient astronomers and philosophers declared eternal. Only circular motion can have a constant

30 Aristotle, *De caelo* I, 3, 270b21 (Leggatt, p. 59). All English translations from Book I and II *On the Heavens* are S. Leggatt's, and from Book III and IV J. L. Stocks'. On how Aristotle demonstrates the existence of real spheres, see M.-P. Lerner, *Le monde des sphères*, vol. 1, pp. 41–8.

31 See Aristotle, *De caelo* I, 5–7, 271b–276a16; especially 274b29–276a16.

32 See *Physica* VIII, 9, 265a15–265b15.

speed, and be both complete and continuous at the same time. This signifies that: (1) ideally, celestial motions are circular or composed of several circles, and (2) that the circular motion of the celestial bodies must proceed at a constant velocity.

Thus the arguments of Eudoxus and Callippus, the only authors mentioned by Copernicus in his *Commentariolus,* perfectly satisfy the requirements for uniform circular motion of celestial bodies. Copernicus acknowledged that the solution drew on correct "principles and assumptions," but it had one shortcoming: its failure to account for certain observable motions of celestial bodies. As Copernicus justly observes, the use of concentric spheres could not "give an account of everything in the planetary motion, that is, not only those motions that appear in connection with the revolutions of the planets, but also that the planets appear to us at times to ascend and at times to descend in altitude, which concentric circles in no way permit."[33] Or, to quote Noel Swerdlow, the concentric spheres furnish an unsatisfactory representation of a "planet's motion in longitude and in no way account for its apparent change of distance."[34]

Eccentrics and epicycles: Ptolemy. Alexandrian astronomy, which culminated in Ptolemy's *Almagest*, attempted to solve this problem with the so-called theory of eccentrics and epicycles.[35] Though purely mathematical in nature, the astronomy of the *Almagest* implies a physical state of the universe. Geometric procedures which Ptolemy uses to account for celestial phenomena build on the premise that the earth remains at rest at the centre of the universe, and that the celestial bodies move in uniform circular motion. In sum, Ptolemy's method as set out in the *Almagest* comprises the following steps:[36] (1) to make the greatest possible number of accurate observations; (2) to determine the anomalies of the observed motions in relation to uniform circular motion; (3) to deduce experimentally laws that would lead to the understanding of how the periods and magnitudes of these anomalies are created; (4) to account for the observed phenomena by combining uniform circular motions with the use of eccentrics

---

33 *Commentariolus* (Swerdlow, pp. 433–4; Zekl, p. 2).

34 N. Swerdlow, "The Derivation and First Draft of Copernicus' Planetary Theory," p. 434.

35 On the development of Greek astronomy which led to eccentric and epicyclical models see J. Evans, *The History and Practice of Ancient Astronomy*, pp. 289–355. For Ptolemy's *Synthaxis mathematica*, see the English translation of G. J. Toomer, *The Almagest.* For a more detailed overview of the *Almagest* in its entirety with the commentary, see O. Pedersen, *A Survey of the Almagest.*

36 I am closely following R. Morelon's concise summary of Ptolemy from his "General Survey of Arabic Astronomy," pp. 4–5.

and epicycles; (5) to calculate the parameters of these motions and on this basis compose tables to calculate the positions of the stars.

The simplest model used by Ptolemy and his followers to account for the apparent motions of the celestial bodies is that of the eccentric, which places the earth outside the centre of its circle, thus rendering it "off-centred." The second model reproduces the same motion with the use of a deferent (a basic circle with the earth at its centre) and an epicycle (a smaller circle added to it). The third model employs combinations of two or more eccentric and epicycle circles. In addition to reproducing a planet's varying distances from the earth according to the principle of uniform and circular motion of the celestial bodies, which are, according to Ptolemy, proper "to the nature of divine beings,"[37] these models furthermore reflect two anomalies observed against the backdrop of the uniform circular motion of the entire heavens from east to west. The planets generally travel in the opposite direction, that is to say, eastward around the zodiac. But those trajectories exhibit two kinds of observed motions that contravene the postulated uniformity and circularity. First, planets do not always appear to move at a constant or uniform speed. Instead there is a point of minimum speed (the apogee) and a point of maximum speed (the perigee).[38] This irregularity is called the first anomaly, or the zodiacal anomaly. And, second, there are motions of planets that appear to take place neither circularly nor in the eastward direction at all times, but are observed to slow down, come to a halt (a station), reverse direction, and go backward for a while or retrograde, and then resume direct or prograde, eastward motion again. The position of each planet has a special relationship to the position of the sun; this is called the second anomaly, or solar anomaly. These two anomalies, can be accounted for by geometric models employing combinations of circles or by the so-called eccentric and epicycle models that are represented as intersecting two-dimensional geometric objects.[39]

---

37 Ptolemy explicitly stated in the *Almagest* IX, 2 (Toomer, p. 420; Heiberg, pars 2, p. 208) that the five planets moved in uniform circular motion: "Now it is our purpose to demonstrate for the five planets, just as we did for the sun and the moon, that all their apparent anomalies can be represented by uniform circular motions, since these are proper to the nature of divine beings, while disorder and non-uniformity are alien [to such beings]."

38 Apogee is a point at which a planet is farthest from the earth, perigee is a point at which a planet is nearest the earth.

39 For a more accurate and still sufficiently concise presentation of kinematic models of individual planets, see, for example, O. Neugebauer, *The Exact Sciences in Antiquity*, pp. 191–207.

As for the sun, Ptolemy uses the eccentric model to explain why it appears to us on earth not to move through the zodiac at a uniform speed. To a terrestrial observer, a celestial body in this model appears to move more swiftly when it comes closer to the earth and more slowly when it moves farther away from it. The eccentric model accounts for the so-called first, or zodiacal anomaly, that is, the motion of a celestial body at a non-uniform speed (see *Figure 2*).

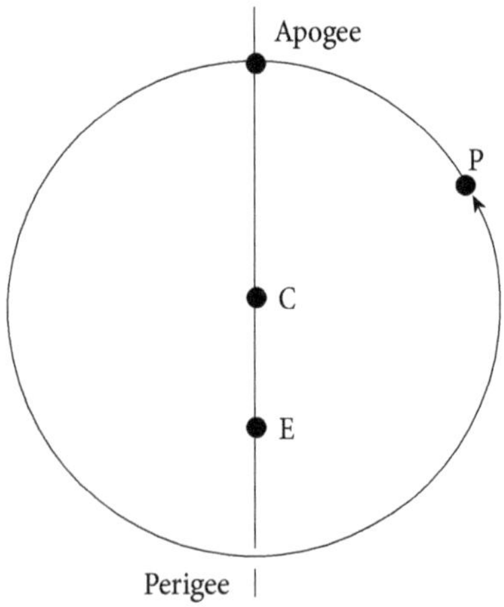

Figure 2.
The eccentric model.

The diagram shows a circle with the centre C and E (= earth) located at some distance away from the centre. P, located on the arc or the circumference, represents a planet. The circle moves at a constant speed around its centre C. When the planet is observed from the earth at E, both its distance and the speed of its motion appear to vary. The celestial body actually moves at a constant speed with respect to its centre, but this is not apparent to the terrestrial observer, to whom the planet appears to be moving more swiftly when it is at its perigee, or the top of its circle, and more slowly when it is at its apogee, or the bottom of its circle.

The second anomaly produces an effect in which a planet appears to come to a halt, move backwards or retrograde and then resume its forward motion again. This is precisely what the term "planet" defines: "one who wanders," hence "wandering stars," as planets were called in Antiquity. Since planets do not always start retrograding at the same zodiacal point, this anomaly does not depend on their relationship to the sphere of the fixed stars but to that of the sun. The ancient astronomers observed that the three superior planets (Mars, Jupiter, and Saturn) reach the top or the centre of their retrograde arcs when they are in opposition to, or 180° from, the sun, and that the two inferior planets (Venus and Mercury), which are never observed too far from the sun, show the similar pattern of retrograde motion.

The second anomaly may be accounted for by means of an epicycle (see *Figure 3a* and *Figure 3b*).

The diagram shows a circle that moves at a constant speed around its centre E. Point EC, located on the circumference of the circle, is the centre of the smaller

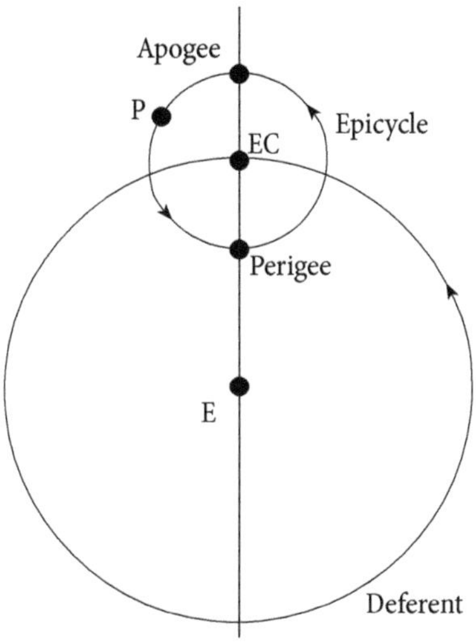

Figure 3a.
The epicycle model.

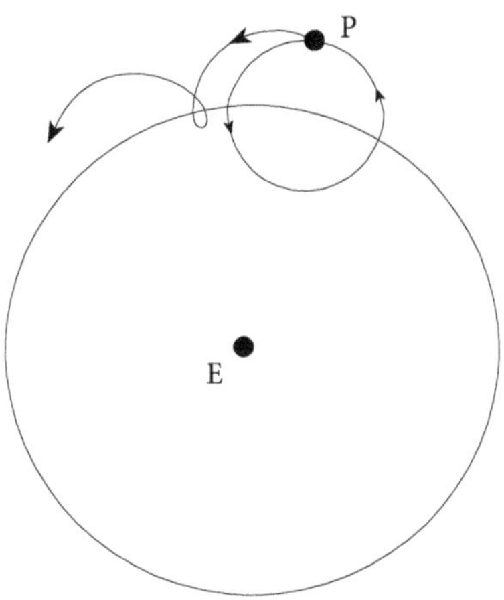

Figure 3b.
The path of a planet in the epicycle model.

circle (epicycle). P, placed on the circumference of the epicycle, represents a planet. The larger circle (the deferent) revolves around the centre E and the smaller circle (the epicycle) revolves at a constant speed around its own movable centre EC. This can account for the varying distance of the planet from the earth, and particularly its apparent retrograde movement. When the planet is at its apogee, or its furthest point from the earth, it appears to be moving backward; contrary observations are made when the planet is at its perigee or the point closest to the earth.

As in other models, the combination of an epicycle and a concentric deferent enables the adjustment of radial distances of both circles, their speeds and

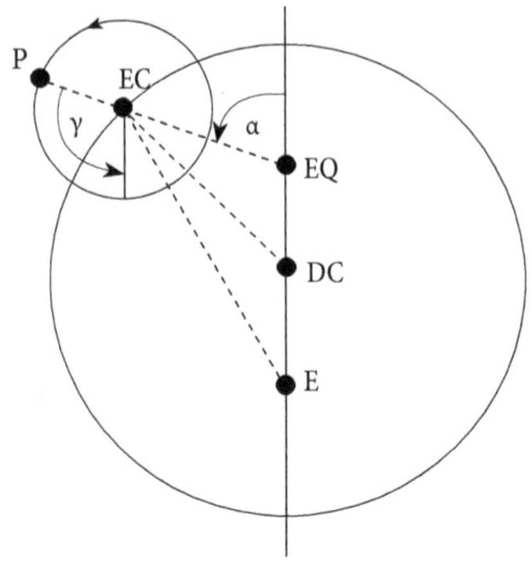

Figure 4.
The equant model.

directions to reproduce all apparent changes in distance and speed, as well as stations and retrograde motion. Models can be made more consistent with actual observed positions by using as a deferent an eccentric circle instead of a concentric one, or by adding another epicycle to the first.

The eccentric and epicycle are therefore two basic mathematical devices employed in Ptolemy's astronomy. Nevertheless, Ptolemy is compelled, so as not to increase the number of circles necessary to account for the motion of a celestial body, to abandon the principle of uniform circular motion – and to do so in a relatively imperceptible way. As Copernicus states in the *Commentariolus*, theories

> concerning these matters that have been put forth far and wide by Ptolemy and most others, although they correspond numerically [with the apparent motions], also seemed quite doubtful, for these theories were inadequate unless they also envisioned certain equant circles, on account of which it appeared that the planet never moves with uniform velocity either in its deferent sphere or with respect to its proper centre.[40]

Ptolemy's solution, presented here in a slightly simplified form is as follows. (see *Figure 4*).[41] A planet (P), that is, the centre of its epicycle, moves uniformly;

---

[40] *Commentariolus* (Swerdlow, p. 434; Zekl, pp. 2–4). For Copernicus' own technical explanation and rejection of the equant, see *De revolutionibus* IV, 2 (Rosen, pp. 175–7; Gansiniec, pp. 175–7), where he discusses lunar theory, and V, 2 (Rosen, p. 240; Gansiniec, p. 246), where he discusses planetary theory. For the quotations, see infra, pp. 73–4, n. 86.

[41] See also N. Swerdlow's reconstruction, "The Derivation and First Draft of Copernicus'

however, its motion is uniform with respect neither to the geometric centre of the deferent (DC) nor to the earth (E), but with respect to a certain eccentric point in the deferent (EQ), in the Middle Ages called the equant. In this model, the earth is eccentric to the geometric centre of the universe, much in the same way as is the actual centre of motion, the so-called *punctum aequans*. The required uniformity of planetary motions is thus achieved with regard to this imaginary, immaterial point.

The planet (P) moves on the epicycle, whose centre (C) moves on the deferent. The centre of the deferent (DC) does not coincide with the earth (E); the deferent is eccentric to the earth. The centre of the epicycle (EC) moves in a constant and uniform motion with respect neither to DC nor to E, but with respect to EQ, i.e. the equant (*punctum aequans*) located on the straight line EDCEQ at an equal distance from DC as E, so that EDC = DCEQ. The EQEC line revolves in a constant and uniform motion around EQ; the motion of the planet on the epicycle is measured from the same line. Ptolemy thus retains constant angular speed but not constant linear speed.

Planetary Theory," pp. 434–5, of the Ptolemaic model on the basis of Peurbach's and Regiomontanus' *Epytoma Almagesti* IX, 6, with "equant circles."

CHAPTER TWO

Celestial Spheres and the Problem of the Equant

Let us once again recapitulate Copernicus' presentation and criticism of astronomy in his *Commentariolus* and *Preface* to *De revolutionibus*.

Our predecessors, says Copernicus in the *Commentariolus*, assumed a large number of celestial orbs for the purpose of explaining the apparent motion of the planets by the principle of uniformity. It was absurd to them that perfectly spherical heavenly bodies would move any way but uniformly. Working from within this principle, however, they discovered that by connecting and combining uniform motions in various ways they could make any celestial body appear to move to any position. Eudoxus and Callippus met this requirement of astronomical science with the theory of concentric orbs, but their model failed to account for the varying distances of the planets from the earth. In Copernicus' reconstruction – and in historical reality – this particular problem with the concentric model eventually leads to the Ptolemaic astronomy of eccentrics and epicycles. The theories of Ptolemy and many other astronomers readily predicted the apparent motions of the planets. But they also presented no small difficulty because, in order to provide more accurate predictions of celestial motions, Ptolemy was compelled to introduce the equant, a device which violated the principle of uniform circular motion. Hence, Copernicus concludes that Ptolemy's "notion (*speculatio*) seemed neither absolute enough nor sufficiently in accordance with reason (*non satis absoluta neque rationi satis concinna*)."[42]

As seen earlier, both the tradition of concentric spheres and that of eccentrics and epicycles are presented as deficient also in Copernicus' *Preface* to *De revolutionibus*. The concentric system is completely unacceptable because it fails to account for the varying distances of the planets from the earth. The Ptolemaic tradition of eccentrics and epicycles, on the other hand, violates the principle of uniform motion. Copernicus' first objection to the Ptolemaic astronomy therefore arises not from its disagreement with observations or its incapacity to calculate the positions of the celestial bodies, but from its violation of the ancient principle of uniform circular motion. That is the reason why the Ptolemaic *speculatio* is not sufficiently "absolute" or "perfect" and "in accordance with reason."

But what does this mean?

The equant model shows *how* the Ptolemaic astronomy violates the principle of uniform motion, but it does not explain *why it should not* do so. Does Copernicus see Ptolemy's violation as philosophical, aesthetic, metaphysical or even theological? Why does he find Ptolemy's concept of the equant – this imaginary reference point located within the deferent but in an eccentric position to the planet on the epicycle moving uniformly with respect to it – not "sufficiently in accordance with reason"?

Copernicus nowhere explicitly states the reasons for his objection. But he is not the first to object to the Ptolemaic astronomy due to its violation of the principle of uniform motion. We may learn something about the nature of his objection by examining the reasons evoked by other astronomers in criticizing this concept.

Ibn al-Haytham and the equant. Ptolemy's concept of the equant was criticized for the first time in the eleventh century by the famous Arabic scholar and astronomer Ibn al-Haytham, known to the Latin West as Alhazen (965–ca. 1040),[43] in one of his later works, *Doubts concerning Ptolemy* (*al-Shukūk 'alā Batlamiyūs; Dubitationes in Ptolemaeum*).[44] The standard interpretation has

---

42 *Commentariolus* (Swerdlow, p. 434; Zekl, p. 4). Compare with a similar formulation in *De revolutionius* IV, 2 (Rosen, p. 175; Gansiniec, p. 175). For quotation, see infra, p. 73, n. 86.

43 In the Latin West Alhazen was better known for his *Optics* and optic theory.

44 The entire work is divided in three parts. In the first, Ibn al-Haytham criticizes Ptolemy's *Almagest*, his *Planetary Hypotheses* in the second, and *Optics* in the third. For more

it that according to Ibn al-Haytham the concept of the equant contravened Ptolemy's own tenets of the uniform motion of the celestial bodies as well as his own understanding of astronomy – though not that presented in the *Almagest*. According to this reading Ibn al-Haytham criticized the concept of the equant on the basis of another characteristic of Ptolemy's astronomy that was developed in his *Planetary Hypotheses* in an effort to construct a physical equivalent to mathematical astronomy of the *Almagest*. According to historic discoveries made over the last fifty years, medieval Latin authors ascribed to Ptolemy the system of distances and sizes of real planetary spheres or the so-called "Ptolemaic system of the world," which, however, could not have been derived from the *Almagest*. And more importantly, they believed that Ptolemy developed the system himself. The discovery of the second part of Book I of *Planetary Hypotheses*, preserved in its entirety only in Arabic translation but not in Greek, has confirmed the existence of such a system.[45] In addition to the mostly pure mathematical astronomy constructed in the *Almagest*, Ptolemy also devised physical astronomy.

In the *Almagest*, Ptolemy focuses on determining planetary positions, usually expressed in longitude and latitude, while disregarding the distances of the planets from the observer. What is more, the model for each planet is constructed independently of theories or models for other planets, so that he does not account

---

on this, see G. Saliba, "Arabic Planetary Theories after Eleventh Century AD," p. 75; *Islamic Science and the Making of the European Renaissance*, pp. 97–109. I shall present al-Haytham's criticism of Ptolemy's equant following Saliba. For a comprehensive overview of Arabic astronomy, see the first volume of *Encyclopedia of the History of Arabic Science*, edited by R. Rashed. After the initial period of assimilation, Ptolemy's observation methods and models were subject to constant criticism and improvement efforts.

45 On how the discovery occurred and on the deficiencies of Heiberg's edition of the Greek text and translation from Arabic, see B. Goldstein, "The Arabic Version of Ptolemy's Planetary Hypotheses," pp. 3–4, and A. Murschel, "The Structure and Function of Ptolemy's Physical Hypotheses of Planetary Motion," pp. 34–5. For the preserved Greek segment of the text, see the third volume of *Claudii Ptolemaei opera quae extant omnia*. It contains the Greek text with a German translation and translation of the Arabic version of the first and second book, but omits the most relevant second part of the first book. For the complete Arabic text and the English translation of Book I, part 2, see the above mentioned article by B. R. Goldstein, pp. 5–9. For more on Ptolemy's *Hypotheses planetarum*, see also O. Pedersen, *A Survey of the Almagest*, pp. 393–7, and M. P. Lerner, *Le monde des sphères*, vol. 1, pp. 67–81. In the first book of the *Planetary Hypotheses*, Ptolemy explains the general order of the spheres, whereas in the second, he describes the mechanism contained in each of these spheres.

for the system of the entire cosmos or the entire *machina mundi*. But the physical reality of Ptolemy's cosmos, often implied but rarely explicated in the *Almagest*,[46] comes to the fore in the *Planetary Hypotheses*. Therein Ptolemy transforms two-dimensional geometric models of planetary positions into real, three-dimensional spheres that do not overlap or intersect each other, and attempts to compute the real size of the planetary system. The computation in the *Planetary Hypotheses* is performed on the basis of the following assumptions:[47] (1) the order of the planets is the same as in the *Almagest*; (2) the model of each planet is a physical mechanism enclosed in a spherical shell concentric with the earth, so that the maximum distance of one planet from the earth equals minimum distance from the earth of the planet above it; (3) the spheres are fitted inside each other, so that the outer surface of each one coincides with the inner surface of the next one, with no intermediate space; there is no vacuum between the spheres, as assured by the postulate of nesting spheres. Thus, the geometric models in the *Almagest* are transformed into physical models in the *Planetary Hypotheses*. In the *Almagest*, the models for planetary motions appear as geometrical models of intersecting circles, but the *Planetary Hypotheses* presents the same models as spheres enclosed one within another, with epicyclic spheres moving between the surfaces of eccentric spheres to prevent any intersecting or overlapping.

Given here is a somewhat simplified variant for Saturn which retains all significant characteristics (see *Figure 5*). A is an aethereal body. Its outer surface is concentric with the earth (T) and its inner surface is eccentric to it. Spherical shell B is placed within A. Its surfaces are concentric with each other but not with respect to T. B encloses a third, eccentric-concentric spherical shell C which encloses another concentric "hollow" that contains the mechanisms for Jupiter and all other planets that follow. Saturn is placed at point P in sphere E, which circulates within the spherical shell in the "hollow" enclosed by B. The kinematics of such a device fully correspond to the geometric model in the *Almagest*. The first sphere A causes the diurnal circular motion of the entire system and of the slower precession motion of the fixed stars. The circular motion of B equals that of the deferent, whereas E and F reproduce epicycle motions.

According to standard interpretation, Ibn al-Haytham's objection to the equant in *al-Shukūk* is based on the established assumption that these motions are the motions of real rather than imaginary bodies. In his own words: "[n]othing

46 Ptolemy sometimes talks about spheres, for instance, when describing the first lunar model. See O. Pedersen, *A Survey of the Almagest*, p. 167 and p. 392.

47 See ibid., p. 393.

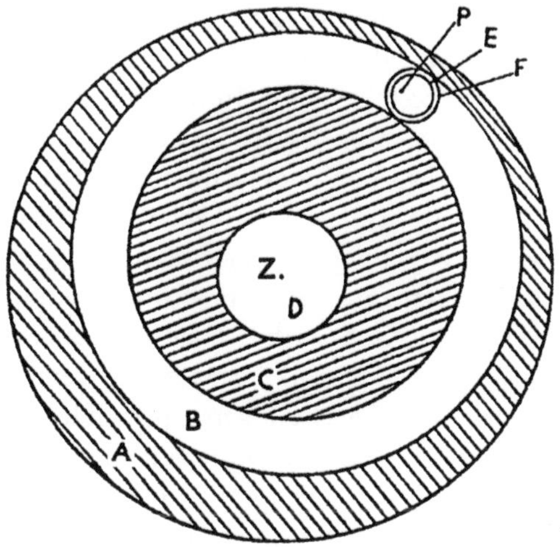

Figure 5.
Ptolemy's model for Saturn
from the Planetary Hypotheses.
Reconstruction by O. Pedersen,
A Survey of the Almagest, *p. 396.*

moves in any perceptible motion that produces an existing entity in this world except the body which [really] exists in this world."[48] Ptolemy's fictitious equant circles are, in Ibn al-Haytham's opinion, in contradiction with the physical reality: "The contradiction in the configuration of the upper planets that is taken against him [i. e. Ptolemy] was due to the fact that he assumed the motions to take place in imaginary lines and circles and not existent bodies. Once those (motions) are assumed in existent bodies, contradictions followed."[49]

According to George Saliba, Ibn al-Haytham makes it clear that if a sphere is to move in a uniform circular motion around a certain point, its distance from that point should always remain the same. If bodies described by Ptolemy were real, physical bodies, then a sphere could rotate uniformly only about an axis passing through its centre. According to Ibn al-Haytham, astronomy should devise mathematical models that are not in contradiction with the physical reality of real spheres, as was done by Ptolemy. To quote Saliba, Ptolemy "assumed the existence of an equant that would force a physical sphere to move uniformly, in place, on an axis that did not pass through its centre."[50] For Ibn al-Haytham the problem of the equant is in the inability of a sphere or spherical shell to uniformly rotate about any axis that does not pass through its centre.[51]

---

[48] Quoted from, G. Saliba, *Islamic Science and the Making of the European Renaissance,* p. 98.

[49] Ibid., p. 101.

[50] Ibid., p. 102.

[51] According to Ibn al-Haytham, Ptolemy must have in fact been aware that he was

Copernicus and the equant: the standard account. Does Copernicus, five centuries later and in a different cultural and scholarly environment, raise the same objection to the equant? In the light of Ibn al-Haytham's criticism I ask more simply: Does Copernicus, too, believe in the existence of real, planet-carrying spheres? If so, how does he become acquainted with this tradition?

The transmission and accuracy of Ptolemy's physical astronomy of the *Planetary Hypotheses* remains a subject of debate, but the translation of one of Ibn al-Haytham's non-critical works unquestionably had an important role in the process. As shown by Abdelhamid Sabra,[52] Ibn Al-Haytham's criticism of the equant from *al-Shukūk*,[53] in a text apparently never translated into Latin, drew on the first two of Aristotle's four principles that were presented in the addendum to Ibn Al-Haytham's earlier work on planetary theory known to the Latin West as *De configuratione mundi*,[54] in which Ibn Al-Haytham accepted Ptolemy's *Planetary Hypotheses* rather uncritically in comparison to his later *al-Shukūk*. The fundamental problem which Ibn al-Haytham treated in *De configuratione mundi* was not the equant but eccentrics and epicycles or, more accurately, the difficulties they posed to the Aristotelian physics. His proposition to develop a coherent physical picture of the Ptolemaic mathematical astronomy was quite similar to the physical-cosmological system devised in Ptolemy's *Planetary Hypotheses*. According to Ibn al-Haytham,[55] the construction of spheres, stars and

---

violating his own principles and that he even admitted that in *Almagest* IX, 2 (Toomer, p. 422; Heiberg, pars 2, p. 212), stating that he was using "some basic assumptions which we arrived at not from some readily apparent principle." See G. Saliba, *Islamic Science and the Making of the European Renaissance*, pp. 98–104.

52 See A. I. Sabra, "An Eleventh-Century Refutation of Ptolemy's Planetary Theory."

53 Ibn al-Haytham's critique of Ptolemy gave rise to the so-called "eastern school of Islamic astronomy," whose fundamental task was to develop models that would preserve Ptolemy's observations, "save the phenomena," and be consistent from the philosophical point of view. Beginning in the first half of the thirteenth century, Islamic astronomers, many of whom worked at the Marāgha observatory, devised new techniques that did not violate the accepted philosophical principles and allowed for the solution to several problems of Ptolemy's system. "Western," Andalusian Islamic thinkers criticized Ptolemy on different grounds.

54 For the Arabic text and English translation, see *Ibn al-Haytham's On the Configuration of the World*. The Latin West disposed of two translations. The more influential was the translation of *De configuratione mundi*, now housed by the library in Toledo. See J. M. Millàs Vallicrosa, *Las traductiones orientales en los manuscritos de la Bibliotheca Catedral de Toledo*, pp. 285–312.

55 For Ibn al-Haytham's principles that are not preserved in Latin translation, see also

all bodies that circle the world has to be based on the following four principles of Aristotelian natural philosophy: (1) a natural celestial body can undergo only one natural motion; (2) a simple celestial body cannot be moved in an irregular (non-uniform) motion: it travels the same distance in the same time; (3) the body of the heavens is incorruptible, and it allows no penetration; (4) there can be no void, no vacuum in the world.

In compliance with peripatetic philosophy, Ibn al-Haytham divides the universe in two major realms: sublunary and superlunary. The sublunary realm contains two kinds of bodies, heavy and light, whereas the bodies that belong to the superlunary realm are neither heavy nor light. The superlunary realm constantly revolves around the other two worlds (those of light and heavy bodies), and it contains all the stars. It is bounded by two equidistant surfaces whose centre coincides with the centre of the world. The entire orb (*orbis*) moves swiftly in a circular motion from east to west, carrying with it all celestial bodies it contains. Ibn al-Haytham draws particular attention to the polysemy of the term *orbis* (*falak* in Arabic): "*Orbis* is a common term used to denote any round quantity, be it a spherical body (*corpus spericum*), a spherical surface, a surface of a circle or a circumference of a circle."[56] It should be borne in mind that *corpus spericum* does not designate a full sphere, or *kura* in Arabic.[57] Rather, it has a specific meaning as it denotes a body "bounded by two parallel spherical surfaces," irrespective of whether or not their centre coincides with the centre of the world. In addition to the term *orbis*, the Latin translator refers to the same concept also by using the term *spera*: "Part [of the world] which revolves about the centre is called an orb, and it is divided [...] into nine parts composed of spherical bodies [...], and the centre of all spheres is the centre of the world, and each of these spheres is called an orb."[58]

M.-P. Lerner, *Le monde des sphères*, vol. 1, pp. 92–3. The outline of the contents is taken from *De configuratione mundi* (manuscript from the cathedral in Toledo), but compare also with M.-P. Lerner's presentation in the above mentioned book., vol. 1, pp. 92–8, and N. Swerdlow, "Pseudoxia Copernicana: or, Inquiries into Very Many Received Tenets and Commonly Presumed Truths, Mostly Concerning Spheres," pp. 117–8.

56 *De configuratione mundi* (Millàs Vallicrosa, p. 287). Ibn al-Haytham, *De configuratione mundi* (Millàs Vallicrosa, p. 285), defines the "sphere" as a "solid figure contained by one surface; at its centre lies a point, so that all the straight lines drawn from it to the circumference are equal to one another; this point is the centre of the sphere." All translations from *De configuratione mundi* are mine.

57 See M.-P. Lerner, *Le monde des sphères*, vol 1, p. 94.

58 *De configuratione mundi* (Millàs Vallicrosa, p. 287).

Ibn al-Haytham further subdivides the superlunary realm into nine contiguous spheres. They include, in an ascending order, seven planetary spheres, the sphere of the fixed stars, and, lastly, the *orbis maximus*, which causes all other spheres to make a complete revolution from east to west in twenty-four hours. Then follow the descriptions of the sun as the "noblest among the wandering stars," the moon, Mercury, Venus, and three superior planets (Mars, Jupiter and Saturn), the heavens of the fixed stars, and the highest sphere. In compliance with the stereotype model, each sphere is placed relative to the sphere encompassing it and the sphere encompassed by it. Ibn al-Haytham's description of the sphere or the spherical shell of the sun is, for instance, as follows: "The solar orb is a spherical body bounded by two equidistant surfaces whose centre is the centre of the world. The upper of these touches the concave surface of the orb of Mars; the lower touches the convex surface of the Venus' orb."[59]

Further on, Ibn al-Haytham lists "partial spheres" within each total planetary sphere of and construes their inherent motions. In the case of the sun, Venus and superior planets, he first mentions the eccentric sphere, which is described as follows: "[The eccentric sphere] is bounded by two equidistant surfaces whose centre is a point outside the centre of the world. The upper surface of this sphere touches the upper surface of the first sphere in one point and similarly the lower surface in one point; this orb is called the eccentric orb."[60]

Planets like Venus, Mars, Jupiter, and Saturn are affixed to the surface of a small total sphere (*spera brevis* or *spera epicili*, or *epiciclus*) whose diameter equals the thickness of the eccentric (also known as the *orbis deferens*, and carries the epicycle (see *Figure* 6). The body of sun, on the other hand, spans the entire thickness of its sphere: "The sun [...] is a spherical body placed inside the body of the orb [...]. The diameter [...] of the solar sphere equals the thickness of the body of its orb."[61]

Regardless of whether or not the Latin West caught the first glimpse of the realistic, physical astronomy through Ibn al-Haytham's treatise *De configuratione mundi*, or in some other way, the fact remains that the astronomy of real, three-dimensional planet-carrying spheres was gradually established and then fully accepted in the fifteenth century by Peurbach's *Theoricae novae planetarum*, one of the most popular works in the entire history of astronomy, which

---

[59] Ibid. (Millàs Vallicrosa, pp. 297–8).

[60] Ibid. (Millàs Vallicrosa, p. 298).

[61] Ibid.

Figure 6.
Ibn al-Haytham's model for Venus and superior planets. From M.-P. Lerner, Le monde des sphères, *vol. 1, p. 97 (reconstruction by W. Hartner,* Oriens-Occidens, *vol. 1, p. 482).*

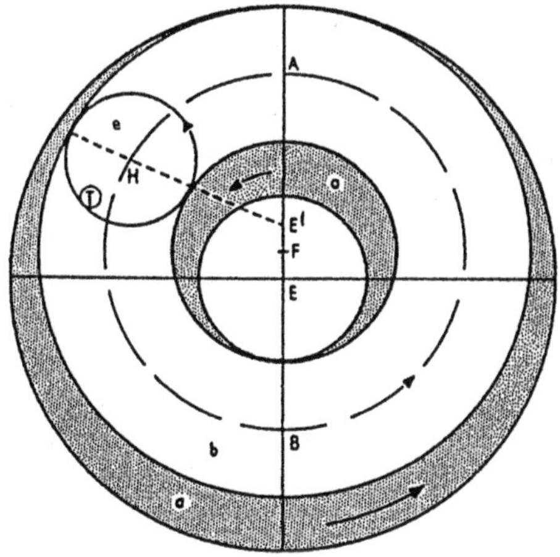

considered *orbes* as combinations of real spherical shells.[62] Peurbach's textbook definitively established the understanding of the celestial order as a series of real, material, three-dimensional orbs or spheres (*orbis totalis*) composed of various combinations of spherical orbs (*orbis partialis*). His materialization of Ptolemaic models was one of the six orthodox astronomical-cosmological beliefs held by most astronomers of the second half of the sixteenth century.[63] Despite some ambiguous considerations regarding the reality of "partial spheres" – but rarely of spheres in general – most authors of the sixteenth century believed, no matter which cosmological system they advocated, that *orbes* were real, three-dimensional, planet-carrying spheres and spherical shells, respectively, or epicycles, which were likewise considered spherical shells.[64]

---

62 For a more comprehensive exposition, see infra, pp. 241–65.

63 According to J. M. Lattis, *Between Copernicus and Galileo*, pp. 65–73, other elements of astronomical orthodoxy include: (1) adopting planetary constructions from Ptolemy's *Almagest* (sometimes with minor variations) as an explanatory system for planetary motions; (2) adopting the order of celestial spheres in accordance with Ptolemy; (3) a consensus on the magnitude of celestial spheres and the general size of the cosmos; (4) harmonizing the Ptolemaic planetary system with Aristotelian physics; (5) the Christian vision of the universe.

64 For the best up-to-date overview and discussion of the published commentaries on *Theoricae novae*, see P. Barker, "The Reality of Peurbach's Orbs: Cosmological Continuity in Fifteenth and Sixteenth Centuries Astronomy," pp. 7–32. Barker considers Albert of Brudzewo (ca. 1445–97), Sylvester Mazzolini (ca. 1456–1527), Jacobus Faber Stapulensis (Jacques Lefèvre d'Etaples; ca. 1450–1536), and Franciscus Capuanus de Manfredonia (Iohannes Baptista Capuanus de Manfredonia; n.d.). "With the exception of Prieras,"

However, does the fact that the astronomy of real, three-dimensional spheres was established in the time of Copernicus and known by him through Peurbach's *Theoricae novae* also mean that it was accepted by Copernicus himself?

Based on the discussion in *De revolutionibus*, (which is, for the most part, with the exception of Book I, devoted exclusively to mathematical astronomy), it would seem at first sight that Copernicus rejects physical astronomy. But taking a closer look is nonetheless required. Just as Ptolemy was believed to have developed mathematical astronomy in the *Almagest* but had actually devised its physical equivalent in the *Planetary Hypotheses*, Copernicus may be assumed to have done the same. And he did – in his *Commentariolus*, which Matthew of Miechów (1452–1523), the Cracow physician and professor of medicine, described in a brief entry in a catalog of his books and manuscripts of May 1, 1514: "Next a six-folio [manuscript] of a *Theorica* maintaining that the earth moves while the sun is at rest (*Item sexternus* Theorice *asserentis Terram moveri, Solem vero quiescere*)."[65] From this it may be concluded that – even though the treatise is not of a rudimentary nature and its present form does not contain any illustrations – Matthew regards it as a text in which *orbes* signify spheres or spherical shells describing real models for planetary motions. It is only through such understanding that it seems reasonable to classify Copernicus' *Commentariolus* as the *theorica* literature, which alongside Peurbach's *Theoricae novae* typically presents planetary models as combinations of real spherical shells. As maintained by Swerdlow, Copernicus assumes that the few readers to whom the brief treatise is dedicated are already familiar with the theory of real spheres, and utilizes an almost identical terminology in the *Commentariolus* to that in the *Theoricae novae*.[66]

But the conclusion that Copernicus regards *orbes* as real, three-dimensional

---

claims Barker, p. 14, "all these commentators follow Peurbach in treating the celestial orbs as real – and there may be historical reasons for Prieras' reservations." According to Barker, the same conclusion may also be drawn from the inspection of commentators on Sacrobosco's *Sphaera* (Wenceslaus Faber de Budweyss, Petrus Ciruellus Darocensis (or Pedro Ciruelo)). His investigation brought him to the conclusion, p. 26, "that a substantial number – perhaps a majority – of European astronomers – accepted the physical reality of celestial orbs in the period from Peurbach to Copernicus, and later." M.-P. Lerner, *Le monde des sphères*, vol. 1, pp. 128–31, is a little bit more reserved.

65 Quoted from N. Swerdlow, "The Derivation and First Draft of Copernicus' Planetary Theory," p. 431. Emphasis added.

66 That the *Commentariolus* was indeed written in the style of the *theorica* literature is demonstrated by N. Swerdlow in "Pseudoxia Copernicana," pp. 116–7. For a more detailed analysis of the text of the *Commentariolus*, supported with the evidence that *orbis* always denotes a real, material sphere, see ibid., pp. 148–56.

spheres – and that his objection to the equant should be therefore identical to Ibn al-Haytham's – seems to be confirmed not only by the context, by the "scientific paradigm" of his time, but also by several passages in Book I of *De revolutionibus*.[67] Let me provide one example.[68] Copernicus makes at least one explicit use

[67] This issue on Copernicus' belief in the existence of real spheres brought about a major contention between E. Rosen and N. Swerdlow. Rosen aspired to make Copernicus a scientific hero, a revolutionary that broke with the tradition with a single stroke and had consequently abandoned real spheres. See E. Rosen, "Copernicus' Spheres and Epicycles"; "Reply to N. Swerdlow," and N. Swerdlow, "Pseudoxia Copernicana: or, Inquiries Into Very Many Received Tenets and Commonly Presumed Truths, Mostly Concerning Spheres"; "The Derivation and First Draft of Copernicus' Planetary Theory: A Translation of the 'Commentariolus' with Commentary"; "Translating Copernicus." On this issue, see also P. Barker's summary of the arguments, "Copernicus, the Orbs, and the Equant," and "The Reality of Peurbach's Orbs." See also M.-P. Lerner, *Tre saggi sulla cosmologia alla fine del Cinquecento*, pp. 38–40. A. Goddu, "Hypotheses, Spheres, and Equants in Copernicus' *De revolutionibus*," maintains that Copernicus only believed in the existence of the large orbs or spheres (composed of partial orbs) and in the reality of the eccentric or deferent orb in which the planets were imbedded, but not in the existence of real epicycle orbs. I am convinced that Copernicus *in principle* also believed in the existence of real epicyclic orbs, but could not explicitly claim their reality because he had several options to organize them in order "to save the phenomena." See *De revolutionibus* III, 15 (Rosen, p. 156; Gansiniec, pp. 154–5): "Whatever is done by an epicycle, however, can be accomplished in the same way by an eccentric. [...] From all of these analyses it is clear that the same apparent non-uniformity always occurs either through an epicycle on a concentric or through an eccentric equal to the concentric. There is no difference between them provided that the distance between their centres is equal to the epicycle's radius. Hence it is not easy to decide which of them exists in the heavens." And *De revolutionibus* III, 20 (Rosen, p. 164; Gansiniec, p. 163): "Since so many arrangements lead to the same result, I would not readily say which one is real, except that the perpetual agreement of the computations and phenomena compels the belief that it is one of them." For more on this problem, see P. Barker, "The Reality of Peurbach's Orbs," pp. 20–3, and P. Barker and M. Vesel, "Essay Review of André Goddu, Copernicus and the Aristotelian Tradition: Education, Reading and Philosophy in Copernicus's Path to Heliocentrism," pp. 321–7. On the subject of orbs, see also E. Grant, "Celestial Orbs in Latin Middle Ages"; N. Jardine, "The Significance of the Copernican Orbs"; E. J. Aiton, "Celestial Spheres and Circles." The history of these fishy objects from Greeks up to their complete abolition in the seventeenth century is nicely treated in M.-P. Lerner, *Le monde de sphères*.

[68] Copernicus' traditional conception of the mechanism of celestial spheres – but not their order – can also be inferred from Chapter 4 of Book I of *De revolutionibus*, in which he explains the principle of uniform circular motion of the celestial bodies. His

of the term *orbis* designating a three-dimensional spherical shell, that is, a sphere with two surfaces. In Chapter 10 of Book I, he shows that the paths of Mercury and Venus – and later also those of Mars, Jupiter and Saturn – are heliocentric, and continues:

> But since all these planets are related to a single centre, the space between Venus' convex orb and Mars' concave orb must be set apart as an orb or sphere, both of whose surfaces are concentric with those orbs (*inter conuexum orbem Veneris et concauum Martis relinquitur spatium, orbem quoque siue sphaeram discerni cum illis homocentrum secundum utranque superficiem*). This [intercalated sphere] receives the earth together with its attendant, the moon, and whatever is contained within the moon's globe.[69]

Therefore, the earth goes around the sun between the spheres of Venus and Mars, not in an empty space, but between the convex surface of Venus' orb and the concave inner surface of Mars' orb, so that it travels in some material, real sphere whose outer and inner surfaces are concentric with the former ones. The terms *orbis* and *sphaera* therefore refer to the same sphere or spherical shell, that is, a sphere which encloses the earth and the sphere of the moon within its surfaces.[70] A combination of Copernicus' description in Chapter 10 of Book I of *De revolutionibus* with purely geometric depictions in Books III and V of the same volume yields a reconstructed model of real spheres that provides the physical basis for the mathematical model (see *Figure 7*).

The answer to whether Copernicus accepted the materialization of astronomical models is therefore affirmative, even though he nowhere explicitly

---

reflection would not make any sense at all if he conceived the *orbes* as circles or imaginary spheres. For a different, very original view, see A. De Pace, "Copernicus against a Rhetorical Approach to the Beauty of the Universe," pp. 90–5, and *Niccolò Copernico e la fondazione del cosmo eliocentrico*, pp. 79–86; for her interpretation of Chapters 4 and 10, see pp. 107–13, 225–6, and n. 669. She claims, *inter alia*, that Chapters 1 to 4 of Book I of *De revolutionibus* do not necessarily represent Copernicus' own beliefs but common assumptions which he accepts only in order to be able to pursue dialectical discussion. I shall analyze Copernicus' Chapter 4 infra, pp. 128–38.

69 *De revolutionibus* I, 10 (Rosen, p. 20; Zekl, p. 132).

70 Compare with Rheticus' paraphrase in the *Narratio prima* X (Rosen, p. 144; Hugonnard-Roche and Verdet, p. 59): "Between the concave surface of Mars' orb and the convex of Venus', where there is ample space, the globe of the earth together with its adjacent elements, surrounded by the moon's orb, is carried around by a great orb (*a magno quodam orbe [...] circumferri*), which encloses within itself, in addition to the sun, the orbs of Mercury and Venus, so that the earth moves among the planets as one of them."

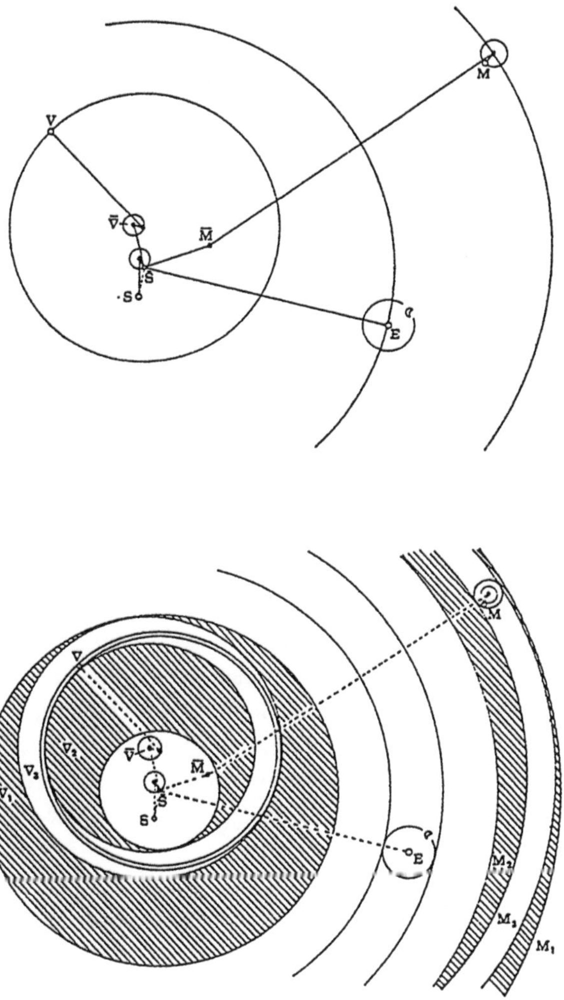

Figure 7. Geometric and physical model for the earth's path between the spheres of Venus and Mars. Reconstruction by N. Swerdlow, "Pseudoxia Copernicana," pp. 124–5.

and unequivocally expressed his conviction about the reality of celestial spheres. There is consensus now among scholars that Copernicus believed in the existence and reality of (at least) the total spheres that moved the planets. In line with the astronomical and cosmological tradition of the late Middle Ages and Renaissance, Copernicus considered the planetary motions dependent on the motions of real, material spheres in the Ptolemaic and Aristotelian traditions.

Therefore, Copernicus' objection to the equant seems identical to that voiced by Ibn al-Haytham. As he states in the *Commentariolus*, Ptolemy's violation of the "principle of uniform circular motion," is caused by the introduction of the equant. However, the equant is not considered in geometric terms, as an imaginary reference point in the deferent that is eccentric with respect to a planet's motion on the epicycle and in relation to which a planet moves in uniform motion. Rather, it is conceived of as a physical, real entity within the framework of the astronomy of real celestial spheres which can only rotate uniformly on axes that pass through their centres. It seems, at least at first sight, that Ptolemy's introduction of the equant or *aequantes quosdam circulos* is neither sufficiently *absoluta* nor sufficiently *rationi concinna*, because, as Swerdlow states, Copernicus "considers the motion of a planet to be directed by the revolution of a material sphere or spheres in which the planet is fixed. The only motion permitted to the sphere is a simple uniform rotation about its diameter; it cannot move uniformly with respect to any other straight line passing through it."[71] Copernicus' objection to the non-uniform motion of spheres as posited by Ptolemy's concept of the equant thus seems to be based on "mechanical" grounds.[72] As I shall explain in due time, Swerdlow's interpretation of the nature of the equant problem, which we followed above, does not exhaust the question. In my view, for Copernicus the problem of the equant is definitively neither mathematical nor just "mechanical"; it is primarily philosophical or, better yet, metaphysical.

This problem of the equant is the reason, Copernicus concludes in the introductory part of the *Commentariolus*, that he "often pondered whether perhaps a more reasonable model composed of circles could be found from which every

71 N. Swerdlow, "The Derivation and First Draft of Copernicus' Planetary Theory," p. 435.

72 Such characterization is given by N. Swerdlow, "The Derivation and First Draft of Copernicus' Planetary Theory," p. 435. P. Barker, "Copernicus, the Orbs, and the Equant," p. 318, however, justifiably warns that such characterization should be understood as ahistorical, since in Copernicus' time, there was no canonical mechanics that would allow such a conclusion to be drawn. A. Goddu, "Hypotheses, Spheres, and Equants in Copernicus' *De revolutionibus*," pp. 90–4, is of a different opinion. His contention, p. 94, is that "Copernicus rejected the equant as a mathematical hypothesis," because "as Copernicus discusses the problem, his objection is expressed in purely geometrical terms referring to circles, not orbs." In my opinion, Goddu contradicts himself in the very next sentence, correctly evoking that Copernicus in the *Commentariolus*, addressing the equant problem, does refer to "a deferent sphere." Copernicus' characterization of his own solution in terms of circles instead of spheres (*modus circulorum*) can be easily transformed into a solution in terms of spheres, provided that we know exactly which one is real. See supra, p. 61, n. 67.

apparent irregularity would follow while everything in itself moved uniformly, just as the principle of perfect motion requires."[73]

[73] *Commentariolus* (Swerdlow, p. 435; Zekl, p. 4).

CHAPTER THREE

A Stationary earth and the *Forma Mundi* Problem

Let us return to the *Preface* and Copernicus' analysis of the state of astronomy. The opening segment of the *Preface* is, in a way, a cross between the introductory sentences of the *Commentariolus* and the last paragraph of the *Introduction*. Astronomy is faced with two problems. The first is a specific astronomical problem: astronomy was so far incapable to "establish and observe a constant length [...] for the tropical year." The second, the more general of the two, is that astronomical theories contradict each other in regard to the principles they use in their attempts to describe the motions of the celestial bodies. There are two astronomical schools, each founded on different principles and assumptions: one which draws on the model of concentric spheres, another which applies the theory of eccentrics and epicycles. Both try to explain the same celestial phenomena by drawing on two conflicting theories. Copernicus writes:

> Secondly, in determining the motions not only of these [celestial] bodies [i.e. the sun and the moon] but also of the other five planets, they do not use the same principles, assumptions, and demonstrations of the apparent revolutions and motions. For while some employ only homocentrics, others utilize eccentrics and epicycles, and yet they do not quite reach their goal.[74]

[74] *Preface* (Rosen, p. 4; Zekl, p. 70).

The astronomical traditions contravene each other in their basic principles and are mutually exclusive by assuming either concentric or eccentric spheres. Moreover, each proves inadequate in itself:

> For although those who put their faith in homocentrics demonstrated that some nonuniform motions could be compounded in this way, nevertheless by this means they were unable to obtain anything certain that would be in absolute agreement with the phenomena. On the other hand, those who devised the eccentrics seem thereby in large measure to have solved the problem of the apparent motions with appropriate calculations. But meanwhile they introduced a good many ideas which apparently contradict the first principles of uniform motion.[75]

The astronomical tradition which tries to explain the motions of celestial bodies by employing the system of homocentric spheres is deficient, because it can only partially account for irregular motions of the celestial bodies. In other words, it fails to provide a solid explanation for the motions of the celestial bodies – an explanation that will be "in agreement with the phenomena." The problem is then addressed by the astronomical tradition of eccentrics and epicycles, which also uses circular motions and spheres carrying the celestial bodies, only in this case the spheres are not concentric with the earth but eccentric to it. This tradition brings theory into agreement with the observed phenomena by solving the problem of "apparent motions with appropriate calculations." However, the problem with this tradition is that it relies on a notion, which apparently contradicts the first principle of uniform motion, that is, it uses the equant. The shortcomings of the first system bring about the construction of the second one, which ultimately provides appropriate computations for the motions of the celestial bodies, but only by violating the principle of uniform motion. The concentric astronomy conforms to the first principles of uniform motion but contradicts the observations, whereas the second astronomy for the most part agrees with the observed phenomena but violates the first principle of astronomy, that is, the principle of the uniform circular motion of celestial bodies.

Copernicus obviously considers the two theories to be related.[76] The first theory sacrifices agreement with the phenomena by insisting on correct principles, taking into account both the uniformity and circularity of celestial motions. But the second theory, while sufficiently congruent with celestial phenomena, does

---

75 Ibid.

76 Compare also with J.-J. Szczeciniarz, *Copernic et la révolution copernicienne*, p. 46.

so at the expense of one of the fundamental postulates of the ancient astronomy, that is, the principle of the uniform motion of the celestial bodies. Copernicus' description of the deficiencies of existing astronomy implicitly reveals one of the goals of his astronomical reform: he wants to compensate for the deficiencies of one theory by using the advantages of the other, but without violating the ancient principle of the uniform motion. He will try to reproduce the observed varying distances of the planets from the observer, which the theory of concentric spheres cannot accomplish, using the deferent-epicycle astronomy, and he will at the same time seek to avoid the use of the already mentioned equant (*punctum aequans* also referred to as *circulus aequans* or bisected eccentricity) by employing only uniform circular motion. Hence, by combining appropriate calculation methods provided by the eccentric astronomy with the physical model of the concentric astronomy Copernicus will attempt to devise a mathematically and physically sound astronomy.

This leads us to another essential objection of Copernicus to the Ptolemaic astronomical tradition, one first explicitly mentioned in the *Preface* to *De revolutionibus*. Astronomers, says Copernicus, failed to determine and demonstrate the "principal consideration": "the structure of the universe" and the certain "symmetry" of its parts. Therefore, the universe of Ptolemaic astronomy was a *monstrum*. Astronomers could not

> elicit or deduce from the eccentrics the principal consideration, that is, the structure of the universe and certain symmetry of its parts (*mundi formam ac partium eius certam symmetriam*). On the contrary, their experience was just like some one taking from various places hands, feet, a head, and other pieces, very well depicted, it may be, but not for the representation of a single person; since these fragments would not respond to one another at all (*nullatenus inuicem sibi respondentibus*), a monster rather than a man would be put together from them.[77]

Copernicus' objection to Ptolemaic astronomy has been rightfully characterized as an aesthetic one, but this aesthetic objection has an astronomical equivalent.[78] Ptolemy's astronomy in the *Almagest* is not a system in the true sense, like the one presented in the *Planetary Hypotheses*.[79] In the *Almagest* Ptolemy built

---

77 *Preface* (Rosen, p. 4; Zekl, p. 70).

78 See O. Gingerich, "'Crisis' versus Aesthetic in the Copernican Revolution."

79 On the concept of the "world system," see M.-P. Lerner, "The Origin and Meaning of 'World System'."

his mathematical astronomy on the computation of angles rather than distances and the actual planetary order may have varied.[80] He develops a suitable model to account for the motions of each individual planet and to predict planetary positions, but fails to devise a *system* that would allow the correlated celestial phenomena to be coherently accounted for as a whole. In other words, Ptolemy investigates "parts" of the universe one after another, producing a sum of relatively unrelated elements. Combinations of the deferent/epicycle units illustrating the paths of individual planets are described in the function of inferior units, without being integrated into the whole.

In the beginning of Book IX of the *Almagest*, Ptolemy briefly addresses the question of planetary arrangement, but he admits that there is no objective criterion to determine planetary distances.[81] The planetary distances and definitive arrangement of the planets would only be possible if planetary parallaxes were detected, but they remained unobserved.[82] Ptolemy acknowledges the agreement among astronomers that the sphere of the fixed stars is the most distant from the earth and the lunar sphere the nearest to it. And he agrees with them that the spheres of the three superior planets are farther away from the earth than the spheres of the other two and the sun. Astronomers, likewise, agreed that the sphere of Saturn was the largest (and hence most distant from the earth), followed by the spheres of Jupiter and Mars. The problem arose with the order of the sun, Venus, and Mercury. While some located Venus and Mercury below the sun, others placed them above it, as neither had ever been seen to pass in front of the solar disk. Ptolemy argues instead that the planets may perhaps be traveling below the sun on a different plane from that of the earth's viewpoint of the sun, which may ultimately explain why a spot on the sun has never been observed. Since no star, fixed or wandering, has shown any detectable parallax, Ptolemy finds the order propounded by ancient astronomers far more credible since it used the sun as a divider between the planets that move away from the sun in any direction and distance and those that always remain near it.

Ptolemy's criterion for ordering the planets is to divide them into two groups: those that reach some maximum elongation from the sun and those that

---

80 See O. Pedersen, *A Survey of the Almagest*, pp. 261–2, and 295–6.

81 See *Almagest* IX, introduction (Toomer, pp. 419–420; Heiberg, pars 2, pp. 206–7).

82 The only absolute distances which Ptolemy determines in the *Almagest* are the distances of the moon and the sun. The former is determined on the basis of the observations of the lunar parallax and the latter on the basis of an argument in which he uses the observations of eclipses. See *Almagest* V, 13–15, and O. Pedersen, *A Survey of the Almagest*, pp. 203–13.

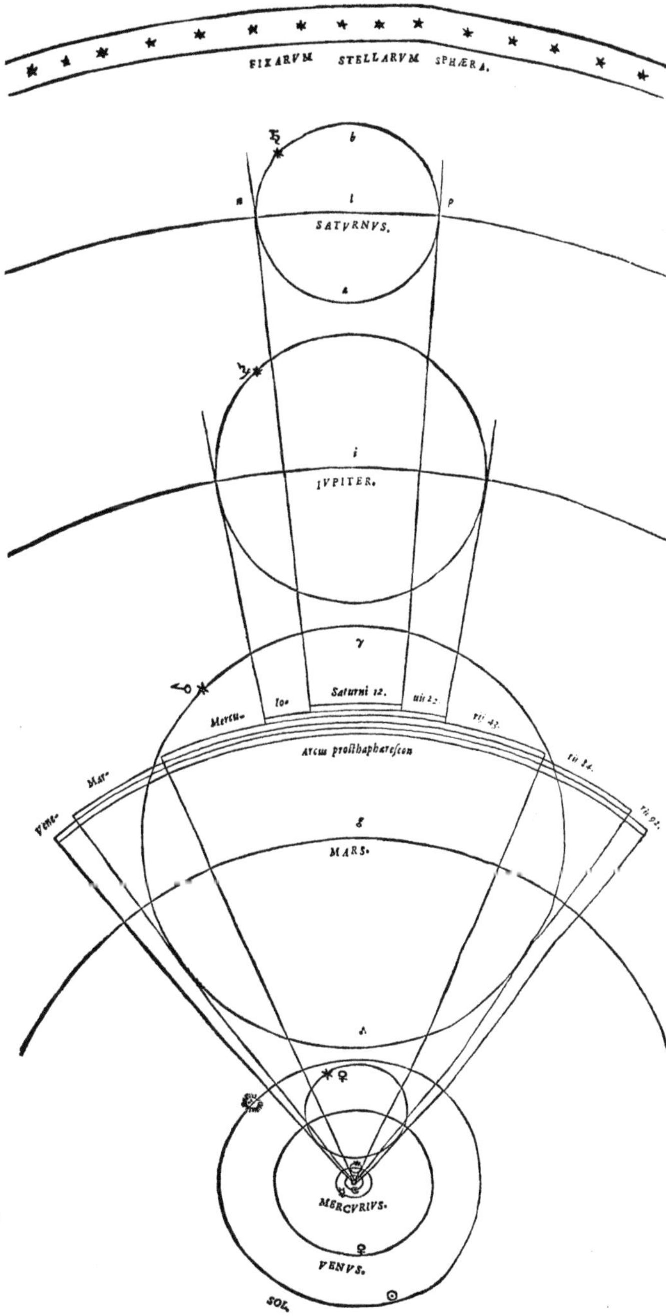

Figure 8. The Ptolemaic universe (drawn to scale) according to Johannes Kepler, Mysterium cosmographicum *(Tübingen, 1596). Image courtesy of the History of Science Collections, University of Oklahoma Libraries.*

do not. Unlike all of the other planets, Mercury and Venus do not travel away from the sun but always remain in its proximity. Mercury's greatest elongations vary between 18° and 28° (usually given as 23°), and Venus' between 45° and 47° (usually given as 45°). On the other hand, the angular distance of the superior planets Mars, Jupiter and Saturn can reach no less than 180°. When at apogee, they are in opposition with the sun, and when at perigee, they are in conjunction to it. But Venus and Mercury have a peculiar relationship to the sun because of the limitations in angular distance with respect to the sun for they can only be in conjunction with the sun and never in opposition to it. Ptolemy thus distinguishes Venus and Mercury from other planets and places them below the sun, closer to the observer on the earth.

Seen from a slightly different perspective, the perspective that Copernicus will adopt, a major problem of Ptolemaic astronomy stemmed from its inability to completely satisfy the ancient requirement to establish a correlation between planetary periods and distances. Put another way, the Ptolemaic system failed to demonstrate that the distance of each planet from the earth (and hence the entire planetary order) depends on its period. Saturn completes its cycle through the zodiac in thirty years, Jupiter in twelve years, and Mars in two, whereas Mercury, Venus and the sun each return to their starting points in one year. Ptolemy arranged the superior planets as the ancient astronomers had: according to period, with the planets with longer periods lying further from the centre of their motion. But he failed to apply the same principle to the sun, Venus and Mercury, thus rendering the arrangement of the inferior planets arbitrary and highly problematic in Copernicus' view.[83]

All of the above clearly reveal Ptolemy's uncertainty about the order of planetary spheres and unsystematic nature of the Ptolemaic universe. So far there has been no objective criterion for determination of the order of the universe, which is the reason that the "parts" of Ptolemy's universes, as Copernicus argues, do not have *certam symmetriam*; they do not "belong [or respond] to one another" when put together.

Copernicus' description of the critical state of astronomy in his *Preface* has an epistemological function as well. To his view, the advocates of the Ptolemaic

---

[83] See also, for example, T. Kuhn, *The Copernican Revolution*, pp. 174–7; F. Hallyn, *The Poetic Structure of the World.*, p. 83 et seq. Ptolemaic astronomy suffered from yet other "monstrous" technical problems. Copernicus mentions, at least in passing, in the *Commentariolus* and *De revolutionibus* at least two or three of them. See infra, p. 272.

astronomical tradition have inadvertently drawn a monstrous picture of the universe, because the fundamental building blocks on which they rely are in themselves inadequate. Their inability to derive a harmonious structure of the world is the consequence of inadequate principles. The postulated *certa symmetria*, or harmony of the parts of the universe, is a result of lack of harmony among the principles of cognition.[84]

> Hence in the process of demonstration or *methodos*, as it they call it, those who employed eccentrics are found either to have omitted something necessary or to have admitted something extraneous and completely impertinent. This would not have happened to them, had they followed certain principles. For if the hypotheses assumed by them were not false, everything which follows from them would be verified beyond any doubt.[85]

Had the Ptolemaists devised a theory that had neither omitted anything necessary nor admitted anything extraneous, they would have obtained a real, determined structure of the universe, whose constitutive parts would have been in fixed proportion. As can be inferred from the *Preface*, and in the parts of *De revolutionibus* where he dismisses the equant, Copernicus regards non-uniform circular motion as "something extraneous and completely impertinent."[86] Likewise,

84 See also J.-J. Szczeciniarz, *Copernic et la révolution copernicienne*, p. 49.

85 *Preface* (Rosen, p. 4; Zekl, p. 72).

86 See also *De revolutionibus* IV, 2 (Rosen, pp. 175–6; Gansiniec, pp. 175–6), where he speaks of "an extraneous cycle": "This combination of circles was assumed by our predecessors to be in agreement with the lunar phenomena. But if we analyze the situation more carefully, we shall find this hypothesis neither suitable enough nor adequate (*non aptam satis nec sufficientem*), as we can prove by reason and by senses (*ratione et sensu*). [...] Therefore, the epicycle's motion on the eccentric described by it is nonuniform. But if this is so, what shall we say about the axiom that the heavenly bodies' motion is uniform and only apparently seems nonuniform, if the epicycle's apparently uniform motion is really nonuniform and its occurrence absolutely contradicts an established principle and assumption? But suppose that you say that the epicycle moves uniformly with respect to the earth's centre, and that this is enough to safeguard uniformity. Then what sort of uniformity will that be on *an extraneous circle* on which the epicycle's motion does not occur, whereas it does occur on the epicycle's own eccentric?" Emphasis added. See also *De revolutionibus* V, 2 (Rosen, p. 240; Gansiniec, p. 246), where he speaks of "an extraneous centre": "The motion on the epicycle, however, clearly should be uniform with respect to E, the centre of its deferent, and the planet's revolution should be uniform with respect to line LME. Here too, then, as they admit, a circular motion can be uniform with respect to *an extraneous centre* not its own, a concept of which Scipio in Cicero would hardly have dreamed.

he regards a variety of natural philosophical principles – the motion of the earth, the stationary sun in the middle of the universe, the stationary sphere of the fixed stars, and the immense distance of the fixed stars – as having been "necessary" and therefore erroneously omitted. The monstrosity of the Ptolemaic universe is thus the consequence of the fact that Ptolemaic astronomers have admitted as part of their principles (assumptions, hypotheses, axioms) something extraneous to the subject, i.e., the equant, which contravenes the principle of the uniformity of motion, and omitted the motion of the earth and other relevant principles of natural-philosophy.

Copernicus' account of the state of astronomy is crushing: astronomers do not agree among themselves in their investigations; they are unable to establish the exact duration of the tropical year, and they do "not use the same principles, assumptions, and explanations of the apparent revolutions and motions." Not only do astronomical theories contradict each other in their basic principles (concentric versus eccentric astronomy), but they are hardly adequate in and of themselves. Whereas the proponents of concentric astronomy fail to provide appropriate calculations for the positions of celestial bodies, those who employ eccentrics and epicycles are largely able to honor this requirement – but only by violating the principles of uniform circular motion. Moreover, they are also unable to deduce the principal consideration, that is, the actual form of the world and the certain *symmetria* of its parts. The major starting point of Copernicus' astronomical reform is thus, beside the equant problem, the *uncertainty* of the existing explanations for the organization of the planetary spheres of the *machina mundi*.

But why should he be so concerned about it? As we shall see, most of the other astronomers were able to do their work without knowing the actual arrangement of planetary spheres. Why is Copernicus bothered by that? Copernicus himself provides us with first indication of his broader, metaphysical concerns. In his view, neither uncertainty nor monstrosity belongs naturally to the universe. The universe was created by the best Artisan of all, God, and has therefore a certain structure and is in itself absolute and perfect. Nor is the uncertainty and

---

And now in the case of Mercury the same thing is permitted, and even more. But (in my opinion) I have already adequately refuted this idea in connection with the moon. These and similar situations gave me the occasion to consider the motion of the earth and other ways of preserving uniform motion and the principles of the science, as well as of making the computation of the apparent nonuniform motion more enduring." Emphasis added.

monstrosity of the universe the result of our inability to comprehend it: because it was made "for our sake," we must be able to learn about it. If the absolute and perfect universe was made "for our sake," we are capable to discover its harmonious structure.

The universe is in principle comprehensible by the human mind and this is the reason that Copernicus undertook the task of "rereading the works of all the philosophers," from which he could "learn whether anyone had ever proposed other motions of the universe's spheres than those expounded by those who teach mathematics in the schools."[87] Given that the prevailing theories were contradictory, unsuitable, and unable to account for the principal consideration, that is, the "structure of the universe," it seems reasonable and logical to investigate the theses which these prevailing theories refute. In other words: since the theory which Copernicus criticizes for disregarding the uniformity of motion and producing a distorted image of the universe rests on the assumption that the earth remains at rest, it is only logical that he begins his investigation among the theories which suggest the contrary.

> And in fact first I found in Cicero that Hicetas supposed the earth to move. Later I also discovered in Plutarch that certain others were of this opinion. I have decided to set his words down here, so that they may be available to everybody: 'Some think that the earth remains at rest. But Philolaus the Pythagorean believes that, like the sun and moon, it revolves around the fire in an oblique circle. Heraclides of Pontus, and Ecphantus the Pythagorean make the earth move, not in a progressive motion, but like a wheel in a rotation from west to east about its own centre.' // Therefore, having obtained the opportunity from these sources, I too began to consider the mobility of the earth. And even though the idea seemed absurd, nevertheless I knew that others before me had been granted the freedom to imagine any circles whatever for the purpose of demonstrating the heavenly phenomena. Hence I thought that I too would be readily permitted to ascertain whether more firm demonstrations than those of my predecessors could be found for the revolution of the celestial orbs on the assumption of some motion of the earth.[88]

87 *Preface* (Rosen, p. 4; Zekl, p. 72).

88 Ibid. (Rosen, p. 5; Zekl, pp. 72–4). Copernicus reference is Cicero, *Quaestiones academicae* II, 39, 123, and second pseudo-Plutarch, *De placitis philosophorum* III, 13, 896a. See infra, pp. 364–5, and n. 362–4.

The concept of the earth's motion is the vital foundation which allows Copernicus not only to determine the planetary positions in the past and future (which was already possible in the traditional geostatic and geocentric Ptolemaic astronomy) but also to build a harmonious and non-arbitrary construction of the entire universe. While it is possible to "save the celestial phenomena" without the concept of the earth's motion, that is, without linking the motions of the planets to the movement of the earth, it is not possible to account for the essential "order and size of all the planets and orbs." Therefore, the motions of the earth allow not only the construction of an arbitrary order of the universe but an accurately firm, necessary structure of the universe. In geokinetic and heliocentric astronomy, "heaven itself is so linked together that in no portion of it can anything be shifted without disrupting the remaining parts and the universe as a whole."[89]

89 Ibid. (Rosen, p. 5; Zekl, p. 74).

CHAPTER FOUR

The Consensus of Many Centuries: The Motion of the earth is Inconceivable

Copernicus' solution to the problematic state of astronomy is the concept of the earth's motion. Nonetheless, it is both a cure and a poison at the same time. The motion of the earth, even though the "consensus of many centuries" dismisses the concept as utterly ridiculous and inconceivable, can provide more certain explanations for the revolution of the celestial spheres than those of Copernicus' predecessors who were unable to provide certainty in establishing the arrangement of the universe. In fact, it departs from the general norm so much that the first reaction to Copernicus' work, as he himself announces in the opening sentence of the dedicatory *Preface*, will certainly be that of repudiation:

> I can readily imagine, Holy Father, that as soon as some people hear that in this volume, which I have written about the revolutions of the spheres of the universe (*de Revolutionibus spharearum mundi*), I ascribe certain motions to the terrestrial globe, they will shout that I must be immediately repudiated together with this belief.[90]

The concept of the earth's motion is crucial for Copernicus' project to restore

[90] *Preface* (Rosen, p. 3; Zekl, p. 66). The wording "de Reuolutionibus sphaerarum mundi" clearly confirms that the title *De revolutionibus* orbium caelestium (emphasis added) was Copernicus' own – and that he believed in the reality of celestial orbs.

astronomy, yet also problematic. The introduction of this concept in scholarly discourse is far from self-evident, despite the remarkable scholarly results it brings forth. Just the opposite: at the same moment that Copernicus presents it as the basis of the entire project of restoring astronomy, it is also regarded as utterly absurd and so inconceivable as to provoke sharp, even extreme reactions. Why?

According to Copernicus, the thesis of the earth's motion contradicts the "consensus of many centuries," which is not confined to any one particular domain or authority but is upheld – as we shall see in a while – by the entire spectrum of the existing articulations of knowledge. The earth's motion is refuted by the established and approved Aristotelian philosophy of nature, theology, and *sensus communis*. Copernicus is perfectly aware of this consensus and it is most likely one of the principal reasons for his hesitation in deciding whether or not to publish the volume. As his disciple Rheticus reports in his "Encomium Prussiae," an addendum to the *Narratio prima*, Copernicus expressed his concern to his friend Bishop Tiedemann Giese (1480–1550), and the same concern was also clearly voiced in Copernicus' correspondence with Andreas Osiander, and in Osiander's letter to Rheticus.

Giese and Copernicus on Copernicus' hypotheses. In his highly illuminating discussion with Giese, Copernicus' basic concern is that his *hypotheses*, that is, primarily, the motion of the earth and sun's position in the centre of the universe, will overturn the commonly accepted opinion on the long disputed order of the spheres, that is, the order of the planets, and that they are in conflict with our senses:

> [...] he [Copernicus] had long been aware that in their own right the observations in a certain way required hypotheses which would overturn the ideas concerning the order of the motions and orbs that had hitherto been discussed and promulgated and that were commonly accepted and believed to be true; moreover, the required hypotheses would contradict our senses.[91]

Giese urges Copernicus, whose accomplishments and insights he has known for many years, to provide an ecclesiastical calendar, as well as a correct theory and explanation of the motions of the celestial bodies.[92] But Copernicus proposes instead to compile a set of astronomical tables with new rules following

---

[91] *Narratio prima,* "Encomium Prussiae" (Rosen, p. 192; Hugonnard-Roche and Verdet, pp. 84–5).

[92] See ibid. 107–13 (Rosen, p. 192; Hugonnard-Roche and Verdet, p. 84).

the example of the "Alfonsine astronomers" rather than deliver demonstrations imitating Ptolemy. In this way, he hopes to sidestep any philosophical controversy and simultaneously provide ordinary astronomers (*vulgares mathematici*) with correct calculus of celestial motions. On the basis of numerical data, true scholars will, in Copernicus' opinion, easily arrive at "the principles and sources from which everything is deduced."[93] Scholars, learned astronomers who so far had to elicit the true hypothesis of the motions of the stellar spheres from the "Alfonsine doctrine," will henceforth be able to do the same with Copernicus' tables. As for ordinary astronomers, who are not so much interested in *scientia*, they will at least have the tables at their disposal. Finally, this will also be in line with the Pythagorean principle that philosophy must be pursued in such a manner that its innermost secrets are reserved for learned men, those trained in mathematical philosophy.[94]

No less interesting is Giese's answer to Copernicus' elusiveness. Giese argues that Copernicus' gift to the world will remain incomplete unless he reveals "the system or theory and the foundations and demonstrations upon which he relied"[95] for his astronomical tables, just as Ptolemy did before him. The omission of such things, Giese argues, has produced many problems and errors in the Alfonsine tables, since their assertions had to be approved on the principle "The Master said so," which is utterly unacceptable in mathematics. Moreover, given that Copernicus' principles and hypotheses are diametrically opposed to those of the ancients, there will be scarcely any scholar to examine the principles of the tables in the future and publish them. Hence, Copernicus himself is to set forth the reasons, the proofs and explanations for the celestial motions, rather than to merely determine them.

Giese also has an answer to Copernicus' concerns regarding the opposition from learned philosophers and ordinary persons who are convinced by the evidence of their senses that the earth stands motionless in the middle of the universe. According to Giese, more erudite and prudent philosophers will study Aristotle's arguments in *De caelo* and carefully examine the fact that having established the immobility of the earth, Aristotle resorted to the following argument: "What the mathematicians say in astronomy also testifies to this, since the apparent facts – that is, the changing of the configurations in terms of which the arrangements of the stars is determined – results from the supposition that the earth lies at the

93 Ibid. 126 (Rosen, p. 192; Hugonnard-Roche and Verdet, p. 85).

94 See ibid., 121–31 (Rosen, pp. 192–3; Hugonnard-Roche and Verdet, p. 85).

95 Ibid., 133–6 (Rosen, p. 193; Hugonnard-Roche and Verdet, p. 85).

centre."[96] To Giese's way of thinking, philosophers will realize that if Aristotle's conclusion cannot be linked to his previous discussion, they should waste no more time and effort, but rather "assume the true basis of astronomy." Then they should work out appropriate solutions to other problems:

> By returning to the principles with greater care and equal assiduity, we must determine whether it has been demonstrated that the centre of the earth is also the centre of the universe. If the earth were raised to the lunar sphere, would loose fragments of earth seek, not the centre of the earth's globe, but the centre of the universe, inasmuch as they fall at right angles to the surface of the earth's globe? Again, since we see that the magnet by its natural motion turns north, would the motion of the daily rotation of the circular motions attributed to the earth necessarily be violent motions? Further, can the three motions, away from the centre, toward the centre, and about the centre, be in fact separated? We must analyze other views which Aristotle used as fundamental propositions with which to refute the opinions of the *Timaeus* and the Pythagoreans. They will ponder the foregoing questions and others of the same kind if they desire to look to the principal end of astronomy and to the power and industry of God and nature.[97]

---

[96] *De caelo* II, 14, 297a2–6 (Leggatt, p. 165).

[97] *Narratio prima*, "Encomium Prussiae" (Rosen, p. 194; Hugonnard-Roche and Verdet, pp. 85–6). Giese's magnetic analogy found supporters in such authors as William Gilbert (*De magnete*, 1600), Johannes Kepler (*Astronomia nova*, 1609, and *Epitome astronomiae Copernicanae*, 1617–1621), and Galileo Galilei (*Dialogo sopra i due massimi sistemi del mondo, tolemaico e copernicano*, 1632). On those who used magnetic philosophy against Copernicus, see M. R. Baldwin, "Magnetism and the Anti-Copernican Polemic." For reference to Plato, see also *De caelo* II, 14, 296a24–26 (Leggatt, p. 161): "Let us say first whether the earth has a movement or rests, since as we said, some people [i. e. Pythagoreans] make it one of the stars, whereas others [i. e. Plato, *Timaeus* 40b–c] put it at the centre and say that it winds, i. e. moves about the central axis." See also ibid. II, 13, 293b30–32 (Leggatt, p. 153): "Some also maintain that the earth lies at the centre and winds, i. e. moves, about the axis extended through the whole universe, as is written in the *Timaeus*." See *Timaeus* 40b–c (Cornford, p. 120): "And earth he designed to be at once our nurse, as she winds round the axis that stretches right through, the guardian and maker of night and day, first and most venerable of all the gods that are within the heaven [...]." This passage incited numerous interpreters to present Plato as a supporter of the rotation of the earth. As Cornford explains, it comes down to the interpretation of the word *hyllomenen* ('winds'). For the ancient authors who interpreted Plato as supporter of the movement of the earth, see D. Knox, "Ficino, Copernicus and Bruno on the Motion of the Earth," p. 335, and n. 10.

There is a problem with Giese's optimistic belief that philosophers will be swayed by astronomic theory, however; this requires philosophers to readily let go of their principles instead of insisting on them. Should they fail to do so, Copernicus will face a situation not unlike that faced by Ptolemy, the "king" of astronomical science. Averroes, otherwise a remarkable philosopher, established that Ptolemy's epicycles and eccentrics could not exist in the nature and stated: "Ptolemaic astronomy is nothing, so far as existence is concerned; but it is convenient for computing the nonexistent."[98] As for the untutored, Giese's advice to Copernicus is simple: the shouting of all people who do not know theory, music, philosophy, and geometry is simply to be ignored.

Osiander and Copernicus on Copernicus' hypotheses. Apart from other compelling facts, such as reference to Plato's *Timaeus*, the discussion between Giese and Copernicus also clearly reveals two categories of potential opponents to the earth's motion: learned philosophers and ordinary, uneducated persons. But there is another category as well: theologians. As evident from the extant fragments of Osiander's correspondence with Copernicus and Rheticus, both written on April 20, 1541, Copernicus expresses in the now lost letters to Osiander his fear of the opposition not only from the Aristotelian philosophers, peripatetics, but also from theologians. In other words, he fears not only the reaction from the established learned community committed to the study of natural truth but also the reaction from the established learned community committed to the study of revealed truth.

In response to Copernicus' fear, Osiander proposes to him to declare his thesis that "the earth moves whereas the sun is at rest in the centre of the universe" as one of many possible astronomical hypotheses, an instrument to determine the exact positions of the celestial bodies in the past and predict them in the future. In this way he will placate peripatetics and theologians, and perhaps even win them over. In a letter to Rheticus, Osiander writes:

> The peripatetics and theologians will be easily placated if they hear that there can be diverse hypotheses about the same apparent motion [of the celestial bodies] and that they are not advanced as being certainly so, but rather as governing the calculation of apparent and composite motion as expediently as possible; that it could happen that someone else should think

[98] *Narratio prima*, "Encomium Prussiae," 173–4 (Rosen, p. 195; Hugonnard-Roche and Verdet, p. 86).

> up appropriate constructions and another more appropriate ones, both giving rise to the same appearance of motion; and that anyone is free and, moreover, is to be congratulated if he thinks up more expedient ones. Thus, called away from severity in condemnation and summoned to the pleasures of inquiry, they will at first be more reasonable and then, seeking in vain, will go over to the author's opinion.[99]

Osiander maintains that there are different hypotheses about the same apparent celestial motion, that these hypotheses are not necessarily true (or do not suit the actual state of matters), that they are "appropriate" so long as they yield accurate calculations of celestial positions, and that it is possible to invent other, perhaps even better hypotheses. In a letter to Copernicus, however, he explains in detail what he means by arguing that the same motion of a celestial body might be accounted for by two different but equally valid hypotheses, and again emphasizes that this is the right way to pacify the peripatetics and the theologians:

> I have always been of the opinion that hypotheses are not articles of faith, but bases for calculation, so that even if they are false it does not matter provided they yield the phenomena of the motions [of the celestial bodies] exactly. For who could make us surer that the unequal motion of the sun is due to an epicycle than that it is due to an eccentric, if we follow Ptolemy's hypotheses, since it could happen in either way. So it would seem to be a good idea for you to say something on this matter in the preface. For thus you would pacify the peripatetics and the theologians whom you fear to be about to raise objections.[100]

In the *Preface* to *De revolutionibus*, all three categories of likely opponents to the earth's motion appear together. The thesis of the earth's motion contravenes the established and accepted *scientia* and, as Copernicus explains a few paragraphs further, aims against "traditional opinion of mathematicians," that is, astronomers, who have, of course, espoused peripatetic cosmology with the motionless earth at the centre of the universe. It challenges certain passages of the Holy Scripture as the divine word of truth, because there might be babblers (*matailogoi*), as Copernicus complains, who will, "badly distorting some passage of the Scripture to their purpose, […] dare to find fault with my undertaking and

---

99 Quoted from, J. Kepler, *Apologia pro Tychone contra Ursum*, in N. Jardine, *The Birth of History and Philosophy of Science*, p. 153 (Latin text, p. 98).

100 Ibid., p. 152 (Latin text, p. 97).

censure it."[101] It is also *contra communem sensum,* against common sensory experience or against the widely held opinion, common sense, in modern connotation as *sensus communis* could also be interpreted.[102] All possible opponents and scoffers of the earth's motion may thus be divided into three categories: theologians, peripatetic philosophers, whose physics is also accepted by mathematicians, and the proponents of sensory experience or common sense. These categories coincide with three types of arguments against the motion of the earth: theological, philosophical and experiential.

Theological reasons against the motion of the earth. Copernicus nowhere mentions any specific theological objection to the movement of the earth; in Chapter 7 of Book I of *De revolutionibus*, he confines himself to presenting philosophical arguments and those based on experience. Nevertheless, as is clear from the correspondence with Osiander and his *Preface*, he is well aware of potential theological rebuttals to the earth's motion.

As a matter of fact, already sometime before 1536 Copernicus' fears of theological objections were answered by his friend Giese, who wrote a now lost treatise *Hyperaspisticon* (*Supershield*) in which he claimed that the Holy Scripture was compatible with the new astronomy. His fears might have grown bigger, if Rheticus had told him about Luther's critique of "a new astrologer" and a "fool" who went against the Holy Scripture, which tells us that "Joshua commanded the sun to stand still, and not the earth," and "tried to prove that the earth revolves, not the heavens or the firmament, the sun and the moon."[103] But there are a number of

---

[101] *Preface* (Rosen, p. 5; Zekl, p. 76).

[102] When Copernicus used the phrase *sensus communis*, he was, of course, referring to Aristotelian faculty psychology and not today's "common sense." This distinction should be borne in mind. However, the Aristotelian *sensus communis* is, in a certain measure, also a "common sense." For a typical account of Aristotle's common sense, see Stobaeus, *Eclogarum libri duo*: "Aristotle did not indeed postulate a sixth sense but rather a common sense, the judge of composite forms to which the individual senses send their images. It comprehends motion and change of shape; is midway between partaking and not partaking of reason; participates in memory and the mind; and is attributed to beasts as well insofar as they have a spark of intelligence. Objects of the common sense are shape, evident to sight and touch, also distance between things, evident to sight and hearing, and, moreover, the motion, size and number of all things." English translation is taken from N. Jardine, *The Birth of History and Philosophy of Science*, p. 155, n. 84.

[103] Quoted from P. Barker, "The Lutheran Contribution to the Astronomical Revolution: Science and Religion in the Sixteenth Century," p. 34, n. 16.

other passages in the Holy Scripture besides Joshua's that speak of the moving celestial bodies while the earth remains at rest, and which provide – without using excessive "distortion" – a splendid starting point for the theological rejection of the earth's motion.[104] Rheticus, obviously very much concerned about this matter himself, wrote sometime shortly after 1540 and before September 1541, most likely with the approval of Copernicus and one can speculate that even on his or even Giese's instigation, a short treatise of compatibility of the Holy Scripture and movement of the earth, *On Holy Scripture and the Motion of the Earth*,[105] in which he, as a convinced Copernican *and* an Orthodox Christian – albeit a Protestant one – showed "very clearly [...] that the motion of the earth does not contradict the Holy Scriptures."[106] His main strategy was the Augustinian doctrine: the Holy Scripture was to teach what was necessary for salvation, and it should accommodate in scientific matters to what had been discovered or proven as true.

Exactly the opposite was the case with the first critic of Copernicus, Dominican Giovanni Maria Tolosani (ca. 1471–1549), who in 1547 or 1548, authored (but never published) *Opusculum quartum: De coelo supremo immobili et terra infima stabili, ceterisque coelis et elementis intermeddis mobilibus*. In the first chapter of his *opusculum* Tolosani mainly refers to Genesis, Isaiah and Proverbs to show that the "supreme heaven" – the empireum – is immobile (and not the sphere of the fixed stars); to Psalms 103, 5 and 92, 1, that the earth is motionless in the centre of the universe; and finally to Ecclesiastes, I, 5–6, in support of the movement of the sun. Tolosani considers Copernicus to be "an expert in mathematics and astronomy" but "very deficient in physics and dialectics." Copernicus "seems

---

[104] It was already in the fourteenth century that Nicole Oresme, discussing the earth's possible rotation in his *Le livre du ciel et du monde* II, 25 (Menut and Denomy, pp. 526–7), enumerated three passages from the Holy Scripture that seemed to be speaking against it: Ecclesiastes 1, 5–6; Psalms 93 (92)); and Joshua 10, 12–13.

[105] The treatise, first published only in 1651 as *Epistola cujusdam Anonymi de terrae motu*, was discovered in the second half of the twentieth century by R. Hooykaas. See R. Hooykaas, *G. J. Rheticus' Treatise On Holy Scripture and the Motion of the Earth*, and "Rheticus's Lost Treatise on Holy Scripture and the Motion of the Earth." For a critical edition, see also *Receptio Copernicana*, pp. 37–73. See also M.-P. Lerner, "The Heliocentric 'Heresy'," pp. 12–3. For a more systematic treatment on this subject, see R. S. Westman, "The Copernicans and the Churches," p. 90, who lists the following four groups of references from the Holy Scripture that are relevant for the theological polemic on Copernicanism: (1) stability of the earth; (2) motion of the sun with respect to the terrestrial horizon; (3) the sun at rest; (4) motion of the earth.

[106] So Giese in his letter to Rheticus, July 26, 1543. Quoted from, M.-P. Lerner, "The Heliocentric 'Heresy'," p. 27, n. 5.

to be unfamiliar with Holy Scripture since he contradicts some of its principles, not without the risk to himself and to the readers of his book of straying from the faith."[107]

Copernicus obviously had reasonable grounds to devote a relatively substantial part of his *Preface* – as we shall see – to defending the view of Rheticus: that the interpretation of the Holy Scripture should be adapted to astronomical discoveries rather than vice versa.

Philosophical and experiential reasons against the motion of the earth. Nonetheless, Rheticus' account in "Encomium Prussiae" also clearly reveals that Copernicus is much less concerned about theologians than he is about peripatetic philosophy and the fact that the motion of the earth contradicts the evidence of the senses.

Several Aristotelian reasons against the motion of the earth were already implicitly evoked by Bishop Giese. Copernicus makes – a relatively passing – reference to some of them later in Chapter 7 of Book I of *De revolutionibus*. He mentions the so-called reason of "heaviness and lightness" and the reason of "nature of motion," and adds to them a few experiential reasons (observational tests, appearances or experiences),[108] which he ascribes to Ptolemy. His formulation of observational tests is – not without reason, as we shall see in due time – sufficiently specific to call for a thorough scrutiny. His formulation of the two above mentioned Aristotelian reasons is more accurate, but still relatively simplistic in

---

107 See *Opusculum quartum*, cap. 2, 340r (Lerner, pp. 701–3). Tolosani's treatise confirms that Copernicus' fear of the "opposition from theologians" was well founded. It shows that the weight of theological implications of the earth's motion was far from negligible, even in the time before Galileo's "first trial" (1615–1616). The text of *Opusculum quartum* was discovered in the second half of the twentieth century by Eugenio Garin. He published it in 1973 in *Studia Copernicana* 6, and reprinted in *Rinascite e rivoluzioni*, pp. 283–94. M.-P. Lerner published it in 2003, together with the French translation, in "Aux origines de la polemique anticopernicienne (I): *L'Opusculum quartum* de Giovanmaria Tolosani." See also his "The Heliocentric 'Heresy'," pp. 14–7. To my knowledge, there is no English translation of this text. Tolosani had a general interest in problems related to the calendar and in astronomy. On Tolosani, see also E. Rosen, "Was Copernicus' *Revolutions* Approved by the Pope?," and especially M. A. Granada, "Giovanni Maria Tolosani e la prima reazione romana di fronte al *De revolutionibus*: la critica di Copernico nell' opusculo *De caelo et elementis*," especially pp. 21–7.

108 So called by F. J. Ragep, "Tūsī and Copernicus: The Earth's Motion in Context," p. 151; J. Buridan, *Quaestiones in Aristotelis De caelo* II, 22 (Patar, pp. 503–4); and N. Oresme, *Le livre du ciel et du monde* II, 25 (Menut and Denomy, pp. 520–1), respectively.

comparison with the Aristotle's account in *De caelo* II, 14 and especially in comparison with the additional arguments used in the Middle Ages and Renaissance. For the sake of brevity, we shall only present Aristotle's arguments as formulated by Copernicus' contemporary and first critic Giovanni Battista Tolosani, and add some others, discussed by John Buridan (1292–1363) and Nicole Oresme (ca. 1320–1382) in their respective treatises, *Questiones in Aristotelis De caelo* and *Le livre du ciel et du monde*.

(i) Philosophical reasons. Let us first take a look at the broader theoretical context of Aristotle's rejection of the earth's motion; his doctrine of natural places, natural motion of the elements, and heaviness and lightness in *De caelo*.[109]

Aristotle argues that each of the simple bodies or elements (earth, water, air, and fire) composing the sublunary realm has its proper place, which scholastic scholars called its natural place (see *Figure 9*). There are four concentric elemental spheres beneath the moon, one for each sublunary element. The natural place of element earth is, so says Aristotle, at the centre of the universe; next is the natural place of the element water occupied by the sphere of water; then follows the natural place of the element air occupied by the sphere of air, and finally the natural place of the element fire occupied by the sphere of fire, which is at the periphery of the sublunary realm, beneath the lunar sphere.

Each element has a natural tendency to move to its natural place, where it comes to rest. The natural movement of fire towards the extremity of the sublunary realm means that fire is an absolutely light body, whereas the tendency of earth to naturally fall towards the centre of the universe means that earth is an absolutely heavy body. Hence, heavy bodies tend to move towards the centre or fall downwards, whereas light bodies tend to rise or move upwards.[110] The remaining two elements, however, have only relative heaviness. When situated below its natural place somewhere in earth, water tends to rise, and when above its natural place, in air or fire, it tends to descend towards its natural place. Air naturally falls when in the region of fire and rises when in earth or water.[111]

---

109 What follows is a very simplified account focusing on ideas relevant to Copernicus' discussion in *De revolutionibus*. For more, see H. S. Lang, *Aristotle's 'Physics' and its Medieval Varieties*, pp. 63–84; A. Falcon, *Corpi e movimenti*, pp. 84–118; B. Morison, *On Location: Aristotle's Concept of Place*, pp. 25–53, 163–4; D. Knox, "Copernicus' Doctrine of Gravity and the Natural Circular Motion of the Elements," pp. 157–64.

110 See *De caelo* IV, 1, 308a29–31.

111 See ibid. IV, 4, 311a16–32.

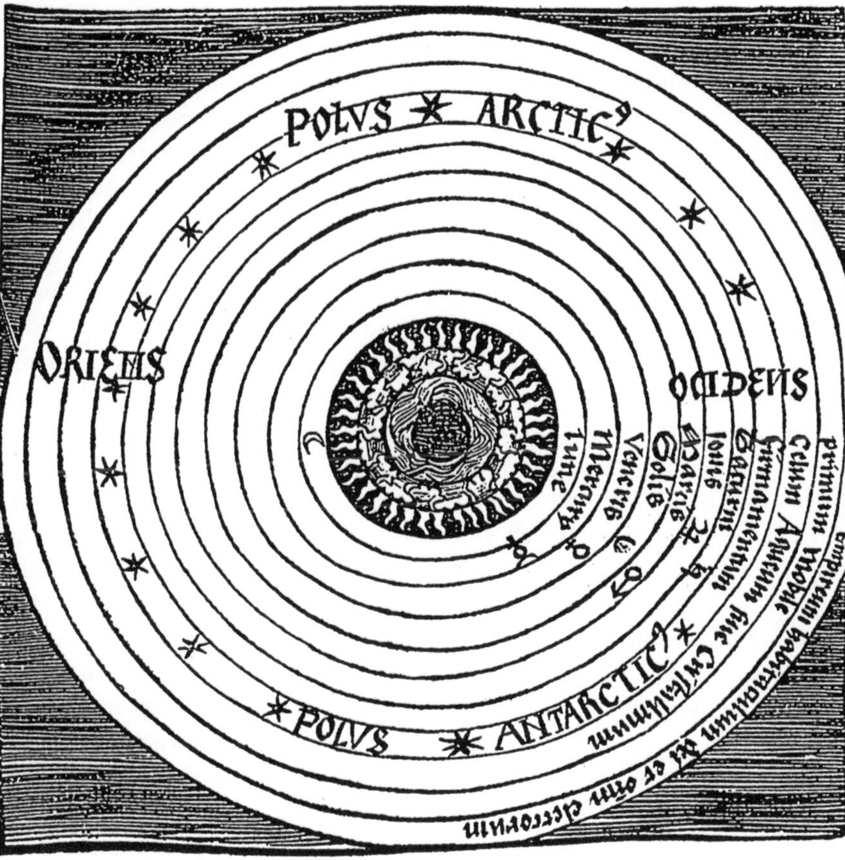

Figure 9.
Aristotelian cosmos from Gregor Reisch, Margarita philosophica nona *(Strasbourg, 1508). This figure illustrates Aristotle's general and abstract scheme. The spheres of earth and are depicted as forming two separated spheres, but as is evident from the text, Reisch believes that earth and water form a single sphere.*

Since the centre of the universe is the natural place of earth as a whole and of its loose fragments, terrestrial bodies tend to move towards the centre of the universe, where each comes to rest once it has reached its natural place.[112] If the

[112] See ibid. II, 13, 295b21–297a1.

earth's globe were placed where the moon is, fragments of earth would try to return to their natural place, that is, to the centre of the universe, not to the earth's globe.[113] Elements move by virtue of an inanimate, but intrinsic principle. When out of their natural place they are in a state of potentiality and acquire their actuality, their proper nature (form), only when they return to their proper places. The "like moves to like."[114] The movement to their proper, natural place, to their sphere, is therefore very like any other process of change, for example, that of healing.[115]

It should be added that, according to Aristotle, each element, each simple body, can have only one natural motion and that all motions are divided into rectilinear and circular or a composite of the two, since only these two motions are simple. Rectilinear motion is further divided into upward-downward motion, with upward motion denoting movement away from the centre of the earth (as the centre of the universe) and downward motion denoting movement towards the centre of the earth. Circular motion, however, takes place around the centre of the earth as the centre of the universe. From this argument it follows that every simple (i.e. non-composite) local motion takes place away from the centre, towards the centre or around the centre.

All simple bodies, all elements contain a principle of natural motion and must move with one simple movement. The first two elements in the sublunary realm, earth and water, have a natural tendency to move rectilinearly downwards, each to its natural place. Air and fire have a natural tendency to move rectilinearly upwards to their respective natural places, away from the centre. The only element occupying the superlunary realm is aether, which naturally moves in a simple circular motion. Given that in the sublunary sphere many bodies are not simple but composed of more elements, the motion of sublunary bodies can be either simple or composite. Motions of simple bodies are simple and motions of composed bodies are composite. Hence, the natural motions of earth, water, fire and air can only be rectilinear, or straight, whereas circular motion is the natural motion of the fifth element, aether, which composes the bodies of the superlunary realm (planet-carrying aethereal spheres and planets).[116]

---

113 See ibid. IV, 3, 310b2–5.

114 See ibid. 310a33–10b1. For an explanation of what exactly "like moves to like" means according to Aristotle, read the passage of the note above and the sequence, ibid., 310b8–15.

115 See ibid., 310b16–19. See also ibid., 310b24–31, and 311a8–9.

116 See ibid. I, 2, 268b14–269a9.

Based mostly on these general considerations and some additional particular assumptions, Aristotle argues in Chapter 14 of Book II of *De caelo* that the earth stands still in the middle of the universe. Here are his arguments from Tolosani's *Opusculum quartum*, who basically copied those found in Thomas Aquinas' (1225–1274) *Commentary on De caelo*, who was relatively faithful to Aristotle's text.[117]

Aristotle's first reason is that if the motion of the earth were circular (at or around the centre of the universe), it would be violent and contrary to its nature.[118] It is obvious that circular motion is against the nature of the earth, as it does not apply to its various parts, which must have the same natural motion as the whole. However, the evidence shows otherwise. All parts of the earth move in rectilinear motion to the centre of the world. If the earth's circular motion were violent and contrary to its nature, it would not be eternal, because nothing violent is eternal. However, if the earth's motion were circular, it should be eternal, because for Aristotle the order of the universe is eternal. The order of the universe governs the mobility or immobility of its elements. Hence it follows that the earth does not move in a circular motion.

We shall omit Aristotle's second, astronomical reason due to its complexity,[119] but the third one relates to the question of earth fragments and the earth as a whole.[120] Assuming that the earth is fixed in the middle of the universe so that the centre of the earth is also the centre of the universe, the question becomes: Towards which centre do heavy bodies naturally move, the earth's centre because it is the centre of the earth or the earth's centre because it is the centre of the universe? According to Aristotle, heavy bodies must move towards the centre of the entire universe. The motion of heavy bodies is contrary to the motion of light bodies. These, particularly fire, move to the boundary of the heavens, that is, towards the lunar sphere, whereas heavy bodies, particularly earth, move to the centre of the world. However, since the centre of the earth happens to coincide with the centre of the universe, Aristotle proves his assumption to be correct: heavy bodies, which fall in natural motion to the earth, are carried neither in opposite directions nor in parallel lines which never meet, but in equal angles, perpendicular to the earth's surface or to the line parallel to it. Therefore, each

---

[117] According to Tolosani, Aquinas divided Aristotle's' arguments into physical and astronomical. See *Opusculum quartum*, cap. 3, 341^r (Lerner, p. 711): "[...] probat physicis rationibus, deinde astronomicis et mathematicis."

[118] See *De caelo* II, 14, 296a27–34.

[119] See ibid., 296a34–296b6.

[120] See ibid., 296b6–296b21.

heavy body, irrespective of where it is coming from, moves to the centre of the earth. In Aquinas' view, this is demonstrated by the fact that a pillar, situated anywhere on earth, and erected not perpendicular to but with a certain inclination to the ground, falls to the side with which it makes an acute angle with the ground. The third book of Euclid's *Elements*, however, demonstrates that a straight line drawn perpendicular to the tangent of the circle passes through the centre of the circle and beyond.[121] Therefore it is obvious that heavy bodies descending from various directions would, if unhindered, meet at the centre of the world, each following a straight line running perpendicular to the tangent at the point of contact. From this it follows that all heavy bodies move towards a single centre, that of the whole universe and of the earth.

On this basis Aristotle draws two conclusions. First: that the earth is at the centre of the universe. This conclusion rests on two premises: all heavy bodies move by themselves at the centre of the universe and, as has been demonstrated, all heavy bodies move towards the centre of the earth, which means that the earth is at the centre of the universe.[122] The second conclusion is that the earth is motionless, and it derives from the following premise. First, it is in the nature of everything to move to the place where it naturally comes to rest. Second, the earth moves to the centre of the universe. Therefore, being at the centre of the universe, the earth does not move in any way.

Finally,[123] says Aquinas – in Tolosani's paraphrase –, Aristotle explains the *cause* of the earth's immobility,[124] which is already evident from what has been said. The element earth moves to the centre from all directions, just as fire, so our senses tell us, naturally moves from the centre of the universe to its periphery. Hence, every part of earth, large or small, can only move away from the centre by constraint. A single body has a single natural motion, and a simple body a simple motion. A simple body cannot have two contrary motions. Any motion away from the centre is contrary to the motion which naturally leads towards it. Therefore, if no single part of earth can be moved away from the centre otherwise

---

121 See Euclid, *Elements* III, prop. 19 (Heath, vol. 2, p. 45): "If a straight line touch a circle, and from the point of contact a straight line be drawn at right angles to the tangent, the centre of the circle will be on the straight line so drawn."

122 Tolosani's text (Lerner, pp. 711–3) is completely faulty and makes no sense. I paraphrase Aquinas' text, *In De caelo* II, lectio 26, n. 7.

123 For the fourth reason, which in accordance with the classification specified above belongs to experiential arguments, see infra, pp. 93–4.

124 See *De caelo* II, 14, 296b–297a2.

than by constraint, it is evidently still less possible for the earth to move away from the centre as a whole. This argument may be subject to objection. Somebody could say that the earth as a whole does not move to the centre. But Aristotle's response is that the earth as a whole moves – or is assumed to move – to the same place as its parts. If one part of the earth naturally moves to the centre, it is also in the nature of the earth as a whole to do so. This means that the earth cannot move. Therefore, the only possible conclusion is that the earth is motionless at the centre of the world.

In the fourteenth century John Buridan and Nicole Oresme add some new arguments against the motion of the earth to the already existing stock in their very original discussions on whether it is possible that the earth rotates about its axis.[125] Buridan and Oresme recognize that the fundamental problem of an

[125] While Oresme, as we have seen (see supra, p. 84, n. 104), included some theological considerations into discussion, Buridan, *Quaestiones in Aristotelis De caelo*, did not present any theological arguments against the earth's motion but limited himself to stating "probable reasons" that stemmed from philosophy and "appearances." Albert of Saxony (ca. 1316–1390), who discussed the rotation of the earth some years later, was less original and more derivative; he more or less repeated Buridan's and Oresme's arguments. See E. Grant, *Planets, Orbs & Stars*, p. 639, n. 79. Albert is important, though, because his philosophy of nature was influential in Cracow, where Copernicus started his studies. In discussing Buridan and Oresme – here and infra, pp. 173–81. – I have no intention of making them precursors of Copernicus as so many scholars have done. See, for example, T. Kuhn, *The Copernican Revolution*, pp. 114–22; A. Crombie, *From Augustine to Galileo*, vol. 2, pp. 75–84; M. Clagett, *Science of Mechanics in the Middle Ages*, p. 614; E. Grant, "Late Medieval Thought: Copernicus, and the Scientific Revolution," pp. 205–12; *The Foundations of Modern Science in The Middle Ages*, pp. 112–7; and *Physical Science in the Middle Ages*, pp. 69–70; M. Kokowski, *Copernicus's Originality*, pp. 82–4, especially n. 3. See also A. De Pace, *Niccolò Copernico e la fondazione del cosmo eliocentrico*, p. 165, n. 485. I do not believe that Copernicus was philosophically or, to be more precise, epistemologically and metaphysically *influenced* by Buridan or Oresme. Nevertheless, I am convinced that he adopted some key elements of their discourse (most probably not directly from their writings, perhaps through Albert of Saxony or more likely through some other source) regarding the argumentation about the possible rotation of the earth and used them for his own purposes. There are many important differences between Copernicus on one hand and Buridan and Oresme on the other. Firstly, Buridan and Oresme were philosophers, whereas Copernicus was an astronomer who believed that it was an astronomer's duty to also answer the questions regarding the constitution of the universe; he was an astronomer-philosopher. Secondly, Copernicus' principal concern was, as we shall demonstrate, not the diurnal rotation of the earth but its revolution around the sun. Thirdly, despite compelling arguments in favor of its motion, Buridan and Oresme rejected the thesis: Buridan

axial rotation of the earth is that it contradicts the Aristotelian theory of natural and simple motion.[126] According to Aristotle a simple body such as earth can only have a single simple motion. The same applies to water, air, and fire. It is in the nature of the earth to move downwards, towards the centre of the cosmos. Therefore, circular motion is not its natural motion, because a single simple body cannot have more than one single simple motion. It would not be reasonable to consider the earth's circular motion as violent and contrary to its nature, because no violent motion can be perpetual.

But Buridan and Oresme further expand the arsenal of the proponents of the earth's immobility. Buridan, for instance, mentions the argument of priority of circular motion: circular motion, classified as primary motion in the ontological hierarchy of motions, would be more appropriate to celestial bodies as primary

---

for experiential reasons based on his own theory of impetus and Oresme for theological reasons. Fourthly, Buridan's and Oresme's treatment of the earth's motion had a completely different epistemological status than that of Copernicus. For Buridan and Oresme the arguments in favor of the earth's motion were regarded merely as an "intellectual game" (*par esbatement*); they discussed it *secundum imaginationem* rather than a real and likely possibility in the actual world. Buridan and Oresme did not consider the arguments in favor of the earth's motion as *demonstrationes,* but merely as *rationes probabiles*, probable reasons. Both – despite their differences (on that, see P. Souffrin, "Oresme, Buridan et le movement de rotation diurne de la terre ou des cieux") – abandoned the argumentation in its strict sense and reduced it to persuasion. The key methodological term used by both authors was *persuasio.* This comes from the fact that the so-called "nominalists" introduced extremely strict demonstration requirements. The only evidence recognized by them were logical truths, self-awareness and direct sensory experience – and even that was ultimately subject to doubt, because God in his *potentia absoluta* (as opposed to *potentia dei ordinata*) could have created in us the direct notion, evidence of actually existing entities. On this, see A. Funkenstein, "The Dialectical Preparation for Scientific Revolutions," p. 192 et seq. On differences between Copernicus and Oresme, see also A. De Pace, *Niccolò Copernico e la fondazione del cosmo eliocentrico*, pp. 184–5, and n. 552, and infra, p. 179, n. 129.

126 Unlike Buridan, Quaestiones *in Aristotelis De caelo* II, 22 (not in Clagett's translation; Patar, p. 504), Oresme, *Le livre du ciel et du monde* II, 25 (Menut and Denomy, p. 527) split Aristotle's first argument into two and presented as the first *raison* Aristotle's assumption that every simple body has a single simple motion. For – an incomplete – English translation of Buridan's *Questions on the Four Books on the Heavens and the World of Aristotle: Book II, Question* 22 by M. Clagett, see *The Science of Mechanics in the Middle Ages*, pp. 594–9. Clagett's translation and commentary is reproduced in E. Grant, ed., *A Source Book in Medieval Science*, pp. 500–3.

bodies than to the earth.[127] Oresme adds three more arguments, two of which may be considered as philosophical or physical.[128] If the heavens did not perform their diurnal motion around the earth, all astronomy and a major part of natural philosophy would, in his opinion, be false by constantly presuming movable heavens and a motionless earth. The argument against the motion of the earth was also made by the authority of Averroes. Based on the assertion that any local motion, any movement from one place to another, exists only in relation to a motionless body, Averroes derived a conclusion that the earth must be at rest in the centre of the heavens.[129] Moreover, the earth is too heavy to be moved circularly by an internal "motive force." If, however, it were moved by some external motive force, its motion would not be natural but violent and against its nature, and as we have seen, no violent motion can be eternal.[130]

(ii) Experiences or appearances against the motion of the earth. The remaining considerations against the earth's motion are based on "experience," as Oresme puts it, or "appearances," in Buridan's words. The starting point of these considerations is Aristotle's argument stated in *De caelo* II, 14, which is by Aristotle intended to be in support of the already examined ones. If the earth moved, objects thrown straight up into the air should not fall back to the same spot from which they were thrown. But this does not happen: "It is evident, therefore, that the earth must be at the centre and motionless, both for the reasons given, and because weights thrown straight upward by force return to the same point, even if the force flings them an unlimited distance."[131]

The argument is further expanded and elaborated by Aquinas and repeated by Tolosani. Let us, says Aquinas through Tolosani, place a stone on the table and then throw it straight upwards. The stone will fall back, retracing the same straight line as when it was thrown. If the table is not moved, the stone will fall back to the same spot on it. However, if the table is moved, the stone will fall on a spot behind the spot from which it was thrown. The distance from the starting point and the spot to which a stone will fall will be proportionate to the height reached by the thrown stone, because more time will elapse between the moment

[127] See *Quaestiones in Aristotelis De caelo* II, 22 (not in Clagett's translation; Patar, p. 504).

[128] Oresme's last argument could be called epistemological. See *Le livre du ciel et du monde* II, 25 (Menut and Denomy, pp. 526–7).

[129] See ibid.

[130] See ibid.

[131] *De caelo* II, 14, 296b21–25 (Leggatt, p. 163).

of the throw and the end of the fall. But we can see that heavy bodies thrown upwards fall down to the same spot on the earth from which they were thrown.[132]

The aforementioned argument, which Buridan finds *magis demonstrativa apparentia*,[133] is also cited by Oresme, although in a slightly different form than that recounted by Buridan and Aristotle: Let us imagine that we are on a ship that moves rapidly eastwards. If we shot an arrow vertically upwards, it would not fall back on the ship but a considerable distance behind the ship, towards the west (with respect to the new position of the ship). The same holds for the earth. If the earth moved with the same speed from west to east, the stone thrown vertically upwards should not fall to the spot from which it was thrown. But the evidence of our experience confirms that this is so. *Ergo…*[134] Further arguments against the earth's rotation can be made on the basis of everyday experiences. Every day we see the sun, the moon, and the stars (planets and the sphere of the fixed stars) move, rise, and set. This, of course, means that the heavens revolve around the motionless earth.[135] Another experience-based argument against the earth's rotation is actually the lack of experience that would have to arise as a result thereof. If the earth rotated about its axis, all things on it (the trees, the houses, and the people) should move so too. The speed of the earth's motion should be extremely high to make a complete course from west to east (i.e. in the opposite direction as we see the heavens move) in a day. This, then, also means that wind should blow continuously and strongly, causing a noise similar to that made when shooting an arrow. But since our experience speaks to the contrary, it is evident that the earth does not rotate about its axis.[136]

---

[132] Tolosani, *Opusculum quartum*, cap. 3, 341^v (Lerner, p. 715).

[133] Buridan's *magis demonstrativa apparentia* will be subject to a more detailed treatment infra, pp. 173–7.

[134] See Oresme, *Le livre du ciel et du monde* II, 25 (Menut and Denomy, pp. 520–1).

[135] See ibid.

[136] See ibid.

CHAPTER FIVE

The Politics of Science: The Dedication to the Pope

Copernicus' central thesis that the earth moves is therefore extremely problematic. It goes, as we have seen, beyond the normal, approved state of existing articulations of knowledge: the *sensus communis*, Aristotelian natural philosophy, and the *regina scientiarum* or the "queen of knowledge," theology. This, in turn, means that the concept of the earth's motion reaches beyond the boundaries of the particular, astronomical or mathematical discourse in which it was generated, and becomes subject to both "learned" (philosophy, theology), and "non-learned" (general or popular) belief. Copernicus is thus pushed into a corner. What can he do? How can he introduce the earth's motion into astronomy without being ridiculed or even condemned?

One possible scenario of how things might happen was, as we have seen, proposed to him by Bishop Giese. Giese maintained that philosophers will be open to changing their minds once they have reflected on Aristotle's conclusion that it is up to astronomy to decide whether the earth moves or not. In short, Giese suggested that Copernicus ignore the negative reactions of people not involved in the creation of astronomical knowledge. On the other hand, Giese had nothing to say with regard to philosophers who are not willing to let go of their principles.

Another solution to his predicament is presented to Copernicus by his correspondent and the author of *Ad lectorem*, Andreas Osiander. As we have seen, he suggests that Copernicus should, in order to avoid the opposition he fears, bring forth the earth's motion as an astronomical hypothesis which need not to true, but may nevertheless enable the astronomical computations of positions of celestial bodies. Like other astronomical hypotheses, Copernicus' hypothesis regarding the movement of the earth should also be understood as a device or an instrument that need not reflect the actual state of things but is nevertheless a useful for astronomical calculation. Osiander's proposal seems feasible. By pronouncing the earth's motion to be a "mathematical hypothesis," in Osiander's understanding of its epistemological status, Copernicus would easily "mollify the peripatetics and theologians."[137]

The problem is that he cannot bring himself to do it. Copernicus' understanding of the epistemological status of astronomical hypotheses is completely different from Osiander's. Copernicus believes that mathematical hypotheses need to be true to fulfill their task, and considers hypotheses upon which his astronomy is built, and especially the hypothesis of the movement of the earth, to reflect the actual state of the universe. As we have seen, he objects to Ptolemaic astronomers and treats their hypotheses as false and considers the real motion(s) of the earth cure for the most serious malady of the Ptolemaic astronomical tradition, its incapacity to account for the true structure of the world. Therefore, his strategy to tackle the opposition from the peripatetics and theologians is by necessity different from that suggested by Osiander.

Finding both Giese's scenario and Osiander's proposal unacceptable, Copernicus is compelled to show that the earth's motion has a sound mathematical or astronomical basis (Chapters 9 and 10 of Book I of *De revolutionibus*), that the existing philosophical and experiential arguments against its motion do not contribute to a critical debate, and that there is some consistent alternative physics that could provide answers to both kinds of objections. This he achieves – or at least so he thinks – in Chapter 8 of Book I of *De revolutionibus*. But this alone does not suffice.

Two years after spring 1540, when Bishop Tiedemann Giese, most likely through Rheticus' intervention, persuades Copernicus to print *De revolutionibus*, at the time when the manuscript is already in the hands of printer Petreius in Nuremberg, Copernicus decides that he has yet to find a powerful patron

---

[137] On Osiander's epistemology, see article of P. Barker and B. R. Goldstein, "Realism and Instrumentalism in Sixteenth Century Astronomy: A Reappraisal." See also P. Barker, "The Role of Religion in the Lutheran Response to Copernicus."

to protect him from the attacks which he anticipates from all directions.[138] At the last moment, in June 1542, he dedicates *De revolutionibus* to Pope Paul III (Alessandro Farnese; 1468–1549), composing the dedication *To His Holiness, Pope Paul III, Nicolaus Copernicus' Preface to his Books On the Revolutions.* One highly probable external reason for the last-ditch decision could be that he received news that Girolamo Fracastoro (ca. 1479–1533) dedicated to the Pope an alternative and competitive proposal for an astronomical reform in his *Homocentrica.*[139] But this does not exhaust all the reasons for Copernicus' change of plan, and one more general and strategic reason can be determined from a close look at the argumentation of the *Preface.*

In the *Preface* itself, which serves as a typical dedicatory letter intended to appeal to the patron as a principal source of protection and favor, he reveals intrinsic, substantial reasons for which he has turned for help and protection to the highest authority (at least symbolically) of the time. With the use of traditional rhetorical and dialectical formulae,[140] Copernicus develops an extremely interesting defense strategy to introduce the concept of the earth's motion into astronomical *scientia* and simultaneously reveal to a careful reader several deeper motives for addressing the Pope. Copernicus' thesis that the earth moves whereas the sun is at rest at the centre of the universe, is – this is the impression Copernicus wants to create – a response to a critical state in astronomic *scientia.* But his scholarly response, the argument that the earth moves, is, as we have seen, paradoxical. Because this idea has been dismissed as utterly absurd and inconceivable by all long-established and approved articulations of knowledge, his proposal is even more problematic than the state of astronomy itself. The sheer universality of the thesis, which was generated within a particular *scientia* but transcended the

---

[138] According to P. Barker and B. R. Goldstein, "Patronage and the Production of *De revolutionibus,*" Rheticus created expectations that book will be dedicated to the Duke of Prussia, but having consulted Bishop Giese, he changed his mind and at the last moment decided to dedicate *De revolutionibus* to the Pope Paul III. For different opinion, see A. Goddu, *Copernicus and the Aristotelian Tradition*, pp. 293–4.

[139] See M. A. Granada and D. Tessicini, "Copernicus and Fracastoro: The Dedicatory Letters to Pope Paul III, the History of Astronomy, and the Quest for Patronage."

[140] Copernicus makes use of the rhetorical formulae typical of sixteenth-century dedicatory letters. On this aspect of the *Preface*, see R. S. Westman, "Proof, Poetics, and Patronage: Copernicus' Preface to *De revolutionibus*"; F. Hallyn, *The Poetic Structure of the World: Copernicus and Kepler*, pp. 7–147; J. D. Moos, *Novelties in the Heaven*, p. 37 et seq. A. Goddu, *Copernicus and the Aristotelian Tradition*, pp. 299–300, characterizes it as "dialectical argumentation in a rhetorically persuasive humanistic style."

boundaries and norms of the approved knowledge, compels Copernicus to step out of the learned discourse, if only for a moment, if he wants to succeed in defending the inherently scholarly solution to the astronomical problems. Given that the concept of the earth's motion is in conflict with established astronomy and philosophy, approved theology and popular belief and regarded by all as absurd and inconceivable, Copernicus must justify his concept before the universal court of all articulations of knowledge, learned and unlearned, scholarly and non-scholarly alike. In other words: the introduction of an absurd and inconceivable concept of the earth's motion into *scientia* is first a matter of the politics of scholarly investigation, and only then a matter of scholarship.

Copernicus' decision to replace the original preface with a new one should therefore not be viewed as just another in a long progression of dedications which Renaissance authors were, for one reason or another, addressing patrons. Rather, it should be recognized as the fruit of his profound realization – which was surely reinforced by Fracastoro's dedication to the same Pope and other external and traditional circumstances of this kind – that scholarship alone cannot provide a sufficient basis for the introduction of the concept of the earth's motion into scholarly discourse; that there certainly is some politics of scholarship that is not controlled by scholarly endeavor itself but by the Church and ultimately the Pope as its symbolic representative.

What is, then, Copernicus' strategy? How, in his mind, is the introduction of the earth's motion into astronomy legitimized? How does he respond to the anticipated objections of *sensus communis*, the peripatetics, and theologians?

The movement of the earth is absurd: a consensus in the name of the people. From the first sentence of the *Preface*, a typical example of *captatio benevolentiae*, Copernicus presents the earth's motion as utterly problematic. "As soon as some people hear that in this volume, which I have written about the revolutions of the spheres of the universe, I ascribe certain motions to the terrestrial globe," he writes, "they will shout that I must be immediately repudiated together with this belief."[141] With this statement Copernicus acknowledges that the concept of the earth's motion is broadly universal in scope and significance. It concerns the All, the universe, everything and everybody. Though formulated within a particular scholarly discourse, astronomy, it is inscribed in a field that transcends the narrow confines of scholarly enterprise. Philosophy (that is, scholarly discourse) is, in principle, thought to be beyond the judgment of "ordinary persons," but

[141] *Preface* (Rosen, p. 3; Zekl, p. 68).

the concept of the earth's motion is subject to their assessment and consensus. Copernicus clearly anticipates a reaction to his thesis, particularly from those whom he refers to as "people" and specifies later as "ordinary persons" or *vulgus*. Even though the thesis of the earth's motion is, in his words, a philosophical one, its universality makes it imperative for him to expect the most general, popular reception and heed it. "I am not so enamored of my own opinions," he writes, "that I disregard what others may think of them. I am aware that a philosopher's ideas are not subject to the judgment of ordinary persons, because it is his endeavor to seek the truth in all things, to the extent permitted to human reason by God. Yet I hold that completely erroneous views should be shunned."[142]

Copernicus, who clearly associates his endeavor with the endeavors of philosophers, can also anticipate the popular reaction to his thesis. Given that the widely held opinion agrees with the "consensus of many centuries" dismissing the thesis of the earth's motion as absurd, and given that the widely acknowledged truth is diametrically opposed to Copernicus' thesis by expounding that the earth remains at rest in the middle of the heaven as its centre, the general reaction to his thesis cannot be other than scorn, rejection, and repudiation.

Therefore Copernicus finds himself in quite a predicament. The motion of the earth is widely held as completely absurd and inconceivable. Hence, the concept of the earth's motion is subjected to criticism which does not understand it and dismisses it, without even trying to understand what it means.[143] He has two options to avoid scorn and condemnation: reserve the learned thesis to circle of select colleagues (i.e. spread it "not by writing but by word of mouth"), or seek refuge in radical silence, complete secrecy, and abandon work altogether. He writes:

> Those who know that the consensus of many centuries has sanctioned the conception that the earth remains at rest in the middle of the heaven as its centre would, I reflected, regard it as an absurd akroama if I made the opposite assertion that the earth moves. Therefore I debated with myself for a long time whether to publish the volume which I wrote to prove the earth's motion or rather to follow the example of the Pythagoreans and certain others, who used to transmit philosophy's secrets only to kinsmen and friends, not in writing but by word of mouth, as is shown by Lysis' letter to Hipparchus.[144]

---

142 Ibid.

143 See also J.-J. Szczeciniarz, *Copernic et la révolution copernicienne*, pp. 38–9. I agree with many points of his analysis of the *Preface*.

144 *Preface* (Rosen, p. 3; Zekl, p. 68). The letter, of course, did not address the famous astronomer Hipparchus (ca. 190–120 BC), who had nothing to do with the Pythagoreans.

Given that the concept of the earth's motion is widely dismissed as absurd and ridiculous, the best defense against the preconceived condemnation is simply not to face those "who are reluctant to exert themselves vigorously in any literary pursuit unless it is lucrative."[145] He must also avoid those who are stimulated to the non-materialistic study of philosophy by the example of others, yet because of their dullness of mind "play the same part among philosophers as drones among bees."[146] In other words, there is a fundamental discrepancy between the formulation of a novel thesis and its reception. Unmanaged, it could breed misunderstanding and scorn, or even hinder the proliferation of the novel scientific thesis. Therefore, it is in the best interest of his scholarly endeavor to prevent the thesis from reaching the public and to keep it a secret, privy only to the circle of the select few. He would transmit his theory in the same way as the Pythagoreans: "not in writing but by word of mouth." In Copernicus' view, the Pythagoreans were not compelled to do so by feelings of possessiveness towards their doctrine. They kept their theories secret not out of concern that they might become available to a broader audience but for fear that once having spread among "ordinary persons," they might produce a distorted, deformed conception of the thesis, thus essentially rendering their promulgation impossible.

Therefore Copernicus concludes that, at least for the time being, he would rather publicly uphold faulty teachings (that the earth is motionless) than expose the correct theory (that it moves) to the public. The *vulgus* is not and cannot be a competent judge of the validity of his argument and would deform it and invalidate in advance. He even goes as far, having considered the possible reactions of the non-learned public to his "novel and unconventional" concept of the earth's motion, as to conclude that even oral transmission of the concept (in imitation of the Pythagoreans) may not be in the best interest of the entire project; he almost decides to relinquish the work as a whole. "When I weighed these considerations," he writes, "the scorn which I had reason to fear on account of the novelty and

---

Copernicus even translated *Lysis' letter to Hipparchus* and initially inserted it at the end of Chapter 11 of Book I of *De revolutionibus*. He later excluded it from the version of the manuscript which was intended for print. The letter is therefore not contained in the Nuremberg edition from 1543. For the English translation, see *On the Revolutions*, pp. 25–6. M.-P. Lerner and A. Segonds edited and translated the letter to French in "Éditer et traduire Copernic," pp. 389–392. See also A. De Pace, *Niccolò Copernico e la fondazione del cosmo eliocentrico*, pp. 283–285; Italian translation pp. 401–4.

145 *Preface* (Rosen, p. 3; Zekl, p. 68).

146 Ibid.

absurdity of my opinion almost induced me to abandon completely the work which I had undertaken."[147]

Friendly encouragement and scientific imperative. Obviously Copernicus does nothing of the sort. Why? Why does he decide to spread his doctrine on the motion of the earth not only by word of mouth but in writing, despite the scorn he expects to receive from the *vulgus*? He cites two important reasons which compelled him to publish the volume, despite the novelty and absurdity of his opinion on the earth's motion: first, the encouragement of his friends and second, the scientific imperative.

Lists of mutual friends are, of course, an inevitable component of dedicatory letters. Copernicus mentions the support and encouragement of Bishop Giese and Cardinal Nicholas Schönberg (1472–1537) in particular to impress the Pope.

Schönberg, elevated to cardinal by Pope Paul III himself, was one of the main political players during times of difficult relations between the Church and the Empire throughout the 1520s and 1530s.[148] He was informed about Copernicus' work on heliocentric astronomy by his secretary Johann Albrecht Widmanstetter (1506–1577) who might even have drafted Schönberg's letter to Copernicus, signed on November 1, 1536, asking him to communicate his discovery to scholars and to send him his writings. Copernicus did not comply. He did, however, but indicated his name in the *Dedication to the Pope* and inserted his letter at the beginning of *De revolutionibus*. By doing so,[149] Copernicus aims at bridging the distance between himself and the addressed audience. The impression he tries to make is "that Schönberg – who had been an influential member of that audience – showed great interest in his work and urged its publication and that *De revolutionibus* and its dedication to the Pope is but the result of an early interest in his research on the part of members of the Church hierarchy."[150]

In the letter, which Copernicus inserts before the *Dedication to the Pope*, Cardinal Schönberg praises his work, adding that it has also earned recognition and admiration from others. For not only has Copernicus "mastered the discoveries of the ancient mathematicians [or astronomers] uncommonly well," but he

[147] Ibid.

[148] See M. A. Granada and D. Tessicini, "Copernicus and Fracastoro: The Dedicatory Letters to Pope Paul III, the History of Astronomy, and the Quest for Patronage," p. 442.

[149] See *De revolutionibus*, (Rosen, p. XXI; Zekl, p. 64).

[150] M. A. Granada and D. Tessicini, "Copernicus and Fracastoro: The Dedicatory Letters to Pope Paul III, the History of Astronomy, and the Quest for Patronage," p. 442.

has devised a *novam mundi rationem*, which Schönberg is eager to learn more about. In return he offers him protection and patronage, stating: "If you gratify my desire in this matter, you will see that you are dealing with a man who is zealous for your reputation and eager to do justice to so fine a talent."[151] Copernicus utilizes Schönberg's letter as the first presentation of his new cosmology because it reveals the fundamental feature of his proposition: a new, heliocentric and geokinetic explanation of the universe. Also, the credibility of Copernicus' system is automatically enhanced by the fact that it has been outlined by an influential member of the Church. Schönberg's letter does not present the heliocentric system as an absurd *akroama* but as the noblest of achievements worthy of the cardinal's attention and protection.

Later on in the *Preface*, Copernicus describes his friends and supporters as eminent authorities in science. He characterizes Cardinal Schönberg as "renowned in every field of learning" and Bishop Giese as "a close student of sacred letters and of all good literature."[152] In this formulation, Copernicus' rhetorical exhortation and encouragement of his friends can also be understood as his construction of a public different from the one that agrees with the consensus of many centuries on the absurdity of the earth's motion, and, *a priori,* ridicules, rejects and condemns such a thesis. So, there is, after all, a public, a learned community that does not regard the concept of the earth's motion as ridiculous; just the opposite. The existence of such a scholarly environment implies that allowing a

---

[151] *Letter of Nicholas Schönberg* (Rosen, p. XXI; Zekl, p. 64).

[152] The mention of Bishop Tiedemann Giese, Copernicus' long-standing friend, and particularly Cardinal Schönberg, who was made cardinal by Paul III, as well as the fact that Copernicus published his letter at the beginning of *De revolutionibus,* are also significant for yet another reason. In this way, Copernicus tried to bridge the abyss between him "living in this very remote corner of the earth" and the courtly audience. Moreover, the cardinal's praise of Copernicus in his letter and his description of Copernicus' system gave the impression of high probability. On this, see M. A. Granada and D. Tessicini, "Copernicus and Fracastoro: The Dedicatory Letters to Pope Paul III, the History of Astronomy, and the Quest for Patronage," pp. 441–7. See also Rheticus' description of Tiedemann Giese as a scholar in *Narratio prima*, "Encomium Prussiae" (Rosen, p. 195; Hugonnard-Roche and Verdet, p. 86): "In addition, the benevolent prelate deeply loves these studies and cultivates them earnestly. He owns a bronze armillary sphere for observing equinoxes, like the two somewhat larger ones which Ptolemy says were at Alexandria and which learned men from everywhere in Greece came to see. He has also arranged that a gnomon truly worthy of a prince should be brought to him from England. I have examined this instrument with the greatest pleasure, for it was made by an excellent workman who knew his mathematics." I owe this point to Peter Barker.

learned thesis to address it has far more beneficial effects on learning than leaving it unpublished.[153] Copernicus references, in addition to Schönberg and Giese, to a few other very eminent scholars who exhorted him "no longer to refuse, on account of the fear which I felt, to make my work available for the general use of students of mathematics."[154] The more absurd Copernicus' theory may appear now, so much the more praise and acclaim it will earn after the demonstrations provided by the scholarly discourse. "My explanations," "my writings," writes Copernicus, can dispel the apparent initial absurdities if we delve deep enough into their arguments and demonstrations: "The more absurd my doctrine of the earth's motion now appeared to most people, the argument ran, so much the more admiration and thanks would it gain after they saw the publication of my writings dispel the fog of absurdity by most luminous demonstrations."[155]

In a nutshell: What Copernicus expects first and foremost from the Pope Paul III is that he will protect the concept of earth's motion from being condemned even before it receives a proper measure of scholarly attention. If allowed to present its demonstrations, true scholarship can transform any thesis into an acceptable, even praised and admirable theory, regardless of how absurd and nonsensical it may appear at first glance. This eventually also persuades Copernicus himself: the encouragement of his friends and the hope that his proofs can dispel the fog of absurdities, even for the non-scholarly, common persons: "Influenced therefore by these persuasive men and by this hope, in the end I allowed my friends to bring out an edition of the volume, as they had long besought me to do."[156]

What leads Copernicus to start contemplating the earth's motion in the first place? Why does he venture "to conceive any motion of the earth against the traditional opinion of mathematicians and almost against common sense"?

Copernicus' answer is already known to us: he is compelled to consider a different system of deducing the motions of the universe's orbs by the realization that astronomers do not agree among themselves in their investigations. They are not certain about the motion of the sun and the moon and do not use the same principles, assumptions, and explanations of the apparent revolutions and motions. Nor do they produce fully adequate astronomical theories: the proponents of the homocentric model are unable to deliver accurate calculations of the positions of the celestial bodies, whereas the proponents of eccentrics and epicycles seem

---

153 See also J.-J. Szczeciniarz, *Copernic et la révolution copernicienne*, p. 40.

154 *Preface* (Rosen, p. 3; Zekl, p. 68).

155 Ibid.

156 Ibid.

to have solved this problem to a considerable measure, but only by contradicting the first principles of uniform motion. Moreover, astronomers are unable to "elicit or deduce the principal consideration, that is, the structure of the universe and the certain *symmetria* of its parts." In other words, they are unable to arrive at any certain conclusions. Copernicus confronts this uncertainty in explaining the motions of the spheres of the *machina mundi* with cognitive optimism stemming from his realization that the *forma mundi* was created "for our sake" and "by the best and the most regular Artisan of all," that is, God:

> For a long time, then, I reflected on this uncertainty in the mathematical traditions concerning the derivation of the motions of the universe's spheres. I began to be annoyed that the movements of the world machine, created for our sake by the best and most regular Artisan of all, were not understood with greater certainty by the philosophers, who otherwise examined so precisely the most insignificant trifles of this world.[157]

The world as God's creation should itself be perfect, regular, orderly and, having been made for our sake, also cognizable. This readability of the world is the reason for which he claims to have undertaken "the task of rereading the works of all philosophers which [he] could obtain to learn whether anyone had ever proposed other motions of the universe's spheres than those expounded by the teachers of mathematics in the schools."[158] Copernicus lists – as we have already seen –[159] Cicero (106–43 BC) and pseudo-Plutarch, who invoked several defenders of the motion of the earth: Hicetas of Syracuse (ca. 400–335 BC), Philolaus of Croton (ca. 470–385 BC), Heraclides of Pontus (ca. 390–310 BC), and Ecphantus the Pythagorean (fourth century BC).

References to ancient proponents of the earth's motion are an extremely telling component of Copernicus' strategy. In the hope that the Pope and humanist clergy will approve his astronomical system as congruent with the principles of Renaissance and humanist culture, he does not present it as a novelty but embellishes it with the authority of a venerable, time-honored tradition. He maintains that the earth's motion is "novel and absurd" only in the eyes of the uninstructed who rely more on their senses than reason and who are not competent to pass judgments in philosophical matters. The earth's motion is, in fact, an age-old doctrine, so old as to have been debated even long before the days of Aristotle. Given the

---

157 Ibid. (Rosen, p. 4; Zekl, p. 72).

158 Ibid.

159 See supra, p. 75.

fact that the earth's motion has the legitimacy of a "time-honored tradition" and that Copernicus' predecessors enjoyed a large measure of freedom in explaining the motions of stars and planets and accordingly used various circles for this purpose, he demands the same measure of freedom discussing the mobility of the earth, regardless of its apparent absurdity.[160]

Mathemata mathematicis scribuntur. Copernicus introduces the concept of the earth's motion into astronomy on the basis of the legitimacy assigned to it by Antiquity, which shed a new light on "the consensus of many centuries" (and people). He realizes that the concept contributes not only to "saving the phenomena" but also to deducing the constitution of the universe, which he explains in his book. The latter is, in Copernicus' construction of events in the *Preface*, thus written and published. What follows next? That which he fears the most and tries to avoid for so long before he finally decides to print his manuscript? Will his work meet with ridicule or even condemnation? What does fate have in store for Copernicus' book and theory?

Copernicus has no doubt that the arguments in his book are sufficiently sound to satisfy the learned community for which they are primarily intended. However, rather than referring to the scholarly community in general, he focuses on the circle of mathematicians, that is astronomers, convinced that his demonstrations will persuade "acute and learned mathematicians." In his opinion, the judgment on whether or not the concept of the earth's motion is appropriate rests in the hands of astronomers who will surely agree with him if, "as this philosophy especially requires, they are willing to examine and consider, not superficially but thoroughly, what I adduce in this volume in demonstration of these matters."[161] Copernicus thus no longer fears that acute and learned mathematicians will condemn the thesis of the earth's motion and oppose it, insofar as they thoroughly examine his theory. To the contrary; as seen earlier, he firmly believes that he will even be able to persuade them to espouse his viewpoint, and that after a careful consideration of the arguments they will find that the apparently absurd thesis is, in fact, worthy of admiration, and will "agree with him." "Acute and learned mathematicians" will eventually advocate precisely what appears to be "against the traditional opinion of mathematicians." In other words: Copernicus firmly believes in the persuasive power of the mathematical discourse among mathematicians.

---

[160] See *Preface* (Rosen, p. 5; Zekl, p. 74).

[161] Ibid. (Rosen, p. 5; Zekl, pp. 74–6).

However, while Copernicus has no doubt about a favorable reception of the thesis of the earth's motion among "acute and learned mathematicians" and its contribution to astronomy, there is still one other community of learned men in which his success is far from guaranteed. It is theologians or, rather, theology as a strain of *scientia* based on the word of God. However, he does not refer to all theologians, but only to those who are ignorant and uninstructed in mathematics, but nevertheless dare to pass judgment on the subject following God's word in the Scripture: "Perhaps there will be *mataiologoi* [babblers, prattlers] who claim to be judges of mathematics although completely ignorant of the subject and, badly distorting some passage of Scripture to their purpose, will dare to find fault with my undertaking and censure it. I disregard them even to the extent of despising their criticism as unfounded."[162] "Mathematics," maintains Copernicus, "is written for mathematicians": *mathemata mathematicis scribuntur.*

Thus, Copernicus gradually shows his hand to the Pope and other readers in the *Preface*. The initial apprehension that his concept of the earth's motion will earn nothing but scorn, rejection and condemnation from "[t]hose who know that the consensus of many centuries has sanctioned the conception that the earth remains at rest in the middle of the heaven as its centre,"[163] ultimately turns into fear of being mocked by theologians and their "distorted" passages from the Holy Scripture. Acute and learned mathematicians and ordinary persons, advocates of *sensus communis* completely disappear from Copernicus' horizon of potential opponents and "schemers" calling for the condemnation of his theory. Why?

The first, "acute and learned mathematicians," disappear because they will agree with him after they examine his work and its mathematical demonstrations, whereas "ordinary persons" are of no consequence to him at all. The fundamental problem facing Copernicus is the reception of his thesis of the earth's motion, which though widely regarded as "novel and unconventional," is, in fact, old. The state of scholarship, religion and general opinion renders its acceptance impossible and allows it no room to spread. Even less does it grant Copernicus the opportunity to present it appropriately, articulate the mode of its promulgation, or have

---

162 Ibid. (Rosen, p. 5; Zekl, p. 76). On the use of the term *matailogoi* in Copernicus' days, see F. Hallyn, *The Poetic Structure of the World: Copernicus and Kepler*, pp. 65–7, and M.-P. Lerner, "The Heliocentric 'Heresy': From Suspicion to Condemnation," p. 28, n. 15: "The Greek word *matailogoi* (rendered by the Vulgate as *vaniloqui*) and here by 'prattlers' is a rare word that Copernicus probably took from the New Testament, Titus 1: 10, where St. Paul denounces false doctors."

163 *Preface* (Rosen, p. 3; Zekl, p. 66).

control over the manner of its reception. Because the destiny of his work depends first on external elements, Copernicus has to garner the support of the factors that control the situation. These, however, are not "ordinary persons," representatives of *sensus communis*, figuratively speaking, but the Church as the ultimate authority with the Pope as its supreme leader and symbolic representative. In this perspective, Copernicus can simply dismiss the reaction of "ordinary persons" as inconsequential. What is relevant is not the reception among the uneducated but among the educated. Certain of a favorable reception among learned mathematicians, he is now primarily concerned with how the thesis will be received among theologians. This is so much truer, as it is theology and theologians who, in fact, control the politics of scholarship. In short, Copernicus knows that the destiny of scholarship depends not (only) on scholarship itself but on some politics of scholarship which is ultimately controlled by the Church, especially when it involves the introduction of what many see as a complete "novelty and unconventionality of his opinion" that the earth moves. Therefore, he is left with no other option but to find allies within the Church who will concur with him, and to discredit *in advance* anyone who will reject, ridicule and condemn his thesis of the earth's motion for theological or biblical considerations.

This is why Copernicus turns to Pope Paul III as an authority that, metaphorically speaking, stands outside the frameworks of Copernicus' time, an authority that can pass judgment unburdened by the existing forms of knowledge. Copernicus places the Pope beyond "the traditional opinion of mathematicians" and "common sense." He releases him from the standard, approved science, universal understanding, common sense, and places him into a position in which the Pope can freely assess what ordinary persons deem a "novel and unconventional thesis" and use his authority to protect the thesis from "calumnious attacks":

> However, in order that the educated and uneducated alike may see that I do not run away from the judgment of anybody at all, I have preferred dedicating my studies to Your Holiness rather than to anyone else. For even in this very remote comer of the earth where I live you are considered the highest authority by virtue of the loftiness of your office and your love for all literature and mathematics too. Hence by your prestige and judgment you can easily suppress calumnious attacks although, as the proverb has it, there is no remedy for a backbite.[164]

The Pope can use his symbolic and real power, which is at once theoretical,

---

[164] Ibid. (Rosen, p. 5; Zekl, p. 76).

The Pope can use his symbolic and real power, which is at once theoretical, political and social, to protect the seemingly absurd thesis of the earth's motion from being condemned and dismissed *in advance*. Copernicus deems the representative of religious orthodoxy adequately well versed "also in the branches of 'mathematics'," so that he, along with other "acute and learned 'mathematicians'," will agree with his thesis of the earth's motion, protect it from "calumnious attacks," and prepare the ground for it to spread.

In the category of "acute and learned mathematicians" within the Church, Copernicus also lists Pope Leo X (1475–1521), Cardinal Schönberg, Bishop Giese, and Paul of Middelburg (Bishop of Fossombrone). As for the other category, those who, "although completely ignorant of the subject [...], would dare to find fault with my undertaking and censure it," it only includes the church father Lactantius. The choice is not accidental. Lactantius, "otherwise an illustrious writer but hardly a mathematician," "speaks quite childishly about the earth's shape" in his *De divinis institutionibus* III, 24, "when he mocks those who declared that the earth has the form of a globe."[165] Much like Lactantius, a theologian completely ignorant in mathematics, talks foolishly about the shape of the earth and mocks those who pronounced it spherical, there are perhaps also contemporary non-mathematicians who will ridicule Copernicus' thesis that the earth moves. Just like Lactantius (ca. 240–320) scorns the fundamental assumption of astronomy, namely, that the earth is a sphere, a thesis that lies at the very foundation of mathematical astronomy, which has been validated by the "traditional opinion of mathematicians," there may also be *matailogoi* within the Church who will, though completely ignorant of the subject, "badly [distort] some passage of Scripture to their purpose," and ridicule the fundamental postulate of Copernicus' astronomy that the earth moves.

The only branch of learning that has the legitimate right to pass judgment on the concept of the earth's motion is mathematics, i.e. astronomy: "Mathematics is written for mathematicians." This maxim, when transposed to the domain of the existing articulations of knowledge, that is, theology, mathematics, and general opinion as a potential arbiter of the earth's motion, excludes *sensus communis* and divides theology into two camps: one that judges mathematical matters by relying on mathematics, and one that is devoid of any knowledge regarding mathematics but nevertheless dares to sit in judgment on mathematical matters (*mathemata*) by drawing on theology. Copernicus expects theologians who are well versed in mathematics to understand his demonstrations and support his thesis of a movable earth (or at least not refute it). But he also expects opposition and censorship from

165 Ibid.

those who are completely ignorant of the subject and invoke certain passages of the Holy Scripture which could be interpreted in favor of the earth's immobility.

Considering all of the above, the assertion *mathemata mathematicis scribuntur* ultimately means that matters of astronomy can, according to Copernicus, be decided by astronomy alone. But in this situation, that is not enough. Since mathematics and mathematicians have no say in the politics of scholarship, which rests in the hands of the Church, they have no control over the reception of their theses. The only legitimate solution for Copernicus' theory to survive is, then, to commit the "matters of mathematics" to the judgment of men within the Church who are competent in the subject. Therefore, in discussing the matters of mathematics even the Church itself or the theological articulation of knowledge, which relies on the word of God, must subject itself to mathematics and not vice versa. When dealing with astronomical questions, astronomy must hold authority over the interpretation of the Scripture rather than the Scripture hold the authority over the interpretation of astronomy. With this thesis Copernicus effectively refutes all particular theological objections to the motion of the earth based on the Biblical quotation,[166] and unequivocally strips theology of the authority to pass judgment on scientific questions "inside its own ambit and challenges its status as the queen of the sciences."[167]

The last group that Copernicus addresses is that of the mathematicians within the Church. They will realize that the concept of the earth's motion will only be ridiculed by non-mathematicians and conclude that his thesis not only does not contradict the Holy Scripture but contributes to the Church by bringing practical solutions to the ongoing calendar reform:

> Mathematics is written for mathematicians. To them my work too will seem, unless I am mistaken, to make some contribution also to the Church, at the head of which Your Holiness now stands. For not so long ago under Leo X the Lateran Council considered the problem of reforming the ecclesiastical calendar. The issue remained undecided then only because the lengths of the year and month and the motions of the sun and moon were regarded as not yet adequately measured.[168]

---

[166] Copernicus' conviction, expressed here in a rather rudimentary fashion, was developed by Galileo in his *Letter to the Grand Duchess Christina* and in some other letters.

[167] M. A. Granada and D. Tessicini, "Copernicus and Fracastoro: The Dedicatory Letters to Pope Paul III, the History of Astronomy, and the Quest for Patronage," p. 457.

[168] *Preface* (Rosen, pp. 5–6; Zekl, p. 76). On the calendar reform within the context of the *Preface*, see supra, pp. 37–8.

To recapitulate: a scholarly response to the critical situation of science requires some external justification. Copernicus, indeed, turns to Pope Paul III. The situation, which is unfavorable to his geokinetic reform of astronomy, has to be changed into a situation which will allow room for the introduction of the "novel and unconventional" concept of the earth's motion into science. This, however, will only be feasible if the concept is also endorsed by the Church – the authority which controls the political situation, including that of learning. Nevertheless, Copernicus only seems to step out of scholarly discourse by providing an external justification, but this stepping out brings him back to the domain of scholarship. His reaction to the critical situation of astronomical *scientia* is a formulation of relations that are or ought to be established between scholarship and the extra-scholarly field (general opinion or "ordinary persons"), as well as a formulation of relations within the learned discourse itself (mathematics–philosophy–theology). These are, as he maintains in the *Preface,* inevitably such that the concept of the earth's motion, which is – at least in his view – inherent to astronomy, must be placed under the authority of the learned discourse which has generated it, i.e. mathematics: "mathematics is written for mathematicians." His tying the concept to the Pope's (or the Church's) support is therefore only apparent. The Pope (along with other theologians and the Church, respectively) as the representative of the authority which controls the politics of scholarship is left with no alternative: if he is a true mathematician, he will, like all other "acute and learned mathematicians," agree with his concept of the earth's motion, once he will have examined the arguments. If not, he has no right to sit in judgment on the matters "written for mathematicians."

An external authority is only allowed to sit in judgment of Copernicus' learned answer to this critical situation in learned discourse insofar as he is learned himself, i.e. insofar as he does not transcend the boundaries of the learned discourse within which the concept under examination has been developed. In such a case, however, that particular scholarship alone is to be strong enough to convince the authority which controls the politics of scholarship of the legitimacy, validity and truthfulness of the new concept. Although Copernicus may give the impression of subordinating *scientia* to the Church, exactly the opposite is taking place: the Church (the Pope, theology) must subordinate to the learned discourse whose politics it controls. Hence, although *scientia*, which transcends the established general norms, has to acquire approval from the authority which controls the politics of scholarship, the latter can only do so by subordinating itself to scholarship. The mid-point between the thesis that "mathematics is written for mathematicians" and the thesis that the politics of scholarship is controlled by the Church rather than mathematicians is that the introduction of the concept of the earth's motion into the learned discourse somehow rests in the hands of the Pope as the supreme mathematician.

PART TWO

ARGUMENTATION IN FAVOR OF THE EARTH'S MOTION

Despite the general consensus that the earth is motionless in the middle of the universe and that terrestial motion is inconceivable,[1] the matter is far from settled for Copernicus. He will call into question the concept of the stationary earth despite its support from theology, the *sensus communis* and almost the entire philosophy of nature until his time. His argument will transpose the inconceivable, absurd and ridiculous concept into a conceivable and scholarly legitimate theory. He believes firmly that the "acute and learned mathematicians" to whom the book is primarily dedicated – *mathemata mathematicis scribuntur* – will agree with him, "if, as this discipline especially requires, they are willing to examine and consider, not superficially but thoroughly," what he adduces in *De revolutionibus*. Let us, then, take a look at what Copernicus offers "in demonstration of these matters." How can he demonstrate the opposite of what is obvious? In what way does he, as he puts it, dispel "the fog of absurdity" of the thesis of the earth's motion?

Anna De Pace has recently proposed a very original and appealing thesis regarding Copernicus' method in the Book I of *De revolutionibus*. According to her,

> Book I of *De revolutionibus* acquires its sense of unity through a particular method that dictates the development of the argumentation. The chapters do

---

[1] See *De revolutionibus* I, 5 (Rosen, p. 11; Zekl, p. 98): "To be sure, there is general agreement among the authors that the earth is at rest in the middle of the universe. They hold the contrary view to be inconceivable or downright ridiculous." Copernicus most probably refers to Chapter 7 of Book I of the *Almagest* (Toomer, p. 45; Heiberg, pars 1, p. 24), in which Ptolemy states: "However, they do not realize that, although there is perhaps nothing in the celestial phenomena which could count against that hypothesis, at least from simpler considerations, nevertheless from what would occur here on earth and in the air, one can see that such a notion is quite ridiculous."

> not follow [...] the procedure of a doctrinal exposition, instead Copernicus's strand of argument finds its justification in a dialectical contrast above all with Ptolemy's *Almagest* (though there are also inclusions into *De caelo*) which brings out contradictions in the answer found there. Step by step Copernicus dismantles the geocentric cosmos, thus making room, step by step, for alternative hypotheses as a starting point for a body of coherent propositions, arriving only at the end of Chapter 10 at God as the foundation of their truth.[2]

This seems to me – at least in its fundamentals – a very adequate description of Book I of *De revolutionibus*. I also agree with De Pace that "a physical or non-physical justification for the world-system squares with a rhetorical or non-rhetorical value of the beauty of the 'visible god': the cosmos."[3] I have no problems regarding her claim that Copernicus here substantially deploys Plato's hypothetical method as described in the *Meno*, the *Republic*, the *Parmenides* and above all in the *Phaedo*.[4] I also agree that Copernicus, by examining Ptolemy's and Aristotle's hypotheses, rejects and supplements some of them as his discourse develops towards its culmination in Chapter 10, but I am less convinced, however, by De Pace's conjecture that this includes Copernicus' rejection of the reality of celestial spheres, very clearly asserted in Chapter 4.[5]

According to Richard Robinson, Plato's hypothetical method consists, *inter alia*, "in paying the utmost attention to the avoidance of contradiction, in valuing at zero any set of opinion that contradicts itself."[6] If one is to follow Plato's hypothetical method one must "avoid both the direct contradiction of saying that this proposition is both true and false, and also the indirect contradiction of asserting a set of propositions such that some of them, by a shorter or longer train of

---

2 "Copernicus against a Rhetorical approach to the Beauty of the Universe," pp. 84–5. See also her *Niccolò Copernico e la fondazione del cosmo eliocentrico*, pp. 95–104. A. Goddu, *Copernicus and the Aristotelian Tradition*, pp. 275–324, also believes that Copernicus' approach is dialectical, but for different reasons. He connects Copernicus' strategy with the dialectical topics, which was between 1490 and 1550 included into the broader and more comprehensive humanistic understanding of rhetoric.

3 "Copernicus against a Rhetorical approach to the Beauty of the Universe," p. 85.

4 See *Meno* 86c–87d; *Republic* 502c–541b; *Parmenides* 128a–137c, *Phaedo* 99b–100e.

5 On the issue of spheres, see *Niccolò Copernico e la fondazione del cosmo eliocentrico*, pp. 55–61, 78–86, 225–6. For her argumentation that in Chapter 4 "l'autore non sta affato esponendo tesi proprie," see pp. 104–14.

6 *Plato's Earlier Dialectic*, p. 106. Copernicus' reliance on Platonist methodology is, in my opinion, even more evident from his *Commentariolus*. See infra, pp. 269–70, n. 62.

implication, imply the falsity of another."[7] Avoiding indirect contradiction means systematizing our opinion into a body of "concordant" beliefs, which is achieved by the exploration of their implications. Exploration of the implications of a set of hypotheses brings to light an indirect contradiction and that is exactly what Copernicus does with Ptolemy and Aristotle in Book I of *De revolutionibus*. This is a negative part of his procedure: he seeks to expose the contradictory outcomes of Aristotle's and Ptolemy's reasoning. His dialectical opponents are shown to be in contradiction to the generally accepted principles or hypotheses, and by doing that Copernicus is up to produce some rhetorical effects. But Copernicus, despite his expressions of occasional doubts and demands for further specifications regarding some of the hypotheses (such as, for example, the spherical shape of the universe in Chapter 8), does nowhere in *De revolutionibus* (or the *Commentariolus*) dismiss celestial orbs and spheres, entities for which there is a very clear affirmation in Chapter 4.

I would change very slightly De Pace's interpretation and argue instead that Copernicus in Chapters 1 to 4 takes widely accepted principles or hypotheses from a variety of sources, mainly Ptolemy, and adds some new ones, based on the new geographical discoveries unavailable before, to support his conclusions: the spherical shape of the earth; the spherical shape of the universe; the earth as a balanced sphere formed from earth and water; the uniform, eternal and circular motion of the heavenly bodies as dependent on heavenly spheres. He presents these commonly accepted principles or hypotheses in a very general form, without any details or specifics, and goes on to show that the exploration of the implications of a set of Ptolemy's (and Aristotle's) hypotheses reveals some serious contradictions. At the same time his goal is to persuade the reader that commonly accepted hypotheses and principles need only slight modification, plus the addition of a few new ones which arise naturally from the accepted principles and hypotheses in dialectical argumentation. These include the hypotheses or principles that the heavens are immense compared to the size of the earth, that the sphere of the fixed stars is immobile and that the revolution in a smaller circle is faster than that in a bigger one. This latter principle implies that the planets should be arranged in accordance with the duration of the revolution, and assumes that of the objects moving equally fast, those further away seem to travel more slowly. And above all, this includes the principle or hypothesis of the movement of the earth in its different forms, which is introduced to produce a "concordant" result, or, in different words, that the hypothesis of the movement of the earth

---

7 *Plato's Earlier Dialectic*, p. 106.

resolves these contradictions. In the process the traditional hypotheses remain more or less intact and slowly incorporated, step by step, into a geokinetic and heliocentric cosmology. Copernicus formulates hypotheses, especially in Chapter 4, in such a way that they can be taken as almost unproblematic starting points and then, in the course of further discussion, smoothly incorporated, without easily detectable changes, into the heliocentric cosmology. Copernicus does not abandon any hypotheses from Chapters 1 to 4, as De Pace seems to believe, but illustrates, when it is to his advantage, that they are perhaps less certain than usually assumed, and that it is more reasonable to build a heliocentric cosmology on these accepted views than a geocentric one. As I will show, his attempts to provide heliocentric cosmology and physical justification for the movement of the earth, if thoroughly examined, are in themselves no less discordant then those of traditional cosmology.

Copernicus begins the dialectical dismantling of geocentric cosmology and simultaneous building of a heliocentric one in Chapter 5 with two questions: the first one regarding the rotation of the earth, which he answers in Chapter 8, and the second one regarding the earth's position in the universe and thus its motion around the sun, which is answered in Chapters 9 and 10. Chapter 6 presents Copernicus' dialectical argumentation with Ptolemy and his refutation of the thesis that the earth must be stationary in the middle of the universe for geometrical reasons. Chapters 7 and 8 are devoted to the dialectical presentation of physical reasons against the movement of the earth (from Aristotle and Ptolemy) and his refutation thereof. Copernicus also presents four more or less philosophical arguments for the rotation of the earth. A comparison of Copernicus' physical proofs in favor of the earth's rotation with standard Aristotelian proofs for the earth's stability reveals that Copernicus certainly does not regard them as crucial ones. He is extremely short, vague and sometimes ambiguous, which makes the interpretation of his actual position extremely difficult. It is my thesis that for Copernicus the decisive argument for the movement of the earth is achieved on purely mathematical reasons in Chapter 9 and 10. In Chapter 9 he discusses the second movement of the earth, its revolution around the sun and thus its position in the universe. Chapter 10 is the culmination of Copernicus' discourse: he explains the harmonious universe that is the result of the second motion of the earth around the stationary sun.

CHAPTER SIX

The earth is a *Terraquea Sphaera*

At the end of Chapter 4 of Book I of *De revolutionibus*, Copernicus sets forth the need to examine "the relation of the earth to the heavens" so as not to "attribute to the celestial bodies what belongs to the earth," that is, not to attribute to the heavens the motion which, he believes, belongs to the earth. He has dedicated the first three chapters of the book to demonstrations to prove the universe is spherical like the earth (one of the arguments for the spherical form of the universe is "that all the separate parts of the universe, I mean the sun, moon, planets and stars, are seen to be of this shape"),[8] and that element earth forms a single sphere with water. Given that the structure of Book I of *De revolutionibus* – very roughly – imitates that of Book I of Ptolemy's *Almagest*,[9] the first two chapters are somewhat expected, addressing the same questions as Ptolemy's Chapters 3 and 4 and furnishing several standard arguments in support of both generally accepted theses. In Chapter 3, entitled *How earth Forms a Single Sphere with Water*, however, Copernicus diverges from Ptolemy's *Almagest* and begins to prepare the ground for setting the earth in motion about its axis and pulling it into circular motion around the sun. The chapter ends with two conclusions: (1) "that land and water together press upon a single centre of gravity; that the

---

8 *De revolutionibus* I, 1 (Rosen, p. 8; Zekl, p. 86).

9 For a chapter-to-chapter comparison between the two books, see, for instance, M. Kokowski, *Copernicus's Originality*, pp. 80–1.

earth has no other centre of magnitude"[10] than the centre of gravity, that is, they share the same centre; (2) that the earth is "perfectly round, as the philosophers hold."[11]

Copernicus' view can be described as identical to the view held by proponents of the so-called *terraquea sphaera*, that is, the elements of earth and water form a single sphere rather than two separate spheres. The centre of gravity of the terraqueous sphere is identical to its centre of magnitude, that is, to the geometrical centre of the common magnitude of earth and water. Hence, "[t]he earth together with its surrounding waters must in fact have such a shape as its shadow reveals, for it eclipses the moon with the arc of a perfect circle."[12] It is therefore "perfectly round" or "spherical." Nonetheless, this brief description conceals more than it reveals. Copernicus' treatment of this question and his two conclusions testify to the medieval revision of the Aristotelian philosophy of nature, which was continuously subjected to critical examinations. It also alludes to changes in the Aristotelian philosophy of nature brought about by geographic discoveries that took place in the sixteenth century. Most importantly, it signals Copernicus' attempt to prepare the earth for its *real* – as opposed to hypothetical (in Osiander's understanding of hypotheses) – axial rotation and planetary revolution around the sun.

Copernicus' belief that earth and water form a single sphere, which was in the seventeenth century called the "terraqueous globe," and that these elements therefore have the same centre of both magnitude and gravity is his response to the ancient and medieval discussions as to what is meant by the "centre" of the universe. Is it the centre of the earth's magnitude (*centrum magnitudinis*) or its centre of gravity or heaviness (*centrum gravitatis*)? And, assuming it is the latter: Is it the centre of gravity of the earth, of the aggregate of earth and water, or of a single terraqueous globe?

Ancient and medieval natural philosophers had serious problems investigating Aristotle's arrangement of the four sublunary spheres of elements, earth, water, air, and fire.[13] They believed that land extended above water only in the northern hemisphere, while the southern hemisphere was, save a few islands,

---

10 *De revolutionibus* I, 3 (Rosen, p. 10; Zekl, p. 92).

11 Ibid.

12 Ibid.

13 On medieval and Renaissance discussions, see E. Grant, *Planets, Stars, & Orbs*, pp. 622–4, 630–7; T. Goldstein, "The Renaissance Concept of the Earth in Its Influence upon Copernicus"; W. G. L. Randles, *De la Terre plate au globe terrestre*, especially pp.

submerged in water and uninhabited. Since the part of the earth not covered by water was dried and rarefied by the sun, it would necessarily be lighter than that in the southern hemisphere. From this it followed that the entire terrestrial globe was rather imbalanced. The questions that naturally arose were: What occupies the centre of the world? Is it the centre of the earth's magnitude or its centre of gravity? Moreover: Should waters covering the earth also be taken into account in determining the centre of gravity?

These questions, as well as answers thereto, are nicely clarified in the following abstract of a description of the sublunary region written by Pierre d'Ailly, also known as Petrus de Alliaco (1350/51–1420), in his *Ymago mundi*:

> Immediately after [or below] the sphere of the moon, the philosophers place the sphere of fire, which is the most pure there and invisible because of its rarity. Just as water is clearer than earth and air than water, so this fire is rarer and clearer than air, and so is the heaven [or sky] rarer or clearer than fire, except for the stars, which are thicker [or denser] parts of the sky so that the stars are lucid and visible. Afterwards is the sphere of air, which encloses water and earth. This is divided into three regions, one of which is the outermost (next to fire) where there is no wind, rain, or thunder, nor any phenomenon of this kind, and where certain mountains, such as Olympus, are said to reach. Aristotle says that starry comets appear and are made there and that the sphere of fire and this supreme region of air with its comets are moved simultaneously with the heaven [or sky] from east to west. The middle region [of air], however, is where the clouds are and where various phenomena occur, since it is always cold. The other [and third] region is the lowest, where the birds and beasts dwell. Then follow water and earth, for water does not surround the whole earth, but it leaves a part of it uncovered for the habitation of animals. Since one part of the earth is less heavy and weighty than another, it is, therefore, higher and more elevated from the centre of the world. The remainder [of the earth], except for islands, is wholly covered by waters according to the common opinion of philosophers. Therefore, the earth, as the heaviest element, is in the centre or the middle of the world, so that the earth's centre is the centre of its gravity; or, according to some, the centre of gravity of the earth and also of water is the centre of the world. And although there are mountains and valleys on the earth, for

41–64. See also A. De Pace, *Niccolò Copernico e la fondazione del cosmo eliocentrico*, pp. 321–31, notes 77–100, for the discussion of the entire Chapter 3.

> which reason it is not perfectly round, it approximates very nearly to roundness. Thus it is that an eclipse of the moon, which is caused by a shadow of the earth, appears round. They say the earth is round, therefore, because it approximates to roundness.[14]

In addition to other specifics (the tripartite division of the region of air, the movement of fire and the highest region of air in a diurnal motion from east to west, the placement of comets into the upper region of air, etc.), d'Ailly's description makes it evident that the spheres of earth and water are not arranged entirely in line with Aristotle's abstract and general scheme, but with quite a large amount of dry earth rising above waters (see *Figure 10*).

According to d'Ailly, the reason why individual parts of the earth rise above water is that "one part of the earth [that is the part on the northern hemisphere] is less heavy and weighty than another," and is therefore "higher and more elevated from the centre of the world." The other part of the earth, the southern hemisphere, is, with the exception of several islands, completely underwater. Therefore, the problem is that the earth is an inhomogeneous body. One part of the earth is relatively free of water and is inhabited; the other part is submerged and therefore uninhabited. The uncovered part of the earth is heated and "rarefied" by the action of the sun and air, while the covered part remains dense and compact. This results in an uneven arrangement of weight over the entire globe. Since one part of the earth is lighter and the other heavier, the earth's centre of gravity is not identical to its centre of magnitude. If the earth were a perfectly homogeneous body, its centre would coincide with the centre of the universe. What is more, if the earth were a perfectly homogeneous body, its centre of magnitude would coincide with its centre of gravity, and both would coincide with the centre of the universe.

According to d'Ailly, the centre of the universe obviously coincides with the earth's centre of gravity and not with that of its magnitude. This answer allows two options: "the earth's centre is the centre of its gravity" or "the centre of gravity of the earth *and* also of water is the centre of the world." D'Ailly assigns a single centre of gravity to the totality of earth and water, and holds that the centre of the universe is reserved for the centre of gravity of the aggregate of earth and water. It is the natural tendency of all heavy bodies – what the aggregate of earth and water is – to seek the centre of the universe and abide there. Hence, the earth does

---

[14] *Tractatus de ymagine mundi,* "Quartum capitulum de quotuor elementis et eorum situ." Quoted from E. Grant, ed., *A Source Book in Medieval Science*, p. 633. Similar, if not identical positions, was also held by Albert of Saxony.

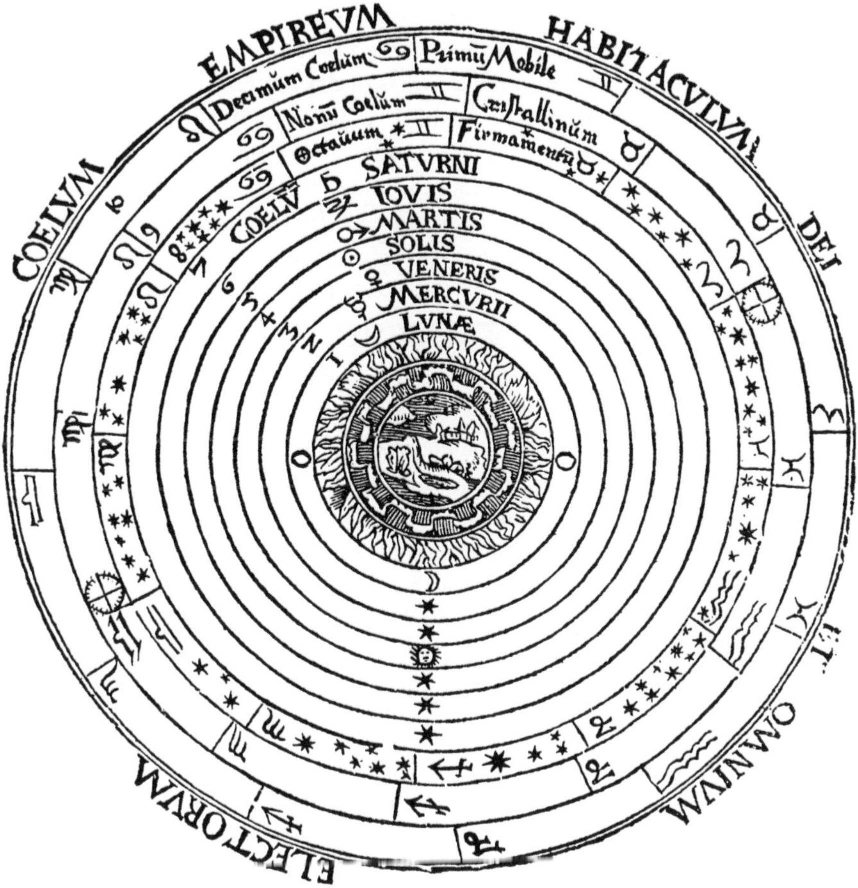

Figure 10.
Aristotelian cosmos from Peter Apian, Cosmographicus liber *(Antwerp, 1584). Earth and water form a composite terraqueous globe. Image courtesy of the History of Science Collections, University of Oklahoma Libraries.*

not lie at the centre of the universe *per se* but only as part of the aggregate, which also includes water on the earth's surface and beneath it. But this does not mean that d'Ailly conceives of earth and water as a single sphere. He most likely thinks of them as two separate spheres that, when taken as a heterogeneous aggregate, constitute a single centre of gravity.

An important shift in conceiving the earth-water relationship and determining the centre of the terrestrial globe occurred in the sixteenth century after a Portuguese expedition discovered vast areas of dry land in the southern hemisphere (modern-day Brazil) in 1501. Joachim Vadianus (1481–1551) wrote in a letter to Rudolf Agricola (1444–1485), published only in 1515, that the earth and water likely formed a single sphere, since over the entire surface of the globe earth was partially submerged in water and partially extending above it.[15] Completely in line with this geographical discoveries and conclusions, drawing in all likelihood on Martin Waldseemüller's (ca. 1470–1520) world map from his *Cosmographiae Introductio*, which appeared in 1507,[16] Copernicus writes the following:

> Moreover, there is no difference between the earth's centres of gravity and magnitude. This can be established by the fact that from the ocean inward the curvature of the land does not mount steadily in a continuous rise. If it did, it would keep the sea water out completely and in no way permit the inland seas and such vast gulfs to intrude. Furthermore, the depth of the abyss would never stop increasing from the shore of the ocean outward, so that no island or reef or any form of land would be encountered by sailors on the longer voyages. But it is well known that almost in the middle of the inhabited lands barely fifteen furlongs remain between the eastern Mediterranean and the Red Sea. On the other hand, in his *Geography* Ptolemy extended the habitable area halfway around the world. Beyond that meridian, where he left unknown land, the modern [...] [voyagers] have added Cathay and territory as vast as sixty degrees of longitude, so that now the earth is inhabited over a greater stretch of longitude than is left for the ocean. To these regions, moreover, should be added the islands discovered in our time under the rulers of Spain and Portugal, and especially America, named after the ship's captain who found it. On account of its still undisclosed size it is thought to be a second group of inhabited countries. There are also many other islands, heretofore unknown. So we have little reason to marvel at the existence of antipodes or antichthones. Indeed, geometrical reasoning about the location of America compels us to believe that it is diametrically opposite the Ganges district of India."[17]

---

15 See W. G. L. Randles, *De la Terre plate au globe terrestre*, pp. 44–5, and E. Grant, *Planets, Stars, & Orbs*, p. 635.

16 See T. Goldstein, "The Renaissance Concept of the Earth in Its Influence upon Copernicus," p. 38.

17 *De revolutionibus* I, 3 (Rosen, pp. 9–10; Zekl, pp. 90–2). According to E. Rosen,

These new geographical discoveries provide a sufficient justification for Copernicus' conclusion

> that land and water together press upon a single centre of gravity; that the earth has no other centre of magnitude; that, since earth is heavier, its gaps are filled with water; and that consequently there is little water in comparison with land, even though more water perhaps appears on the surface. The earth together with its surrounding waters must in fact have such a shape as its shadow reveals, for it eclipses the moon with the arc of a perfect circle. Therefore the earth is not flat, as Empedocles and Anaximenes thought; nor drum-shaped, as Leucippus; nor bowl-shaped, as Heraclitus; nor hollow in another way, as Democritus; nor again cylindrical, as Anaximander; nor does its lower side extend infinitely downward, the thickness diminishing toward the bottom, as Xenophanes taught; but it is perfectly round, as the philosophers hold.[18]

In direct opposition to the premises of the traditional "peripatetics" who argued that earth and water maintain separate spheres, Copernicus uses contemporary geographical knowledge to confirm the globe to be a solid geological mass whose chasms are filled with water. The whole has to be conceived of as a solid, balanced globe whose centre of gravity is also its centre of magnitude. earth and water form a single globe with one single centre of gravity and of – relatively – perfect *rotunditas*. As a single, homogeneous and balanced globe, Copernicus' earth is thus prepared for the uniform circular movement around its axis and ready to be launched on its planetary voyage.

Commentary, p. 346, Copernicus' inspiration here was – at least partly – Pliny the Elder, *Historia naturalium* II, 68, 173 (Rackham, vol. 1, pp. 306–7). D. Knox, "Copernicus's Doctrine of Gravity and the Natural Circular Motion of the Elements," p. 190, n. 185, believes that Pliny's arguments may have encouraged Copernicus' conclusion that the earth and water shared a common centre of gravity. Copernicus' reference to Ptolemy is to his *Cosmographia* VII, 5, 14. According to A. Goddu, "Copernicus's Annotations – Revisions of Czartoryski's 'Copernicana'," pp. 204–26, and 222, Copernicus annotated *Cosmographia* translated by Jacopo d'Angiolo (fl. 1470).

18 *De revolutionibus* I, 3 (Rosen, p. 10; Zekl, p. 92). Here are the dates of the cited philosophers: Empedocles (ca. 490–430 BC), Anaximenes (585–528 BC), Leucippus (fifth century BC), Heraclitus (ca. 535–ca. 475 BC), Democritus (ca. 460–ca. 370 BC); Anaximander (ca. 610–ca. 546 BC), and Xenophanes (ca. 570–ca. 475 BC).

CHAPTER SEVEN

The Postulate of Uniform Circular Motion of Celestial Spheres and Celestial Phenomena

Copernicus' terrestrial globe, ready to be launched on its planetary voyage, will perform its voyage pursuant to predetermined rules. When Copernicus elevates it to a position of a planet, the earth is able to move in the same way as other celestial bodies, that is, in uniform circular motion. Celestial motion is, also in Copernicus' view, "uniform and circular, perpetuous, or compounded of several circular motions."[19]

Circularity. In the *Commentariolus*, Copernicus derives the postulate of uniform circular celestial motion from the form or shape of a celestial body.[20] He does the same in Chapter 4 of Book I of *De revolutionibus* by linking uniform circular motion to the sphere and its shape, which carries individual celestial bodies:

> I shall now recall that the motion of the heavenly bodies is circular, since the motion appropriate to a Sphere is rotation in a circle. By this very act the sphere expresses its form in the simplest body, where neither beginning nor

---

[19] *De revolutionibus* I, 4 (Rosen, p. 11; Zekl, p. 94).

[20] See supra, p. 38, and n. 14.

> end can be found, nor can the one be distinguished from the other, while itself traverses the same points to return upon itself.[21]

This passage, which scholars find very difficult to interpret, becomes more understandable if we take into account that in it Copernicus exposes four common suppositions of Ptolemaic astronomy. These assumptions found support in a variety of philosophical and astronomical sources, but Copernicus presents them in a way that enables him to use them as starting points for his own cosmology, that is to say, without any disturbing particularities and details that would compromise his heliocentric and geokinetic project.

The first assumption Copernicus shared with Ptolemaic theory was his assertion that circular motion of the celestial bodies (i.e. stars and planets) is dependent on the sphere that carries them around. He seems to be speaking of the whole heavenly sphere, but he does it in such a way that it can be extrapolated to any single sphere.

Copernicus also held with Ptolemaic theorists that the movement appropriate to the sphere is circular. It is an astronomical as well as philosophical common assumption. Regiomontanus, for example, in his (and Peurbach's) *Epytoma Almagesti*, published in 1496, says the following in reference to the heavens: "Finally, they attributed to such a noble body the most dignified and to the circular movement most appropriate spherical figure."[22] Similar assertions connecting spherical shape with circular movement can be also found in different philosophical traditions. John Buridan claims in his *Quaestiones in Aristotelis De caelo* that "the spherical figure is an aptitude for the spherical or circular movement."[23] In his *Theologia platonica* Marsilio Ficino writes that "the natural movement of the elements is what does not trespass beyond the bounds of nature

21 *De revolutionibus* I, 4 (Rosen, p. 10; Zekl, p. 94).

22 Johannes Regiomontanus, *Epytoma Almagesti* I, 1: "Demum corpori tam nobili dignissimam, et motui circulari accomodatissimam attribuebant figuram sphericam." Compare also with the end of the "Conclusio prima": "Celo igitur cuncta reliqua comprehensuro figuram impressit sphericam omnium capacissimam. Ad velocitatem quoque motus, quae in hoc corpore reperitur maxima et regularissima, decuit eligere figuram sphericam. Sphaera enim […] circa centrum suum mota nihil penitus habet resistentiae, similitudine partium superficiei sphaericae id efficientis. Nam unaquaeque earum locum sibi vicinae partis subintrat, nullo extrinseco corpore aut cedente aut resistente, quod profecto nullis alterius figurae corporibus accidere constat." See also D. Knox, "Ficino, Copernicus and Bruno on the Motion of the Earth," p. 358, n. 86.

23 Buridan, *Quaestiones in Aristotelis De caelo* II, 22 (not in Clagett's translation; Patar, p. 500).

and of natural place; in other words, it is an everlasting circular movement in its place and sphere perfectly resembling the sphere's natural shape"[24] and in his *Compendium in Timaeum* he comes to the conclusion that "every moving sphere, whether elemental or celestial, moves naturally in a circuit in accordance with its own form, so that it may take fuller advantage of its own position."[25]

Another common assumption is that there is neither beginning nor end to circular movement and that neither beginning nor end can be distinguished from the other since the sphere "traverses the same points to return upon itself." Copernicus' assertion is highly reminiscent of Aristotle's argument that the heavens undergo circular movement which has "no beginning or end."[26] Also similar is Pliny's reflection in *Historia naturalium* on the nature of the spherical shape of the universe and its circular movement; indeed, this passage references almost everything mentioned so far:

> Its [i. e. universe's] shape has the rounded appearance of a perfect sphere. This is shown first of all by the name of 'orb' which is bestowed upon it by

24 *Theologia Platonica* IV, 1 (Allen and Hankins, pp. 284–5).

25 *Compendium in Timaeum*, 25 (Farndell, p. 41; *Opera omnia*, p. 1449). See also D. Knox, "Ficino, Copernicus and Bruno on the Motion of the Earth," pp. 358–9, and 337. Knox warns, however, that the evidence of Copernicus' use of Ficino as a source is inconclusive. Compare also with Plato, who claimed that the world's body is a sphere, rotating on its axis. *Timaeus* 33b–34a (Cornford, pp. 54–5): "And for the shape he gave it that which is fitting and akin to its nature. For the living creature that was to embrace all living creatures within itself, the fitting shape would be the figure that comprehends in itself all the figures that are; accordingly, he turned its shape rounded and spherical, equidistant every way from centre to extremity – a figure the most perfect and uniform of all; for he judged uniformity to be immeasurably better than its opposite. [...] For he assigned to it the motion proper to its bodily form, namely that one of the seven which above all belongs to reason and intelligence; accordingly, he caused it to turn about uniformly in the same place and within its own limits and made it revolve round and round [...]." Alcinous, *The Handbook of Platonism* 12, 3 (Dillon, p. 21), summarized Plato's account in a few lines: "By leaving nothing outside it, he made the world both unique of its kind, and likened numerically to its form, which was itself one. [...] By way of shape, he bestowed on it sphericity, seeing as that is the fairest of shapes and the most capacious and mobile. [...] and removing from it all other types of motion, he granted it circular motion, this being that proper to intellect and thought." Alcinous took this passage almost verbatim from Eusebius of Caesarea, *Preparatio evangelica* XI, 23.

26 See *De caelo* I, 9, 279b1–3 (Leggatt, p. 91): "Thus, that it undergoes an unceasing movement is in accordance with reason; for all things cease moving when they reach their own place, but, in the case of the body that moves in a circle, the place it begins from and ends at is the same." See also ibid. II, 6, 288a17–27.

> the general consent of mankind. It is also shown by the evidence of the facts: not only does such a figure in all its parts converge upon itself; not only must it sustain itself, enclosing and holding itself together without the need of any fastenings and without experiencing an end or a beginning at any part of itself; not only is that shape the one best fitted for the motion with which, as will shortly appear, it must repeatedly revolve [...].[27]

The final suppositional commonality seems to be that the sphere in question moves in aether or some other "simplest body." Since in Aristotle's *De caelo*, and Ptolemy's *Almagest* (as well as Regiomontanus' *Epytoma Almagesti*), the circular movement of the heaven (or in the heaven) is inherently linked to its spherical form and aethereal nature,[28] it seems that when Copernicus says that by rotation in a circle the sphere expresses its form *in simplicisimo corpore*, he actually means to say that the sphere is moving in the simplest body, that is, *in* aether (or in some other unspecified simplest body), and not that the sphere *itself* is the simplest (geometrical) body. If this interpretation is correct, then Copernicus is not saying that the sphere "expresses its form as the simplest body" (that the sphere is the simplest of the geometrical bodies); rather, he is saying that the sphere "expresses its form in the simplest body" (that the sphere moves in the simplest substance, that is, in aether).

---

27 Pliny the Elder, *Historia naturalium* II, 2, 5 (Rackham, vol. 1, pp. 172–3).

28 For Aristotle's conception, see *De caelo* I, 3, 270a12–22; I, 4, 271a20–28; II, 3, 286a8–12 (Leggatt, p. 121): "Each thing that has a function is for the sake of the function. Now, the actuality of god is immortality, which is everlasting life. In consequence, everlasting movement must belong to the divine. Since the heaven is such (for it is a divine body), it thereby possesses the circular body which by nature always moves in a circle." See also ibid. II, 6, 288a34–b1. For Ptolemy, see *Almagest* I, 3 (Toomer, p. 40; Heiberg, pars 1, pp. 13–4): "Furthermore, one can reach this kind of notion from certain physical considerations. E. g., the aether is, of all bodies, the one with constituent parts which are finest and most like each other; now bodies with parts like each other have surfaces with parts like each other; but the only surfaces with parts like each other are the circular; among planes, and the spherical, among three-dimensional surfaces. And since the aether is not plane, but three-dimensional, it follows that it is spherical in shape. Similarly, nature formed all earthly and corruptible bodies out of shapes which are round but of unlike parts, but all aethereal and divine bodies out of shapes which are of like parts and spherical. [...] For this reason it is plausible that the aether surrounding them, too, being of the same nature, is spherical, and because of the likeness of its parts, moves in a circular and uniform fashion." See also A. De Pace, *Niccolò Copernico e la fondazione del cosmo eliocentrico*, pp. 108–9, who also refers to Ptolemy and Aristotle as Copernicus' dialectical opponents here.

But one needs to be careful. It seems to me that Copernicus here and in subsequent chapters of Book I of *De revolutionibus* does not connect the circular movement with the aethereal or with any other simple substance of the heaven. He makes no mention of aether, nor does he explicitly specify what the simplest body is, but he seems to be conceiving of a sphere as the simplest body, and defining the simplicity of a body by its form (or shape or figure), and maintaining that the simplest shape is spherical. In the above quoted passage, he is speaking of a sphere or an orb *in abstracto*: movement "appropriate to [*any*] sphere" – not just to (a) celestial sphere(s) that is (are) moving in aether – "is rotation in a circle." From this it follows that Copernicus geometrizes the motion of the spheres by transferring the question of motion into the field of "geometric reasoning." He transforms Aristotle's definition of simple bodies as primary elements or bodies (fire, earth, air, water, and aether) which cannot be broken down into simpler substances.

Where he explains his own position more clearly in the subsequent chapters of Book I Copernicus always begins his argument for rotations with the form (or shape or figure) of the body in question though he never claims – and this is a point that needs to be stressed – that shape is *the cause* of any motion. In the passage above, Copernicus claims that the appropriate motion of the sphere is "rotation in a circle"; he does not claim that that all spheres rotate due to their sphericity or that sphericity in itself sufficiently accounts for their rotation. Circular motion is just the most naturally "appropriate" to the spherical shape of the sphere; the "form of a sphere" can "entail" the motion;[29] circular motion is "appropriate by nature to its form."[30] Nonetheless, we may infer from the above that it is more natural for the sphere to move than to be at rest. Copernicus' use of the *forma/actus* pair in his formulation *ipso actu formam suam exprimentis* ("by this very act the sphere expresses its form") is highly indicative. According to Aristotle, things fully realize themselves by actualizing their natures, their inner *formae*. In the case of geometrical solids, their natures, their *formae*, are their visible shapes, their visible forms. Thus Copernicus might be implying that movement is a more natural state for the sphere than rest, because by rotational movement

---

29 See *De revolutionibus* I, 5 (Rosen, p. 11; Zekl, p. 98): "Now that the earth too has been demonstrated to have the form of a globe, we must in my opinion see whether also in this case the form entails the motion [...]."

30 See ibid. I, 8 (Rosen, p. 16; Zekl, p. 114): "We regard it as a certainty that the earth [...] is bounded by a spherical surface. Why then do we still hesitate to grant it the motion appropriate by nature to its form [...]?"

its *forma* expresses itself and in a way actualizes, realizes itself.[31] But this still does not mean that *all* spheres rotate.

Uniformity. Copernicus begins his explanation on the uniformity of celestial motions by stating simply that "[t]here are many motions" in the heavens "in connection with the numerous [celestial] orbs."[32] Basically, celestial orbs are responsible for all celestial motions. Hence, all celestial motions or motions of the celestial bodies are circular and uniform. If celestial orbs cause the movements of visible celestial bodies and if the latter perform these movements by virtue of their own rotation, the motion of each celestial body must by necessity be circular (involving a single orb) or compounded of several circular motions (involving several orbs). Therefore, there are many spheres or orbs in the heavens that "actualize" themselves, so to speak, by performing axial rotation, which expresses their spherical form. However, there are also celestial phenomena or apparent celestial motions.

The most visible of these is daily rotation. To the earth-bound observer it appears as though the whole sky makes its daily rotation around the celestial axis. "The most conspicuous of all is the daily rotation," Copernicus writes, "which the Greeks call *nuchthemeron*, that is, the interval of a day and a night. The entire universe, with the exception of the earth, is conceived as whirling from east to west in this rotation. It is recognized as the common measure of all motions, since we even compute time itself chiefly by the number of days." [33] The second category of celestial revolutions comprises the motions of the sun, moon, and five planets, all of which take place in the opposite direction, that is, from west to east. "Secondly, we see other revolutions as advancing in the opposite direction, that is, from west to east; I refer to those of the sun, moon, and five planets. The sun thus regulates the year for us, and the moon the month, which are also very familiar periods of time. In like manner each of the other five planets completes its own circuit."[34]

The second category of celestial phenomena differs from the former in the following ways: (1) the sun, the moon, and the five planets revolve around poles different from those of the "first heaven"; the plane of the ecliptic, in which the

---

31 Compare with M. Wolff, "Impetus Mechanics as a Physical Argument for Copernicanism," p. 219.

32 *De revolutionibus* I, 4 (Rosen, p. 10; Zekl, p. 94).

33 Ibid.

34 Ibid.

sun, the moon, and the planets move, is inclined with respect to the plane of the celestial equator; (2) the sun, the moon and the other five planets also do not move at a constant speed; sometimes they move more swiftly and at other times more slowly; (3) in comparison with the sun, which always moves forward, the other five planets, the errants, also move backwards, stop and resume forward motion again; (4) the planets sometimes appear closer to the earth and sometimes farther away from it.

> Yet [these motions] differ in many ways [from the daily rotation or first motion]. In the first place, they do not swing around the same poles as the first motion, but run obliquely through the zodiac. Secondly, these bodies are not seen moving uniformly in their circuits, since the sun and moon are observed to be sometimes slow, at other times faster in their course. Moreover, we see the other five planets also retrograde at times, and stationary at either end [of the regression]. And whereas the sun always advances along its own direct path, they wander in various ways, straying sometimes to the south and sometimes to the north; that is why they are called 'planets'. Furthermore, they are at times nearer to the earth, when they are said to be in perigee; at other times they are farther away, when they are said to be in apogee.[35]

In this way, Copernicus grounds the geometrical-physical thesis on the rotation of the spheres, "in connection with which" there are many motions of the celestial bodies, in opposition to the apparent celestial motions.[36] The task before him – and every single astronomer – is to bring the phenomena (as seen through the eyes of a geocentric and geostatic observer) in line with autonomous principles of celestial motions. In other words, it is necessary to make observable phenomena explicable on the basis of geometric-physical principles. Irregular, non-uniform motion of planets and an increasing or decreasing distance from the earth, particularly with regard to the motions of the sun, the moon, and the planets, raise some difficulties which can nevertheless be resolved. The non-uniformities of their motions occur pursuant to certain regularity. Therefore, although they may appear unequal, non-uniform "to us," the motions "in themselves" are equal, uniform and circular:

> We must acknowledge, nevertheless, that their motions are circular or compounded of several circles, because these nonuniformities recur regularly

---

35 Ibid. (Rosen, pp. 10–1; Zekl, pp. 94–6).

36 See also J.-J. Szczeciniarz, *Copernic et la révolution copernicienne*, p. 154 et seq.

> according to a constant law. This could not happen unless the motions were circular, since only the circle can bring back the past. Thus, for example, by a composite motion of circles the sun restores to us the inequality of days and nights as well as the four seasons of the year. Several motions are discerned herein, because a simple heavenly body cannot be moved by a single orb non-uniformly.[37]

Thus, the postulated principles of celestial motions are, as Copernicus maintains, also confirmed empirically. Since the inequalities, or non-uniformities of the apparent motions occur in cycles, the actual motions must by necessity be circular or compounded of several circular motions. Every kind of inequality or irregularity is subjected to some higher-level equality, regularity, or constancy over a longer period of time. Therefore, inequality, irregularity itself subsumes equality, regularity. It can only be based on a combination of circular motions. "Orderly," "regular" irregularity can only result from a combination of circular motions that account for it. He summarizes: "This could not happen unless the motions were circular, since only the circle can bring back the past."[38]

The necessity for non-uniformity to arise from a combination of a larger number of uniform elements is also confirmed by the analysis of a body that causes circular motion. Copernicus demonstrates the uniformity of celestial motions by a *reductio ad absurdum* argument. The apparent non-uniformity, which, in the long term, proves to be "orderly disorderliness," can only come from a larger number of regular circular motions, "because a simple heavenly body cannot be moved by a single sphere nonuniformly."[39] Non-uniform circular motion can in no way result from the motion of one sphere or orb, since both possible causes of non-uniformity are inconceivable, that is to say, completely incongruent with the nature of the sphere. A simple celestial body – a body of a spherical shape – could move non-uniformly for two reasons: because of an inconstancy in the *virtus movens* (be it something external or an inner nature),[40] or a variation or inequality (*disparitas*) of the body revolved: "Id enim euenire oporteret vel propter virtutis mouentis inconstantiam, siue asciticia sit siue intima natura, vel propter reuoluti

---

[37] *De revolutionibus* I, 4 (Rosen, p. 11; Zekl, p. 96). It is not clear to me why Copernicus ascribes to the sun "a composite motion of circles."

[38] Ibid.

[39] Ibid.

[40] This *virtus movens* can be identified as a separate Intelligence (and ultimately God as *primus movens*), or as the inner soul of a body moved.

corporis disparitatem."[41] What the potential sources of non-uniform motions are remains irrelevant; Copernicus rejects both, because "from either alternative [...], the intellect shrinks," and because "it is improper to conceive any such defect in objects constituted in the best order." What is of decided importance is that on the basis of this consideration it should seem reasonable "that their uniform motions appear non-uniform to us."[42]

Let me recapitulate. It is an axiomatic truth that only regular, uniform, and circular motions are possible in the heavens. Being the realm of "the best order," the heavens do not allow any kind of inequality or non-uniformity to cause a sphere that moves in a uniform circular motion, made possible by its shape, to move "inconstantly," that is, in any other type of motion than uniform and circular: "[A] simple heavenly body cannot be moved by a single sphere nonuniformly." The only remaining alternative is that the apparent "inequality" or "non-uniformity" derives from a combination of several equal or uniform motions. The nature of the heavens and celestial spheres, which cause the motions of the celestial bodies, does not allow unequal apparent motions of a planet to arise from unequal or nonuniform motion of a sphere; inequality or non-uniformity should be explained by a system of spheres that move in uniform and circular motion. There are two possible explanations, both presupposing uniform motions of spheres, for those celestial inequalities performed by heavenly bodies and observed in the heavens: that their circles "have different poles" or "that the earth is not at the centre of the circles on which they revolve":

---

41 Translators and interpreters disagree yet again on what exactly Copernicus had in mind here, since expression *siue asciticia sit siue intima natura* may refer to *inconstantia* or to *virtus mouens*. E. Rosen translates as if *siue ascitia sit siue intima natura* refers to *inconstantia*: "For this nonuniformity would have to be caused either by an inconstancy, whether imposed from without or generated from within, in the moving force or by an alteration in the revolving body." N. Jardine, "The Copernican Orbs," p. 192, n. 65, and 176 et seq., also argues that it refers to *inconstantia*. His translation of the passage is as follows: "[...] it is impossible for a simple body to be moved non-uniformly by a single orb, for that would have to come about either because of an inconstancy in the motive power, whether due to something adventitious or due to inner nature, or because of a disparity of the body revolved [...]." His view is also shared by M.-P. Lerner, *Tre saggi sulla cosmologia*, p. 39, n. 3, and *Le monde des sphères*, vol. 1, p. 134. According to A. Koyré, *Des révolutions des orbes* célestes, p. 70, and A. De Pace, "Copernicus against a Rhetorical approach to the Beauty of the Universe," pp. 93–4, n. 61, and *Niccolò Copernico e la fondazione del cosmo eliocentrico*, p. 110, n. 322, the expression *siue ascitia sit siue intima natura* referes to *virtus mouens*. I agree with this interpretation.

42 *De revolutionibus* I, 4 (Rosen, p. 11; Zekl, p. 96).

> From either alternative, however, the intellect shrinks. It is improper to conceive any such defect in objects constituted in the best order. It stands to reason, therefore, that their uniform motions appear nonuniform to us. The cause may be either that their circles have different poles [from the earth's] or that the earth is not at the centre of the circles on which they revolve. To us who watch the course of these planets from the earth, it happens that our eye does not keep the same distance from every part of their transit, but on account of their varying distances these bodies appear larger when nearer than when farther away (as has been proved in optics). Likewise, in equal arcs of their orbs their motions will appear unequal in equal times on account of the observer's varying distance.[43]

Copernicus' reasoning seems to be the following. Ptolemaic astronomy explains phenomenal irregularities or inequalities by the establishment of the "second, different motion of the whole" or the "second revolution," preserving the postulate of their equal motion.[44] The movement of the planets is irregular from

---

43 Ibid. The reference to "optics" is according to E. Rosen, *Commentary*, p. 349, to Euclid's *Optica ex recensione Theonis*, prop. 5 (Menge, pp. 158–9), in translation of Bartolomeo Zamberti. Here is Rosen's translation: "Equal magnitudes at unequal distances look unequal, and the one nearer to the eye always seems larger." The claim that "in equal arcs of their orbs their motions will appear unequal in equal times on account of the observer's varying distance" is, as Rosen explains, ibid., p. 349, a special case of the optical principle cited by Copernicus in *De revolutionibus* I, 10 (Rosen, p. 18; Zekl, p. 124), "that of objects moving equally fast, those further away seem to travel more slowly as is proved in Euclid's *Optics*."

44 Compare with *Almagest* I, 8 (Toomer, p. 46; Heiberg, pars 1, pp. 26–7): "The greatest of these circles is called the 'equator', because it is the only [such parallel circle], which is always bisected by the horizon (which is a great circle), and because the revolution which the sun makes when located on it produces equinox everywhere, to the senses. The other motion is that by which the spheres of the stars perform movement in the opposite sense to the first motion, about another pair of poles, which are different from those of first rotation. [...] Now if this motion of the planets too took place along circles parallel to the equator, that is, about the poles which produce the first kind of revolution, it would be sufficient to assign a single kind of revolution to all alike, analogous to the first. For in that case it would have seemed plausible that the movements which they undergo are caused by various retardations, and not by a motion in the opposite direction. But as it is, in addition to their movement towards the east, they are seen to *deviate continuously to the north and south [of the equator]*. Moreover the amount of this deviation cannot be explained as the result of a uniformly-acting force pushing them to the side: *from that point of view it is irregular, but it is regular if considered as a result of [motion on] circle inclined to the equator.* Hence we get the concept of such a

the equator-point-of-view, but Ptolemy finds it "regular if considered as a result of [motion on] circle inclined to the equator."[45] On the other hand, Copernicus may also refer to the homocentric astronomy, which is, according to Averroes based on "two different poles and more."[46] Or he may be deliberately ambiguous. Whatever the case, these apparent irregularities, inequalities of the planetary motions can *also* be explained (Copernicus says: *vel ... sive etiam*) with the use of epicycles on the deferents (by any combination of the circles) in which "the earth is not at the centre of the circles on which they revolve." In the astronomy of eccentrics and epicycles, the cause of the apparent variation in the magnitudes and speeds of heavenly bodies is explained by the fact that the earth is not the centre of their circuits meaning that the distance between the moving object and the observer varies. "[O]n account of their varying distances these bodies seem larger when nearer than when farther away," he writes, and of the apparent inequality of their speed he reasons that "in equal arcs of their orbs their motions will appear unequal in equal times on account of the observer's varying distance." In the former case, the inequality is that of magnitude and in the latter, that of velocity. To us situated on the earth and away from the centre of the observed motion, the angle the planet is observed to make in the course of its journey against the background of the sphere of the fixed stars is smaller when the planet is farther away and larger when it draws nearer. Since motion "in itself" is in both instances equal or uniform, it will necessary appear swifter to us when the planet approaches us than when it recedes from us (see *Figure 3a*, supra, p. 47).

This, in turn, also means that the apparent inequality of motion, to which Copernicus narrows his investigation, stems not from motion "in itself" but from the *relationship* between the observer and the body in motion. It cannot be otherwise. If, for the reasons described above, motions "in themselves" are necessarily

circle, which is one and the same for all planets, and particular to them. It is precisely defined and, so to speak, drawn by the motion of the sun, but it is also traveled by the moon and the planets, which always move in its vicinity, and do not randomly pass outside a zone on either side of it which is determined for each body. Now since this too is shown to be a great circle, since the sun goes to the north and south of the equator by an equal amount, and since, as we said, the eastward motion of all of the planets takes place on one and the same circle, it became necessary to suppose that this second motion of the whole takes place about the poles of the inclined circle we have defined [i. e. ecliptic], in the opposite direction to the first motion." Emphasis added.

45 See the note above.

46 See *In Metaphysicam* XII, 8, comm. 45 (Genequand, p. 179). See infra, p. 246, n. 11, for the quotation.

equal, the "cause" of their apparent inequality can only be the observer and his position in respect to the planetary motions. Inequality or non-uniformity as it appears "to us" also occurs *because of us*. In the Ptolemaic astronomy of eccentrics and epicycles, it arises from the fact that the earth and consequently the terrestrial observers of celestial inequalities are not at the centre of the celestial ballet. The Ptolemaic explanation of the apparent celestial inequalities depends heavily on the relationship between the observer on the earth and the observed moving body, as well as on the observer's displacement from the centre of the observed movement. To rectify the geometrical-physical principle of equal, uniform rotation of the orbs which carry the sun, the moon, and the five planets with their apparent unequal, non-uniform revolutions, it suffices to move the observer "away from the centre" of their motions.

Copernicus emphasizes the need to carefully examine the Ptolemaic situation of the earthly observer's "off-centredness" and the impact of his changing relationship to the phenomena of inequality in celestial motion. These alone is sufficient indicator that all celestial phenomena rest on "the relation of the earth to the heavens," and he concludes "Hence I deem it above all necessary that we should carefully scrutinize the relation of the earth to the heavens lest, in our desire to examine the loftiest objects, we remain ignorant of things nearest to us, and by the same error attribute to the celestial bodies what belongs to the earth."[47]

[47] *De revolutionibus* I, 4 (Rosen, p. 11; Zekl, pp. 96–8). Compare with *De revolutionibus* I, 5 (Rosen, p. 12; Zekl, p. 100): "For the fact that the same planets are observed nearer to the earth and farther away necessarily proves that the centre of the earth is not the centre of their circles. It is less clear whether the approach and withdrawal are executed by the earth or the planets."

CHAPTER EIGHT

The First Motion of the earth and the "Optical Argument"

Copernicus presents in Chapter 5 of Book I of *De revolutionibus* the earth's possible motion as a logical consequence of its spherical form or shape, and once again underlines the necessity to rethink its position in the universe: "Now that the earth too has been demonstrated to have the form of a globe, we must in my opinion see whether also in this case the form entails the motion, and what place in the universe is occupied by the earth. Without the answers to these questions it is impossible to find certain explanation of the appearances in the heavens."[48] He has shown that the earth is a sphere and that earth and water make up a single sphere with a single centre of gravity and magnitude. He has also asserted that the sphere expresses – and in a way actualizes – its form or shape by moving in a circular motion around its axis. The question that now naturally arises is whether the earth itself should also be assigned circular motion, regardless of the fact that "there is general agreement among the authors that the earth is at rest in the middle of the universe" and that "they hold the contrary view to be inconceivable or downright silly." Despite the general consensus that the earth is motionless, the preliminary investigations have led Copernicus to conclude that this question "has not yet been solved" at all.[49]

48 *De revolutionibus* I, 5 (Rosen, p 11; Zekl, p. 98).

49 Ibid.

In what follows I compare how Copernicus and Ptolemy use the so-called optical argument, that is, the relativity of movement or the phenomenological (or optical) equivalence of the earth's rotation and heavens' rest and vice versa. During his argumentation in Chapter 5 and especially in Chapter 8 Copernicus leads us almost to believe that the optical argument is a positive argument in favor of the earth's rotation. On the other hand Ptolemy has basically identical theory of relativity of motion as Copernicus and also evokes it in his discussion in Chapter 7 of Book I of the *Almagest* but concludes that the earth is at rest in the centre of the universe. How is that possible? How do Copernicus and Ptolemy differ in their application of the principle of the relativity of motion? And, if the relativity of motion is a universal principle, what are the criteria for establishing what is true and what is a mere appearance of the truth? And how does this help Copernicus to affirm motion of the earth?

The "optical argument": Copernicus. Copernicus addresses the question of the earth's motion by tacitly drawing on a paragraph in Euclid's *Optics*. Euclid wrote that "if several objects are moving at different speeds in the same direction as that in which the eye is also being carried, any object moving at equal speed with the eye seems to be stationary."[50] Similarly, Copernicus writes: "Every observed change of place is caused by a motion of either the observed object or the observer or, of course, by an unequal displacement of each. For when things move with equal speed in the same direction, the motion is not perceived, as between the observed object and the observer, I mean."[51] The question of the possibility of the earth's motion is therefore subsumed under the more universal question regarding motion in general. The discussion whether the earth moves or is at rest is only relevant if it is based on an established universal theory of motion and rest. Or, to put it more accurately: if we know the unstated assumptions of every motion or rest, then we know what it actually means "to move." Copernicus' paraphrase reveals the following: firstly, he is interested in local movement (*omnis enim quae videtur* secundum locum mutatio); secondly, he speaks of every single local movement (omnis *enim quae videtur secundum locum mutatio*); and thirdly, he is concerned with local movement that is visible, seen (*omnis enim* quae videtur *secundum locum mutatio*), or perceived. What Copernicus considers most essential in dealing with the general nature of locomotion, which is relevant

---

[50] See E. Rosen, *Commentary*, p. 349, and Euclid, *Optica ex recensione Theonis*, prop. 50 (Menge, pp. 238–9).

[51] *De revolutionibus* I, 5 (Rosen, pp. 11–2; Zekl, p. 98).

for the problem of the earth's motion, is what happens when *we* see something, an object or a body to move or change its position. For Copernicus, motion "in itself," motion that takes place independently of us (observers, spectators), is in a way completely irrelevant to the problem under scrutiny.

In this context, Copernicus recognizes three situations that give rise to the appearance of motion. In the first possible situation the observed object moves while the observer is at rest. This situation is the most "natural" and therefore accepted by *sensus communis* as the only real one. It provides the basis on which the observer decides on each apparent motion. If something appears to our *sensus communis* to be in motion, it moves. The second possible situation in which one perceives motion is diametrically opposed to the first one. According to Copernicus, motion is also perceived when the observer or spectator moves, while the observed object is at rest. This instance, too, involves apparent motion, but it is not the motion of the observer or spectator that is apparent. Copernicus asserts that in this instance, too, the observed object appears to be moving, even though it is, in fact, motionless. The third possible situation of motion perception is when both the observer and the observed object move, but only if they move unequally. As can be inferred from the last sentence of the excerpt under examination – "For when things move with equal speed in the same direction, the motion is not perceived, as between the observed object and the observer, I mean [...]" – the inequality can manifest itself as any kind of disparity between the motions of the observer and the observed or, if they move in the same direction, as the unequal or non-uniform speed of their motion.

To put it in more general terms: apparent motion can only be perceived when there is inequality, which can be "absolute" (one element in the relationship is at rest and the other is in motion) or "relative" (the two elements in the relationship move in different directions or, if both move in the same direction, one moves faster than the other). This means that every motion is crucially determined by the relationship between the observer and the object in motion – or at rest. Every single motion is therefore based on a relationship. From the relative nature of motion it follows that locomotion is first and foremost a phenomenon, an appearance. It also follows that it is possible on the level of phenomena to mistake motion for rest. For instance, when we move, we may perceive objects to be moving as well, although they are, in fact, stationary. This means that when we deal with apparent motion on the level of perception, we cannot decide in advance what is actually in motion and what is at rest. The fundamental characteristic of apparent motion is *a priori* incapacity to discern between what is in motion and what is motionless; between what serves as a point of reference and what this point refers to.

And this holds for every motion. Relativity of motion is universal; it governs the entire universe. Therefore, the second situation of motion perception, in which the observer moves and perceives the observed object to move as well, although it is in fact motionless, can be observed in the universe as well. This applies to any kind of motion that may be ascribed to the earth. Whether the earth moves from one place to another as in the earth's revolution around the sun, or rotates in the same place, its motion would have effects on the objects in relationship with it. More specifically, and applied to its assumed rotational movement: even if the earth moved and the heavens were at rest, the observed phenomena would be identical to those observed under the assumption that the heavens move and the earth does not.

> It is the earth, however, from which the celestial circuit is beheld in its repeated performances before our eyes. Therefore, if any motion is ascribed to the earth, in all things outside it the same motion will appear, but in the opposite direction, as though they were moving past it. Such in particular is the daily rotation, since it seems to involve the entire universe except the earth and what is around it. However, if you grant that the heavens have no part in this motion but that the earth rotates from west to east, upon earnest consideration you will find that this is the actual situation concerning the apparent rising and setting of the sun, moon, stars and planets.[52]

In applying the abstract rule of determining the apparent motion (and rest) in the relationship between the heavens and the earth, Copernicus does not employ all three possible situations but only the contrast between the first two, in which the "inequality" between the observer and the observed is "absolute."[53] The

---

52 Ibid. (Rosen, p. 12; Zekl, p. 98). By saying that "upon earnest consideration you will find that this is the actual situation," Copernicus is not – yet – affirming the reality of the rotation of the earth. He is saying that "upon earnest consideration," one can realize that by attributing rotation to the earth and rest to the heavens, there is no consequences as far as the celestial phenomena are concerned.

53 Copernicus could also take account of relative inequality: both the earth and the heavens move although not equally but in different directions and at different speeds. This possibility is mentioned by Ptolemy, *Almagest* I, 7 (Toomer, pp. 44–5; Heiberg, pars 1, p. 24): "But certain people [propounding] what they consider a more persuasive view, agree with the above, since they have no argument to bring against it, but think that there could be no evidence to oppose their view if, for instance, they supposed

observer situated on the earth sees the movement of the heavens. At first glance, the apparent motion of the heavens is the outcome of the first possible situation, in which the observed object moves and the observer is stationary. Yet this, as we know, is not the only possible situation. Apparent circular motion can also arise from the second situation, the one in which the observer of the apparent motion moves and the observed object is stationary. If the earth moves (carrying the observer with it), "in all things outside it the same motion will appear, but in the opposite direction." The movement we perceive is the result of our own motion. The motion of the observer causes the apparent motion of the observed object. The rotation of the earth causes the apparent motion of the celestial vault in opposite direction.

In simplified terms, this relationship yields two possible situations (the earth is at rest while the heavens move; the earth moves while the heavens are at rest), which are completely equivalent from the "optical" point of view; they produce the same visual effect. Regardless of whether we assume that the earth moves and the celestial vault is at rest or vice versa, the observed celestial phenomena are the same. Therefore, since the celestial phenomena can be accounted for by the motion of both the earth and the heavens, the situation remains undecided on the level of phenomena.

Nonetheless, the phenomenological equivalence of these two possibilities does not at all constitute a positive proof of the earth's motion. What does Copernicus benefit from introducing the "optical argument"?

Relativity of motion also has an important role in Copernicus' Chapter 8 of Book I of *De revolutionibus*, in which he compares the diurnal revolution of the heavens and the earth (i.e. the motion of the heavens and the earth) to a ship sailing out of the harbor carrying sailors to whom it appears that the shore and the harbor are moving away. In this presentation, too, Copernicus applies the abstract rules of the "optical argument" to a concrete situation: the apparent motion is produced when the observed object moves and the observer is stationary or vice versa. In both cases it is the observed object that appears to be in motion. Here, Copernicus extends the comparison further by stating that the apparent movement of the town is an image of the movement of the ship. In the sailors'

the heavens to remain motionless, and the earth to revolve from west to east about the same axis [as the heavens], making approximately one revolution each year; *or if they made both heaven and earth move by any amount whatever, provided, as we said, it is about the same axis, and in such a way as to preserve the overtaking of one by the other.*" Emphasis added.

perception, the ship's movement is transposed or mirrored onto the moving shore and the harbor, from which they have set sail. The ship's movement has its image: the movement of the town and the harbor. It is on the basis of this image that the movement of the town is held as true or real. In other words: appearance, mirrored image takes the place of reality. Appearance becomes truth or reality and truth or reality becomes appearance. The same holds true, Copernicus argues, in the case of the moving heavens and a motionless earth. The motion of the heavens is a mirror image of the motion of the earth. However, the motion of the heavens is only an appearance rather than the truth or reality. In fact, just the opposite is true; what in fact takes place is the diurnal rotation of the earth around its axis, while the heavens are at rest.

Since the earth is "bounded by a spherical surface" and since "the motion appropriate by nature to its form" is circular movement,

> [w]hy should we not admit, with regard to the daily rotation, that the appearance is in the heavens and the truth [or reality] in the earth? This situation closely resembles what Virgil's Aeneas says: 'Forth from the harbor we sail, and the land and the cities slip backward.' For when a ship is floating calmly along, the sailors see its motion mirrored in everything outside, while on the other hand they suppose that they are stationary, together with everything on board. In the same way, the motion of the earth can unquestionably produce the impression [or image] that the entire universe is rotating.[54]

The difference in Copernicus' application of the "optical argument" in Chapters 5 and 8 is slight but nonetheless significant. In Chapter 5, he employs the "optical argument" solely for the purpose of demonstrating the phenomenological equivalence between the assumption that the heavens move and the earth is stationary and the assumption that the heavens are stationary and the earth moves. In Chapter 8 he becomes rather more categorical: the earth's motion is reality mirrored in the apparent motion of the heavens. Just as the sailors who have set sail from the harbor perceive the shoreline and the harbor to be drifting away from them, so it appears to us "the inhabitants of the earth" that the heavens move. But in both instances, the apparent motion is only a consequence and image of our own motion. A slight shift in the use of the "optical argument" in relation to Chapter 5 is that Copernicus no longer uses it to point to the general phenomenological equivalence of both possibilities. Rather, the general equivalence is now presented as a platform that has enabled him to deduce – his

54 *De revolutionibus* I, 8 (Rosen, p. 16; Zekl, pp. 114–6).

deduction is first formulated as a rhetorical question and at the end of Chapter 8 as something more probable then the opposite alternative – what truly moves and what is merely an image thereof, and to call for the affirmation of the moving earth and the motionless heavens.

The question naturally arises: What makes Copernicus think he is entitled to assert what the truth is, what reality is and what an image thereof? On what grounds is he able to call for the affirmation of the earth's motion as real and the motion of the heavens as an image of the former? Which is the motion of the boat as seen from the harbor, and which is the motion of the shore as seen from the ship?

The "optical argument": Ptolemy. These questions are all the more pertinent given the fact that the relativity of motion was fully appreciated by Ptolemy, who nonetheless insisted on a geostatic and geocentric cosmos.

In Book II of his *Optics*, Ptolemy maintains that the perception of change in place, that is to say, locomotion, ought to be discussed independently of quantitative and qualitative change since "[v]isual perception of the phenomena associated with locomotion depends primarily on the visual faculty itself [rather than on the object undergoing the change [of place]]."[55] From this it is evident that, like Copernicus, Ptolemy is not interested in motion "in itself" but in apparent motion, motion in relation to the observer, motion as we see it or perceive it. Ptolemy's theory of motion (and rest) is therefore essentially identical to those developed by Copernicus.[56] Ptolemy's conception of motion as visually perceptible is, as Mark Smith puts it, "relative rather than absolute because it is ultimately dependent on a changing relationship between external objects and the vertex of the visual cone."[57]

That Ptolemy's theory is fundamentally identical to Copernicus' is further confirmed and illuminated by his discussion on perceptual *illusions* of motion. Just as Copernicus gives the example of a ship sailing from the harbor and moving away from the shore to draw a sharp distinction between an image (the moving harbor – the stationary ship) and the original (the moving ship – the stationary harbor), between illusion and actuality, so does Ptolemy treat similar examples as illusions. For, regardless of whether the observed object is at rest (and the observer moves) or the observed object moves (and the observer is at rest), the

---

[55] Ptolemy, *Optics* II, 76. All quotations from Ptolemy's *Optics* are taken from the English translation of A. M. Smith, in *Ptolemy's Theory of Visual Perception: An English Translation of the Optics.*

[56] This can be derived from his exposition in *Optics* II, 76 (Smith, p. 103).

[57] A. M. Smith, *Ptolemy's Theory of Visual Perception*, p. 103, n. 94.

optical result is the same: in both cases, it is the observed object that appears to be moving, although this is not the real and actual state of the matter. In the first case, the motion of the observed object is apparent, whereas in the second, the motion of the observed object is real.

Later on in Book II of the *Optics*, Ptolemy provides two examples of apparent motion in which the observer is deceived by perceiving the illusory motion as real and the real motion as illusory according to the already known principles. One of these examples is similar to that used by Copernicus in Chapter 8 of Book I of *De revolutionibus*.[58] If we sail on a ship along the shore during twilight, Ptolemy maintains, and we do not sense the movement of that which is carrying us, we perceive the trees and topographic features of the shoreline to be moving. However, it is also possible to experience the opposite effects of illusory perception,[59] which make us believe that we are moving when in fact we are stationary. This example involves a ship or boat that is (not to our knowledge) anchored in the middle of a calm river. If we do not look at the shoreline but instead focus on the river, it will appear to us that the boat is moving swiftly up the river, while the water is standing still.

Therefore, in aspects relevant to our problem, Ptolemy's theory of motion perception is completely identical to Copernicus'. Motion perception is a matter of relationship. The same apparent motion of an observed object can be perceived both when the observer moves and the observed object is at rest and when the observer is stationary and the observed object moves. Ptolemy explicitly asserts that the latter phenomena of motion are illusions.

Moreover he concludes, just like Copernicus, that the relativity of perceived motion can also have cosmological consequences. Ptolemy, too, explicitly argues that celestial phenomena can be explained by assuming that the earth is at rest and the heavens move as they are by assuming that the earth moves and the heavens do not. In Chapter 7 of Book I of the *Almagest*, in which he demonstrates the earth's immobility, Ptolemy challenges the opinion of "certain people" (most likely referring to Heraclides of Pontus) who hold that the heavens are at rest and the earth makes a revolution from west to east about its celestial axis in approximately one day. This is, roughly speaking, identical to what will be proposed by Copernicus. Such a proposition, Ptolemy argues, contains nothing that would conflict the celestial phenomena. Or, as he puts it: "there is perhaps nothing in

---

58 See Ptolemy, *Optics* II, 132 (Smith, p. 124).

59 See ibid. II, 131 (Smith, p. 123).

the celestial phenomena which would count against that hypothesis."[60] As far as the celestial appearances are concerned, Ptolemy admits, both hypotheses are completely equivalent. Irrespective of whether we assume that the earth moves and the heavens stand still or vice versa, the observed phenomena are the same.

Hence, we have two almost identical "optical" theories of motion, two identical applications of theory to the question concerning the motion of the earth, and two identical opinions that the celestial phenomena can, as a rule, be equally well accounted for by both theses. However they at the same time exhibit radically opposing views of the actual state of the universe. Where Ptolemy insists on a geostatic universe, Copernicus regards the motion of the heavens as appearance and the motion of the earth as reality.

The "optical argument": appearance and truth. How is it possible to arrive at diametrically opposed conclusions on the basis of the same argument and reasoning? Why does Ptolemy, notwithstanding the "optical argument," believe that the earth stands still in the middle of the world? How does the "optical argument" lead Copernicus to the affirmation that with regard to the daily rotation, "the appearance [or illusion] is in the heavens and the truth in the earth"? Are there any – and if so, which – non-explicit criteria that enable us to determine which motion is true and which an illusion? It is easiest to find an answer to these questions by drawing on the analysis of certain examples given by Ptolemy.

Ptolemy's first example of *illusory* motion is identical to that which Copernicus gives in his Chapter 8. If we sail in a boat along the shore and if we do not sense its movement, we perceive the shoreline and the trees to be moving. This argument uses the same logic as Copernicus' example of the sailors who are on board a ship moving away from the shore but nevertheless perceive the harbor and the shore to be drifting away from them. The illusory motion of the shore and the impression that the ship is standing still stem from the fact that we attribute our own motion to a stationary object. How do sailors know what is actually moving? The answer is simply because they know that the shore cannot move. The first possible criterion to distinguish between appearance (an illusion, image) and truth or reality when dealing with apparent motion is therefore banal. *Sensus communis* buttressed by our own and others' experience tells us what can and cannot move. Since we already know in advance that the shore cannot move,

[60] *Almagest* I, 7 (Toomer, p. 45; Heiberg, pars 1, p. 24).

the fact that it so appears can only be the result of an illusion whereby the ship's motion is transferred to the shoreline.

Ptolemy's second example is somewhat more complex. We are in a boat that is, unbeknownst to us, anchored in the middle of a wide river. If we fix our gaze on the water, the movement of its surface may give us the impression that our boat is moving upstream. Since we do not sense the motion of the water, it appears to us that the boat is moving. How can we know what is actually in motion? Do we, together with the boat, move upstream, or does the water alone move? In this case, prior knowledge cannot be used as a guide. In the first example, those who perceive motion are aware that the shore is incapable of any motion of its own, but in this example, both the boat and the water on which it sits are movable. How can we determine whether we are moving or not? Or to put it differently: How can we distinguish between reality and illusion?? In this case, the question can only be answered by reference to "something other," a point external to the relationship between the observer and the observed. This other reference point is the shore, which, as we have learned (from experience or the *sensus communis*) is by nature motionless, and which can ultimately enable us to establish whether we are moving or not. Ptolemy proposes the exact same solution:

> On the other hand, if we look at the water, the shoreline, and the boat all at the same time, and if we take cognizance of the fact that the shoreline is stationary, then we will see that the boat is stationary, since the boat is seen by the same rays that see the shoreline. We will also see the water moving since we will have realized that the boat and the shoreline are stationary.[61]

There is another example that is found neither in Ptolemy nor in Copernicus but is nonetheless extremely interesting, because it is set forth in reference to the problem under our examination and because it transcends both examples presented above. It is treated by Oresme in the aforementioned *Le Livre du ciel et du monde*. Like Ptolemy and Copernicus, Oresme, is also interested in motion as perceived. In his view, locomotion can be perceived only when we can see that "one body assumes a different position relative to another body."[62] We can therefore perceive motion only when we position two bodies or objects relative to one another and determine which body changes its position relative to the other. Oresme supports the thesis with the following example. A man is on board a moving ship, Ship A, and sees nothing except another ship. In this situation,

---

61 Ptolemy, *Optics* II, 131 (Smith, p. 123).

62 Oresme, *Le Livre du ciel et du monde* II, 25 (Menut and Denomy, pp. 522–3).

vision is limited by a frame or screen containing only two elements: the observer (together with Ship A, on which he is standing) and the observed object (Ship B). Nothing outside this relationship exists. It is on this basis that the observer must determine what is moving.

Oresme, too, lists three possibilities of perceiving or not perceiving motion. In the first possibility Ship B moves just like Ship A and therefore motion is imperceptible. In this case, Oresme maintains, it will appear to the observer that neither ship is moving. The second possibility is that ship A is stationary and ship B is moving. In this case, it will appear to the sailor on board Ship A that Ship B is moving. The third example is the most interesting, both in its own right and in connection with the second. If Ship A moves and Ship B is at rest, Oresme posits, it will appear to the observer on Ship A, as before, that Ship A is stationary and Ship B is moving. Thus, Oresme maintains, if "*a* rested an hour and *b* moved, and during the next hour it happened conversely that *a* moved and *b* rested, this man would not be able to sense this change or variation; it would seem to him that all this time *b* was moving. This fact is evident from experience [...]."[63]

Oresme explains: If two bodies, A and B, have a "continual relationship to each other so that, when *a* moves, *b* rests, and, conversely, when *b* moves, *a* rests,"[64] an observer on Object A will perceive movement but may not be able to determine the true source of the motion; he or she will, however, interpret the change as the result of the movement of the other body, Object B. Invoking Witelo's *Optica* or *Perspectiva*, Oresme explains that "to move" means to be now in a different position than before in respect to something else. Motion perception is achieved by "comparing one visible object in motion to some other visible object at rest."[65] Motion is understood by perceiving the change in position of one moving object in relation to another visible object. We do not perceive motion unless we notice that one body is in the process of assuming a different position relative to another: "Now, I take as a fact that local motion can be perceived only if we can see that one body assumes a different position relative to another body."[66] The sailor on

---

63 Ibid.

64 Ibid.

65 Ibid. Erasmus Ciolek Witelo (ca. 1230–after 1280, before 1314). The entire passage from *Optica* IV, 110, reads in Latin as follows: "Quoniam enim moveri est aliter se habere nunc, quam prius: palam quod facilitas huius comprehensionis motus fit ex comparatione rei motae visae ad aliud visibile quiescens non motum. Quando enim comprehenditur situs unius rei mobilis, respectu alterius rei visibilis, tunc etiam comprehenditur diversitas situs eius respectu illius visibilis, et tunc comprehenditur motus."

66 Oresme, *Le Livre du ciel et du monde* II, 25 (Menut and Denomy, pp. 522–3).

Ship A has no "other object" to compare his position to but Ship B. Therefore, even if he is moving on Ship A while Ship B is at rest, it will still appear to him – due to the continual change in the position of Ship B, which, however, is an image of his own motion – that all this time Ship B is moving.

How can one establish reality, truth and its image on the basis of Oresme's third example?So far we have relied on two supporting factors which usually help us resolve the enigma whether our motion is a reality or illusion: advance, prior knowledge and a stationary third point of reference which is external to the relationship and as such enables us to judge movement. But in this particular example, neither advance knowledge nor a third point can be of much help. Both ships are movable objects, and the sailor on Ship A lacks a third element which he knows for certain is motionless and which will enable him to distinguish between what actually moves and what does not, between reality and its mirror image. This example offers no solution: the sailor cannot determine whether it is he who is moving or the observed ship.

Ptolemy, Copernicus, and the enigma of motion of the heavens or the earth. The same situation ensues from the dilemma whether it is the earth or the heavens that rotates. By admitting, albeit at the moment only as a possibility, that the earth, too, is capable of motion, we may find ourselves unable to determine solely on the basis of what we see whether it is the earth that moves or the heavens. Such is also the intention of Oresme's reasoning: to show that no matter whether the motion belongs to the earth or the heavens, our visual experience will be the same, and that it is impossible to determine what actually moves on the basis of what is seen. The relationship between the earth and the heavens leads to similar conclusions as the relationship between the two ships. If the higher part of the universe moved today in a diurnal motion and the lower part remained at rest, and if tomorrow, to the contrary, the lower part moved in a diurnal motion and the higher (the heavens, the sphere of fire, the higher region of air) remained at rest,[67] we would not perceive this change, and everything would appear the same both today and tomorrow. To us observing from the earth it would still appear that the part in which we are is at rest, while the other part is in motion, just as it appears to a man on a ship that the trees on the shore and the shoreline itself are moving. But if an observer were placed in the heavens and moved in

67 Here Oresme follows Aristotle's division of the universe into the sublunary and superlunary realms from *Meteorologica* I, 3, 340b10–12; I, 7, 344a5–23; 345b32–346a6. See d'Ailly's summary supra, pp. 121–2.

a diurnal motion, and if he could see distinctly the earth and its mountains, valleys, rivers, towns and castles, it would appear to him that the earth moves in a diurnal motion, just as it appears to us on the earth that the heavens move. Moreover, if the earth moved in a diurnal motion and the heavens did not, it would likewise appear to us that the earth stands still and the heavens move – "and this can be easily imagined by anyone with clear understanding."[68]

Copernicus, too, is in the position of Oresme's sailor on a ship situated in the middle of the open sea, with no reference point to help him determine whether it is he who is moving or the observed ship. Like the sailor on the ship, a man on the earth cannot rely on any reference point that would help him determine whether he is carried along with the earth's motion and the heavens are at rest or vice versa. In the absence of a third object which he or she knows for certain is motionless, and which he could therefore use a reference to determine whether the earth moves or the heavens, he is unable to legitimately distinguish between reality and illusion. Hence, he is unable to determine legitimately whether the earth moves or not. The "optical argument" can by no means be a decisive reason for Copernicus' implied assertion of the earth's motion in his rhetorical question in Chapter 8 of Book I of *De revolutionibus*: "Why should we not admit, with regard to the daily rotation, that the appearance is in the heavens and the truth [or reality] in the earth,"[69] and his claim that "[t]his situation closely resembles what Virgil's Aeneas says: 'Forth from the harbor we sail, and the land and the cities slip backward.'"[70]

Both Ptolemy and Copernicus propose the same relativistic theory of motion perception, both provide examples of illusory motion or rest that are based on the same logic, and both distinguish between appearance and reality in these examples. Likewise, with regard to the moving earth and the motionless heavens and vice versa, they agree that the answer cannot be sought through "optics" or on the basis of apparent motions. Yet their respective cosmologies are diametrically opposed to each other. Why does Ptolemy believe that the earth stands still and why does Copernicus contend that it moves? On what basis, we can ask again, does Copernicus maintain that the rotation of the universe around the earth is, in fact, a mirror image of the earth's own rotation? What makes him think he is entitled to compare the motion of the heavens with the apparent motion of the shore from which the ship is moving away?

---

68 Oresme, *Le Livre du ciel et du monde* II, 25 (Menut and Denomy, pp. 522–3).

69 *De revolutionibus* I, 8 (Rosen, p. 16; Zekl, p. 114).

70 Ibid.

Such a comparison would, when performed in strict compliance with the "optical" theory, only be justified if Copernicus availed himself of a third object which he knew in advance was stationary and which would enable him to determine whether it is the earth or the heavens that moves. But there is no such object in the heavens. As already said, Copernicus is a sailor on a ship called earth which he cannot abandon and which determines his point of view, that is, the frame or screen of his vision. Therefore, when in Chapter 8 he calls for the affirmation of the earth's motion within the context of relativity of motion, he obviously does so with some prior knowledge that the earth moves and the heavens do not. The same holds for the sailors departing from the harbor, fully cognizant of their motion, despite the contrary appearance that the shore and the harbor are moving away from them. But on what basis does Copernicus arrive at such a conclusion?

The sailors in his example form their conviction by relying on *sensus communis*. The shore, the harbor, and the earth simply do not move and cannot be moved other than by constraint of an external force. Ptolemy uses similar reasoning. His conviction that the earth stands motionless in the middle of a rotating universe is ultimately justified by *sensus communis*, as all of his arguments derived from the accepted Aristotelian philosophy of nature can be reduced to arguments based on what is seen around us "on earth and in the air."[71] In this context, Ptolemy's evocation of the argument for phenomenological indifference has no positive epistemological function whatsoever. Ptolemy mentions it with only one goal in mind: to refute possibility of rotation of the earth by relying on sensory evidence, on what occurs on earth and in the air. While the Aristotelian-Ptolemaic thesis of the motion of celestial bodies may indeed not have any "optical" advantages over the motion of the earth, it at least does not contravene both accepted physics as well as what we see "here on earth and in the air." Ptolemy thinks of the relationship between the earth and the heavens as completely equivalent to that between the boat and the shore. Since he draws on Aristotle's physics and cosmology he assumes in advance that the earth cannot move. Therefore, the motion must necessarily belong to the heavens, even though it could, from the "optical" point of view, belong to the earth. To Ptolemy, the "optical argument" is of secondary importance to the already devised and established cosmological system (the theory of natural motion, heaviness and lightness, etc.).

Copernicus, on the other hand, employs the "optical argument" to overturn the seemingly solid, pre-conviction or "knowledge" that the earth stands

---

[71] See *Almagest* I, 7 (Toomer, p. 45; Heiberg, pars 1, p. 24). Ptolemy's arguments will be subject to a more detailed discussion infra, p. 158 et seq.

motionless while the heavens move. The application of the general rules of motion perception to the relationship between the heavens and the earth eliminates the existence of a pre-established structure to the motion of the universe. For Copernicus, the "optical argument" is not a positive argument in support of the earth's motion but an epistemological premise that makes discussion of the motion of the earth possible in the first place. Copernicus uses it with the sole purpose of demonstrating that the question whether the earth moves or not "has not yet been solved."[72] Like Oresme before him, he begins to consider relative motion perception as the zero point of discussion. Rather than being set forth *post festum*, after the assumption has been made that the earth is motionless, the relativity of motions is incorporated into the very foundation of the discussion. Being universal and applying to everything that moves in the universe, it constitutes the founding principle from which one must start when determining the celestial motions. Copernicus insists from the very beginning on the equivalence of both possibilities and refuses to admit that any cosmological or physical principle could conflict these two possibilities.

In Chapter 5, where he makes the first mention of the relativity of motion, Copernicus therefore only indicates that it would perhaps be more appropriate to ascribe the motion to the earth than to the heavens, since it is more appropriate to ascribe motion to what is "enclosed" than to what is "enclosing."[73] He then buttresses this thesis by referring to the time-honored ancient authors who defended this view: "This opinion was indeed maintained by Heraclides and Ecphantus, the Pythagoreans, and by Hicetas of Syracuse, according to Cicero. They rotated the earth in the middle of the universe, for they ascribed the setting of the stars to the earth's interposition, and their rising to its withdrawal."[74] However, later on, in the final sentence of Chapter 8, this suggestion is expressed a shade more affirmatively. Copernicus considers it "more probable" that the earth moves than that it is at rest, by drawing on the universal relativity of motion, which essentially

72 *De revolutionibus* I, 5 (Rosen, p. 11; Zekl, p. 98).

73 Ibid. (Rosen, p. 12; Zekl, p. 100): "Moreover since the heavens, which enclose and provide the setting for everything, constitute the space common to all things (*caelum sit, quod continet et caelat omnia, communis vniuersorum locus*), it is not at first blush clear why motion should not be attributed rather to the enclosed than to the enclosing, to the thing located in space rather than to the framework of space (*cur non magis contento quam continenti, locato quam locanti motus attribuatur*)." For more on this argument, see infra, p. 191.

74 Ibid.

underlines the optical equivalence of the two possibilities. What convinces him of the earth's motion?

As I have stressed earlier, the scales cannot be tipped in favor of the earth's motion by the celestial phenomena, for these become marginal to the discussion on the earth's motion with Copernicus' universalization of relativity of motion. Using the aforementioned argument of the "enclosing" and "the enclosed" in Chapter 5, Copernicus is expected to defend the earth's motion with natural philosophical arguments similar to those on which Aristotle and Ptolemy draw their contrary belief that the earth is at rest. And he does. This expectation is confirmed by the entire Chapter 8 of Book I of *De revolutionibus* where he states in conclusion that it is "more probable that the earth moves than that it is at rest" and that it "is especially true of the daily rotation, as particularly appropriate to the earth."[75] This conclusion is made dependent on another two natural philosophical arguments, apart from the above mentioned one. In support of earth's rotation Copernicus evokes an argument from the nature of motion, an argument from the nobility of immobility, and one astronomical argument based on the apparent fact that planets approach closer to the earth and recede from it. The last one is, however, only an argument for earth's revolution and not for its rotation. But as has been already stressed by many scholars, these arguments are, save for the relatively complex argument concerning the nature of motion, extremely weak, and belong more to the domain of dialectical argumentation and therefore rhetorical persuasion than to that of necessary conclusion. And Copernicus himself does not disguise it, for he concludes Chapter 8 stating only that it is only *probabilior*, or "more probable," that the earth moves than that it is at rest.

---

[75] Ibid. I, 8 (Rosen, p. 17; Zekl, p. 120).

CHAPTER NINE

The Dialectics and Physics of the Earth's First Motion

Copernicus presents some fundamental traditional philosophical arguments against the moving earth (or, in other words, in favor of its rest), in Chapter 7 of Book I of *De revolutionibus*. He refutes these arguments to show that it is more probable that the earth moves than that it is at rest and that this is especially true of the daily rotation as particularly appropriate to the earth in Chapter 8. This is also his response to Giese's questions presented by Rheticus in the "Encomium Prussiae" of the *Narratio prima*: Is the centre of the earth really the true centre of the universe?; Is circular motion attributed to the earth necessarily violent?; Can Aristotle's three motions – away from the centre, towards the centre, and around the centre – actually be separated?; and finally: Are other reasons that led Aristotle to refute the opinions of Plato's *Timaeus* and the Pythagoreans sound?[76] Arguments presented in Chapter 8 have often been categorized as dialectical or rhetorical, or at least not very demonstrative.[77] However, before I turn to Copernicus' response to Aristotelian, philosophical objections to the rotation of the earth, that is, to Copernicus' own, alternative theory of gravity and

---

[76] See supra, p. 80.

[77] See, for instance, J.-J. Szczeciniarz, *Copernic et la révolution copernicienne*, pp. 59–186, and D. K. Hill, "The Projection Argument in Galileo and Copernicus: Rhetorical Strategy in the Defense of the New System."

elemental motion, as well as to another three positive arguments in favor of the earth's rotation from Chapter 8, I will first shed some additional light on his discourse in Chapters 7 and 8 by analyzing his *formulation* of the third, experiential argument (or rather arguments, because it consists of three, loosely connected issues: centrifugal force, falling bodies, and clouds) against the motion of the earth, the one which he ascribes to Ptolemy, and his response to it.

In Chapter 7 Copernicus first cites two traditional reasons against the earth's motion which were developed in the Aristotelian physics and to some extent also adopted by Ptolemy. The first one is already mentioned by Giese in his conversation with Copernicus, as reported by Rheticus in his *Narratio prima*, and concerns heaviness and lightness of the elements, that is, the theory of gravity and levity:

> Accordingly, the ancient philosophers sought to establish that the earth remains at rest in the middle of the universe by certain other arguments [beside the geometrical one discussed and refuted in Chapter 6 of *De revolutionibus*]. As their main cause, however, they adduce heaviness [or gravity] and lightness [or levity] (*gravitas et levitas*). Earth is in fact the heaviest element, and everything that has weight is borne toward it in an effort to reach its inmost centre. The earth being spherical, by their own nature heavy objects (*gravia*) are carried to it from all directions at right angles to its surface. Hence, if they were not checked at its surface, they would collide at its centre, since a straight line perpendicular to a horizontal plane at its point of tangency with a sphere leads to the (sphere's) centre. But things brought to the middle, it seems to follow, come to rest at the middle. All the more, then, will the entire earth be at rest in the middle, and as the recipient of every falling body (*cadentia*) it will remain motionless thanks to its weight.[78]

The second argument against the motion of the earth, also mentioned by Giese, is founded on Aristotle's tripartite division of motion (away from the centre, towards the centre, and around the centre):

> In like manner, the ancient philosophers analyze motion and its nature in a further attempt to confirm their conclusion. Thus, according to Aristotle, the motion of a single simple body is simple; of the simple motions, one is straight and the other is circular; of the straight motions, one is upward

78 *De revolutionibus* I, 7 (Rosen, pp. 14–5; Zekl, p. 108–10). It seems that Copernicus summarizes Aristotle, *De caelo* II, 13, 295b21–23. Compare also with Ptolemy, *Almagest* I, 7 (Toomer, pp. 43–4; Heiberg, pars 1, pp. 21–6), partly quoted infra, pp. 160–1, and my discussion there.

> and the other is downward. Hence every simple motion is either toward the middle, that is, downward; or away from the middle, that is, upward; or around the middle, that is, circular. To be carried downward, that is, to seek the middle, is a property only of earth and water, which are considered heavy; on the other hand, air and fire, which are endowed with lightness, move upward and away from the middle. To these four elements it seems reasonable to assign rectilinear motion, but to the heavenly bodies, to revolve in an orb around the middle. This is what Aristotle says.[79]

Copernicus' formulation of both arguments – that from "heaviness and lightness" and that from "nature of motion" – is more or less faithful to Aristotle's. But this cannot be said for the next passage where Copernicus presents reasons that he explicitly ascribes to Ptolemy and that belong to the category of experiential arguments (or *sensus communis* or observational tests). "Summarizing" Ptolemy, Copernicus points to the devastating effects of the rotation on the rotating earth itself and which would also occur to things "on earth and in the air," but have not been observed. What consequences would the earth's rotation on its axis have, according to Copernicus' presentation of Ptolemy?

> Therefore, remarks Ptolemy of Alexandria, if the earth were to move, merely in a daily rotation, the opposite of what was said above would have to occur, since a motion would have to be exceedingly violent (*concitatissimum*) and its speed unsurpassable to carry the entire circumference of the earth around in twenty-four hours. But things which undergo an abrupt rotation seem utterly unsuited to gather [bodies to themselves], and seem more likely, if they have been produced by combination, to fly apart unless they are held together by some bond.[80] The earth would long ago have burst asunder, he says, and dropped out of the skies (a quite ridiculous notion); and, what is more, living creatures and any other loose weights would by no means remain unshaken. Nor would objects falling in a straight line (*cadentia in directum*) descend perpendicularly to their appointed place, which would meantime have been

79 *De revolutionibus* I, 7 (Rosen, pp. 14–5; Zekl, p. 110).

80 Compare with Aristotle, *De caelo* I, 2, 269b6–10 (Leggatt, p. 55): "If those bodies that move in a circle move around contrary to nature, it would be amazing and completely absurd for this movement alone which is contrary to nature to be continuous and everlasting, for in other cases *those bodies that move contrary to nature are observed to perish quickly.*" Emphasis added.

> withdrawn by so rapid a movement. Moreover, clouds and anything else floating in the air would be seen drifting always westward.[81]

Copernicus ascribes authorship of these arguments to Ptolemy. But as already noted by several scholars, his formulation differs drastically from the argument in Chapter 7 of Book I of the *Almagest*. One only has to look at his text to see that his "summary" is quite liberal, that he, in words of De Pace, "resumes somewhat freely," or, rather, that he to a large extent distorts Ptolemy.[82] Why would he want to do something like that?

1. Copernicus, Ptolemy and the centrifugal force argument

Copernicus' distortion is most evident from the first, so-called centrifugal force or projection argument, stating that a rotating earth would burst asunder and "dropped out of the skies." According to Copernicus, Ptolemy fears that rotation of the earth about its axis would produce effects contrary to the accepted Aristotelian philosophy concerning the movement of the four sublunar elements. The speed of the earth's motion would have to be exceedingly high to complete its axial rotation in twenty-four hours. But such abrupt rotation usually produces disastrous consequences for things that move in this way. Things that move at great speed cannot keep their parts to themselves; quite the opposite. Even if uniform, gathered into a single whole, or conjoined in one way or another, they will fly apart unless held together by a firm bond. The earth, rotating around its axis would disintegrate, according to "certain people," and fall beyond the last heaven, the limits of the universe. Moreover, everything that inhabits the earth but is not fixed to it (living creatures and everything possessing weight) would likewise be thrown off of it.

Ptolemy's argumentation. Ptolemy, however, does not discuss the earth's rotation in connection with these disastrous effects. In the first part of Chapter 7 of

[81] *De revolutionibus* I, 7 (Rosen, p. 15; Zekl, pp. 110–2)

[82] A. De Pace, "Copernicus against a Rhetorical Approach to the Beauty of the Universe," p. 98. See also E. Rosen, *Commentary*, p. 351; M.-P. Lerner, *Le monde des sphères*, vol. 2, pp. 96–7; D. Knox, "Copernicus's Doctrine of Gravity and the Natural Circular Motion of the Elements," pp. 173–4, and p. 174, n. 81; M. Wolff, "Impetus Mechanics as a Physical Argument for Copernicanism: Copernicus, Benedetti, Galileo," p. 229. A. De Pace, *Niccolò Copernico e la fondazione del cosmo eliocentrico*, discusses this issue on pp. 121–39.

the *Almagest*, which contains a reference to the falling of the earth "out of the heavens," he only mentions the possibility of its displacement from the centre of the universe under the pressure of bodies descending towards it.

Here is the broader context of Ptolemy's argument. Ptolemy maintains that the earth does not move "downwards" under the pressure of heavy objects. Rather, it rests at the centre of the universe, while all heavy objects are carried down to it. From this it can be inferred that all heavy objects fall towards its centre, as we can see that in all parts of the earth, which is of a spherical shape and situated in the middle of the universe, heavy objects are carried by their proper, natural motion at right angles to the tangent plane drawn through the point of contact. From this fact it is clear that if these objects were not stopped by the surface of the earth, they would reach the centre of the earth itself, since the line, which intersects the tangent drawn on the earth's surface, leads to the centre of the earth.[83]

Ptolemy bases this argument on his theory of natural motion in the universe: light and rarefied bodies are carried towards the circumference and thus seem to move upwards. Heavy, dense and compact bodies are carried towards the middle and the centre; therefore, they seem to fall downwards, to the centre of the earth, where they come to rest due to their mutual pressure and resistance, which is equal and uniform from all directions. Hence, it is perfectly plausible that the earth, whose total mass is extraordinarily huge in comparison with the bodies falling towards it, remains motionless under the impact of these small weights (because they press on it from all sides), receiving the objects falling on it. Ptolemy's argument for the immobility of the earth is therefore different from Aristotle's. It is based on the notion of pressure. According to Ptolemy, all matter consists of molecules. The earth is extremely small compared to the celestial spheres and is therefore, as Pedersen says, "subjected to an immense pressure by the aether molecules. Since these are all alike and surround the earth evenly on all sides, the pressure will act uniformly everywhere on the earth, and keep it in place at the centre, the pressure at one point of the surface being equal and opposite to that at the opposite point."[84]

---

[83] See *Almagest* I, 7 (Toomer, pp. 43–4; Heiberg, pars 1, pp. 21–2). According to O. Pedersen, *A Survey of the Almagest*, p. 43, Ptolemy here provides "no proof of the immobility of the earth, which might still have diurnal motion around the centre. It would have been necessary to underline the principle that heavy bodies move towards the centre in order to be at rest there, but this is not stressed by Ptolemy."

[84] *A Survey of the Almagest*, pp. 43–4.

Nevertheless, Ptolemy continues, there are certain people who think that the earth, not being supported by anything, should move under the impact of these falling bodies. After he recapitulates the traditional classification of "up" and "down" in the spherical universe, Ptolemy challenges such reasoning. He contends that such motion would lead to strange consequences: the earth would be carried "down" faster than any other heavy body because of its much greater bulk, so that "living things and individual heavy objects would be left behind, riding on the air, and the earth itself would very soon have fallen completely out of the heavens,"[85] or be pulled out of the universe. But such things are, in his opinion, "utterly ridiculous merely to think of." Ptolemy is obviously trying to reject the possibility of the earth's "downward" motion, as it can be called from our perspective, for there is no such thing as "up" and "down" or "upwards" and "downwards" in the universe itself.

> Those who think it paradoxical that the earth, having such great weight, is not supported by anything and yet does not move, seem to me to be making the mistake of judging on the basis of their own experience instead of taking into account the peculiar nature of the universe. They would not, I think, consider such a thing strange once they realised that this great bulk of the earth, when compared with the whole surrounding mass [of the universe], has the ratio of a point to it. For when one looks at it in that way, it will seem quite possible that that which is relatively smallest should be overpowered and pressed in equally from all directions to a position of equilibrium by that which is the greatest of all and of uniform nature. For there is no up and down in the universe with respect to itself, any more than one could imagine such a thing in a sphere: instead the proper and natural motion of the compound bodies in it is as follows: light and rarefied bodies drift outwards towards the circumference, but seem to move in the direction which is 'up' for each observer, since the overhead direction for all of us, which is also called 'up', points towards the surrounding surface; heavy and dense bodies, on the other hand, are carried towards the middle and the centre, but seem to fall downwards, because, again, the direction which is for all of us towards our feet, called 'down', also points towards the centre of the earth. These heavy bodies, as one would expect, settle about the centre because of their mutual pressure and resistance, which is equal and uniform from all directions. Hence, too, one can see that it is plausible that the earth, since its total mass is

85 *Almagest* I, 7 (Toomer, p. 144; Heiberg, pars 1, pp. 23–4).

> so great compared with the bodies which fall towards it, can remain motionless under the impact of these very small weights (for they strike it from all sides), and receive, as it were, the objects falling on it. If the earth had a single motion in common with other heavy objects, it is obvious that it would be carried down faster than all of them because of its much greater size; living things and individual heavy objects would be left behind, riding on the air, and the earth itself would very soon have fallen completely out of the heavens. But such things are utterly ridiculous merely to think of.[86]

Not a single word in this passage refers to the earth's rotation or to centrifugal effects arising from it. Ptolemy argues against those people, in Pedersen's words, "who cannot imagine so great and heavy body as the earth to be at rest without its inclining to one side or another," and not its rotation.[87] It is only later on in Chapter 7 of Book I of the *Almagest* that Ptolemy recites various implausible consequences of the earth's rotation – none of them is that the "the earth itself would very soon have fallen completely out of the heavens" – making even more obvious the difference between its rotation and moving away from the centre under the impact of heavy bodies falling towards it. Why, then, does Copernicus distort Ptolemy?

Copernicus' (ab)use of centrifugal force argument. One possible explanation could be that he quotes Ptolemy from memory.[88] Another possible explanation may be that he has borrowed the mistaken interpretation from some commentator. But even if such an account did exist, (still) remaining unknown to us, Copernicus must have certainly been aware of its fallacy. Ptolemy's argument is presented faithfully and unequivocally in both translations of the *Almagest*, on which Copernicus bases his discussion.[89] For this reason, he would hardly have

---

86 Ibid. (Toomer, p. 44 ; Heiberg, pars 1, pp. 23–4).

87 O. Pedersen, *A Survey of the Almagest*, p. 43.

88 Such is the opinion of E. Rosen, *Commentary*, p. 351.

89 Copernicus owned and annotated a Greek copy of Ptolemy's *Syntaxis*, with Theon of Alexandria's commentary, published in Basel in 1538, and annotated a copy of Ptolemy's work in Gerard of Cremona's Latin translation from the Arabic, published in Venice in 1515. See P. Czartoryski, "The Library of Copernicus," pp. 367–8, and 372. Copernicus did not annotate or mark *Almagest* I, 7. It may be assumed from terminological similarities that Copernicus, when writing Book I of *De revolutionibus*, used George of Trebizond's translation of the *Almagest* from the Greek, first published in 1528. See E. Rosen, *Commentary*, pp. 341, and 350–1; D. Knox, "Copernicus's Doctrine of Gravity and the Natural Circular Motion of the Elements," pp. 164–5, n. 39; and D. K. Hill, "The Projection Argument in Galileo and Copernicus," p. 111.

failed to notice that the *Almagest* contains anything similar to what he ascribes to Ptolemy.[90] Why, then, does Copernicus attribute the effects of the centrifugal force to the earth's rotation, when Ptolemy ascribes the very same effects to its possible displacement from the centre?

According to Koyré, it is highly likely that Copernicus so interpreted Ptolemy's text "either because he believed that this disclosed the true meaning of the great astronomer, or because he felt that he ought to make his objection in a better, i.e. stronger way."[91] Copernicus' motive for change was therefore dialectical: he wanted to set forth what he believed was a stronger argument that would later enable him "to reveal the conceptual differences between the two cosmologies in a way that was advantageous to himself."[92] He wanted to present the argument against the earth's rotation in such a way as to provide the easiest possible response to it in Chapter 8 of Book I of *De revolutionibus*, and he indeed used the centrifugal force argument very persuasively to bring out contradictions implied in "Ptolemy's" reasoning.

Copernicus develops his argument in Chapter 8, firstly, by claiming that the rotation of the earth is its natural, not violent motion, and by showing that some basic assumptions of the traditional geocentric cosmology contradict each other when we derive their implied consequences, secondly, by decomposing traditional arguments against the earth's motion that are based on our perception of what occurs "on earth and in the air"; and finally, by establishing some rudimentary positive physical theory that should provide physical explanation of its rotation. Virgil's example of the boat leaving the shore, with which we started the analysis of the critical part of Copernicus' arguments in favor of the earth's motion, is set in the context of these mostly dialectical counter-arguments to the Aristotelian-Ptolemaic position against the earth's motion.

---

90 This option is discussed and rejected by D. K. Hill, "The Projection Argument in Galileo and Copernicus," pp. 111–2.

91 A. Koyré, *The Astronomical Revolution*, p. 112, n. 9. A. De Pace, *Niccolò Copernico e la fondazione del cosmo eliocentrico*, pp. 121–39, proposes a solution according to which Copernicus believed that he was explaining, if we quote Koyré, "the true meaning of the great astronomer." According to her, Copernicus had, during his stay in Italy, access to a Greek codex of *Almagest* from which he copied some passages of importance. If that was the case, I am wondering why he did not compare them with the Latin translation(s) that he used later. He would realize in no time that they differed significantly.

92 D. K. Hill, "The Projection Argument in Galileo and Copernicus: Rhetorical Strategy in the Defense of the New System," p. 112; for the elaboration of this point see pp. 112–5. M.- P. Lerner, *Le monde des sphères*, vol. 2, pp. 96–8, agrees with Hill and Koyré.

The first step Copernicus takes is to pronounce the earth's rotation a natural rather than violent motion.[93] It therefore cannot be expected to have the devastating effects assigned to it by Ptolemy, who, in Copernicus' interpretation, fears that "everything earthly will be disrupted" by some kind of a centrifugal force resulting from the earth's rotation. Copernicus dismisses this possibility, contending that the rotation of the earth is a natural and therefore not violent motion, and that what is created through nature's work cannot have the same effects as what results from violence. Things to which violence, or "force or impetus," is applied "cannot long endure" and must disintegrate. On the other hand, whatever is brought into existence naturally is "well-ordered" and preserved "in its best state":

> Yet if anyone believes that the earth rotates, surely he will hold that its motion is natural, not violent. But what is in accordance with nature produces effects contrary to those resulting from violence, since things to which force or impetus (*vis vel impetus*) is applied must disintegrate and cannot long endure. On the other hand, things which are brought into existence by nature are well-ordered (*recte se habent*) and preserved in their best state (*in optima sua compositione*). Ptolemy has no cause, then, to fear that the earth and everything earthly will be disrupted by a rotation created through nature's handiwork (*per efficaciam naturae*), which is quite different from what art or human intelligence can accomplish.[94]

Moreover, if Ptolemy assumes that the earth would be destroyed, disrupted by its rotation, the same should also hold for the motion of the heavens. Copernicus faces Ptolemy with the following alternative: either rotation of the earth is natural and therefore in harmony with the eternal order of the universe, or we need to apply the same devastating "mechanical" reasoning that Ptolemy applies to the earth in order to prove its rest to the motion of the heavens as well. If the rotation of the earth results in its dissipation, the same should apply to the heavens. Or is the immensity of the heavens, asks Copernicus, a result of the "centrifugal" force and their perpetual movement a necessity that prevents it from collapsing? The heavens that, according to Ptolemy complete their circular movement in twenty-four hours, should, as Copernicus contends, increase with each and every turn. But is that physically possible at all?

---

93 For an interesting analysis of Copernicus' reasoning as implying a confutation of Aristotle, *De caelo* II, 14, 296a24–296b3, see A. De Pace, *Niccolò Copernico e la fondazione del cosmo eliocentrico*, pp. 139–41.

94 *De revolutionibus* I, 8 (Rosen, p. 15; Zekl, p. 112).

There are, in Copernicus' view, two explanations regarding the extension of the universe that should be taken into consideration. The first one, held by Stoics, is that beyond the last celestial sphere there is an infinite void space, which can accept the alleged extension of the universe resulting from the "centrifugal" force of its circular motion.[95] In this case, the heavens, completing their circuit every twenty-four hours, should be expanded by this force and enlarged to infinity by the continuous increase in the speed of the heavenly motion. According to Aristotle, however, the infinite cannot be traversed and moved. So, the heavens should come to a halt.

> But why does he not feel this apprehension even more for the universe, whose motion must be the swifter, the bigger the heavens are than the earth? Have the heavens become immense because the indescribable violence of their motion (*ineffabili motus vehementia*) drives them away from the centre and would they fall apart if they came to a halt? Were this reasoning sound, surely the size of the heavens would likewise grow to infinity. For the higher they are driven by the impetus of their motion (*ipso motus impetu rapiatur in sublime*), the faster that motion will be, since the circumference of which it must make the circuit in the period of twenty-four hours is constantly expanding; and, in turn, as the velocity of the motion mounts, the immensity of the heavens is enlarged. In this way the speed will increase the size, and the size the speed, to infinity. Yet according to the familiar axiom of Physics that "the infinite cannot be traversed or moved in any way," the heavens will therefore necessarily remain stationary.[96]

The other possibility, held by Aristotle, is that the universe is finite and that there is nothing, not even void, beyond the last sphere of the universe.[97] In this

---

95 Copernicus probably learned about this Stoic conception of an infinite extra-cosmic void space in Giorgio Valla's paraphrase or translation of Cleomedes' *De motu circulari corporum coelestium* I, 1 in Valla's *Miscellanea grecae* (= Bowen and Todd, p. 22): "So since the cosmos has Nature administering [it] throughout, it is itself necessarily limited, whereas what is outside it is a void that extends without limit in every direction." On Copernicus and Valla, see E. Rosen, "Nicholas Copernicus and Giorgio Valla," and A. De Pace, *Niccolò Copernico e la fondazione del cosmo eliocentrico*, p. 145, n. 426. For Stoic teaching and scholastic background, see E. Grant, *Much Ado about Nothing*, pp. 103–258.

96 *De revolutionibus* I, 8 (Rosen, p. 15; Zekl, pp. 112–4). Reference to *Physics* is to *Phyisica* IV, 4–8; especially 204a3–7; see also *De caelo* I, 5–7.

97 See Aristotle, *De caelo* I, 9, 279a11–279a18 (Leggatt, p. 91): "At the same time it is clear

case, the alleged extension of the heavens to infinity is impossible. "But beyond the heavens there is said to be no body, no space, no void, absolutely nothing, so that there is nowhere the heavens can go."[98] But this would also mean that "nothing" contains or holds in check "something," which is unreasonable: "In that case it is really astonishing if something can be held in check by nothing."[99] How can "nothing" contain anything at all?

To amend this, Copernicus explains that the expression that "there is [...] absolutely nothing" beyond the heavens could be true if understood as designating an infinite heaven, which is finite at their inner concavity, "inwardly," so to speak, because in that way *everything* would be inside the heavens and *nothing* outside. But this option, exactly like the first one, results in an infinite universe, and consequently in the immobility of the heavens.

> If the heavens were infinite, however, and finite at their inner concavity only, there will perhaps be more reason to believe (*magis forsitan verificabitur*) that beyond the heavens there is nothing. For, every single thing, no matter what size it attains, will be inside them, but the heavens will abide motionless. For, the chief contention by which it is sought to prove that the universe is finite is motion [of the simple body, that is, (according to Aristotle) aether].[100]

---

that neither place nor void nor time is outside of the world. For in every place body can exist; and void, they say, is that by which body does not exist, but in which it can; and time is a number of movement – but there is no movement without physical body. Now, it has been shown that outside of the world there neither is nor can be body. Therefore it is clear that neither place nor void nor time is outside."

98 *De revolutionibus* I, 8 (Rosen, pp. 15–16; Zekl, p. 114).

99 Ibid. (Rosen, p. 16; Zekl, p. 114).

100 Ibid. See Aristotle, *De caelo* I, 5, 271b26–272a7 (Leggatt, p. 63): "That, therefore, the body which moves in a circle has to be limited as a whole, is clear from the following. For if the body which moves in a circle is unlimited, the radii will be unlimited. But the interval of unlimited lines is unlimited; by 'interval of the lines' I mean that outside of which no magnitude in contact with the lines can be found. This, then, must be unlimited; for the interval of unlimited lines will always be unlimited. Further, it is always possible to take an amount greater than that given, and so just as we say that number is unlimited, because there is not a greatest, the same argument holds also concerning the interval. If, then, it is not possible to traverse the unlimited, and if, when the body is unlimited, the interval must be unlimited, then it could not possibly move in a circle; but we see the heavens turning about in a circle, and we have determined by argument as well that movement in a circle belongs to some particular body." See

In Copernicus' hands, traditionally accepted cosmological notions lead to absurd results which devastate Aristotelian-Ptolemaic cosmology. Copernicus deliberately ignores the fact that his main opponents, Aristotle and Ptolemy, conceive of the heavenly region as made of aether, an element of completely different nature than earthly matter and therefore subject to different rules.[101] In this way, Copernicus ties the question of the earth's rotation to the question of whether the universe is infinite or finite. If Ptolemy's reasoning is valid for the rotation of the earth, Copernicus argues, then it should apply to the rotation of the heavens as well, and the heavens should, according to Ptolemy's initial premise (rotation = dissipation), grow to infinity. If the heavens by their rotational movement gradually increase beyond the last celestial sphere and become infinite (rotation = dissipation = infinity), then they should come to a stop, since, according to the accepted Aristotelian physics, infinity cannot move (rotation = dissipation = infinity = rest).

Copernicus has thus skillfully transformed the initial question, "What rotates around its axis, the heavens or the earth?" to the question: "Are the heavens (= universe) infinite or finite?." The question of the rotation of the heavens (and consequently that of the earth) is thus linked to the question of the infinity of the universe. Copernicus' implicit argument seems to be that we can approach the question of the earth's rotation from the perspective of the heavens. If the heavens

---

also following arguments, up to 273a5–7: "That, then, the body that moves in a circle is neither endless nor unlimited, but has an end, is clear."

101 There is some difference of opinion between Ptolemy and Aristotle as far as the nature of aether is concerned. Ptolemy's conception was very likely influenced by Stoics. See A. De Pace, *Niccolò Copernico e la fondazione del cosmo eliocentrico*, p. 143; M.-P. Lerner, *Tre saggi sulla cosmologia alla fine del Cinquecento*, pp. 20–1, and *Le monde des sphères*, vol. 1, p. 66. For Ptolemy's conception of aether, see *Almagest* I, 3 (Toomer, p. 40; Heiberg, pars 1, pp. 13–4): "Furthermore, one can reach this kind of notion from certain physical considerations. E. g., the aether is, of all bodies, the one with constituent parts which are finest and most like each other; now bodies with parts like each other have surfaces with parts like each other; but the only surfaces with parts like each other are the circular; among planes, and the spherical, among three-dimensional surfaces. And since the aether is not plane, but three-dimensional, it follows that it is spherical in shape. Similarly, nature formed all earthly and corruptible bodies out of shapes which are round but of unlike parts, but all aethereal and divine bodies out of shapes which are of like parts and spherical. [...] For this reason it is plausible that the aether surrounding them, too, being of the same nature, is spherical, and because of the likeness of its parts, moves in a circular and uniform fashion." See also *Almagest* XIII, 2 (Toomer, pp. 600–1; Heiberg, pars 1, pp. 43–4).

do not revolve, because they are increased to infinity by the dissipating effect of the "centrifugal force," which is the effect of the rotation, then the earth does not rotate either, since the centrifugal force would dissipate it. But if the universe is finite (and of spherical shape, of course), then it rotates, and consequently the earth as well (the centrifugal force dissipates neither the earth nor the heavens).

But something else needs to be borne in mind as well. In Chapters 5 and 8, when addressing the possibility of the earth's rotation, Copernicus places the earth in juxtaposition to the heavens and claims that as far as the celestial phenomena are concerned, it does not matter whether it is the heavens or the earth that rotates – but one of them must. This is another – implicit – counter-argument to Ptolemy. If my reconstruction of Copernicus' logic of producing contradiction in the traditional cosmology above is correct (if the universe is infinite, the earth and the heavens stand still; if the universe is finite and rotates, the earth rotates as well), it shows another impasse of Ptolemy's reasoning. The heavens and the earth cannot – at least in the way how Copernicus presents the matter – simultaneously stand still or move. According to Copernicus, we are faced with a dilemma: either the earth rotates and the heavens stand still or the earth stands still and the heavens rotate. And as it can be inferred from Ptolemy's reasoning – as presented by Copernicus –, the universe must stand still.

Copernicus claims in Chapter 6 that the heavens, i.e. the universe, are, if not actually infinite, at least "immense by comparison with the earth and present the aspect of an infinite magnitude, while on the testimony of the senses the earth is related to the heavens as a point to a body, and a finite to an infinite magnitude."[102] The heavens are of "unlimited [or indefinite] (*indefinitam*) size in relation to the earth's magnitude," but "how far this immensity extends is not at all clear."[103] Now, Copernicus needs (almost) an infinite universe not solely for the explanation of the absence of the parallax, but also in order to prove that the earth rotates and the heavens stand still. If the heavens were "infinite [...] and finite at their inner concavity," the earth would necessarily have to rotate (with the natural, non-violent movement, of course). Another reason for Copernicus to accept the infinity of the universe is that it would prevent him to move because of its form or shape. In Chapter 1 of Book I of *De revolutionibus* Copernicus demonstrates that the universe is spherical. If the universe has a form, a shape, it can only be finite; but in that case, it should move, being attributed – at least in principle – "a motion appropriate by nature to its form," which is "rotation in a circle." In such

[102] *De revolutionibus* I, 6 (Rosen, p. 13; Zekl, p. 104).

[103] Ibid. (Rosen, p. 14; Zekl, p. 108).

a case, the earth should be still. The infinite universe has no spherical form and consequently cannot move. In Chapter 8, Copernicus seems to be pondering the possibility of a universe that is "inwardly" finite and "outwardly" infinite, which would prevent it from rotating, which, in turn, would mean that it is the earth that rotates.

Whatever his position on the issue of the infinity of the universe, in Chapter 8 Copernicus claims that it is yet to be solved – "Let us therefore leave the question whether the universe is finite or infinite to be discussed by the natural philosophers (*physiologi*) [...]"[104] – and thus cannot help determine whether it is the earth that

---

[104] Ibid. I, 8 (Rosen, p. 16; Zekl, p. 114). According to Rheticus in his *Narratio prima* (Rosen, pp. 144–5; Hugonnard-Roche and Verdet, p. 59), Copernicus' universe is immense but limited by starry vault. It is "immense up to its concave surface, and truly similar to the infinite." The reasons Rheticus gives for this conclusion are: (1) the fact that "all the heavenly bodies twinkle, with the exception of the planets including Saturn, which, being the nearest of them to the firmament, revolves on the greatest circle"; (2) deduction from Copernicus' hypotheses (*De revolutionibus* I, 6), which shows, "that the orb of the stars is, to the highest degree, similar to the infinite, since by comparison with it the great orb vanishes, and all the phenomena are observed exactly as if the earth were at rest in the centre of the universe." The question remains of what is beyond the universe. Rheticus' answer, founded on theological reasons, is that we do not know. See ibid. (Rosen, p. 144; Hugonnard-Roche and Verdet, p. 59): "As I carefully consider this arrangement of the entire universe according to the opinion of my teacher, I realize that Pliny set down an excellent and accurate statement when he wrote: 'To inquire what is beyond the universe or heaven, by which all things are overarched, is no concern of man, nor can the human mind form any conjecture concerning this question.' And he continues: 'The universe is sacred, immense, all in all; indeed, it is the totality, finite yet similar to the infinite, etc.' [see *Historia naturalium* II, 1, 1–2 (Rackham, vol. 1, p. 170–1)] For if we follow my teacher, there will be nothing beyond the concave surface of the starry orb for us to investigate, except insofar as Holy Writ has vouchsafed us knowledge, in which case again the road will be closed to placing anything beyond this concave surface. We will therefore gratefully admire and regard as sacrosanct all the rest of nature, enclosed by God within the starry heaven. In many ways and with innumerable instruments and gifts He has endowed us, and enabled us to study and know nature; we will advance to the point to which He desired us to advance, and we will not attempt to transgress the limits imposed by Him." According to A. Goddu, *Copernicus and the Aristotelian Tradition*, pp. 355–8, details on pp. 89–136, especially 103–4, Copernicus here evokes debates about the nature of an extra-cosmic void that were very intense in the fifteenth and sixteenth centuries and did not produce any major consensus. These doctrines were also thought in the second half of the fifteenth century in Cracow; some scholars there held that extra-cosmic space was actually infinite. A. De Pace, *Niccolò Copernico e la fondazione del cosmo eliocentrico*, pp. 159–64, proposes a different – and in my opinion more to the point – explanation in terms of Plutarch's

rotates or the universe. The question of the finiteness or infinity of the universe, which he makes crucial for determining whether the earth rotates, is yet to be answered. There is no certainty as far it this question is concerned. Therefore, Copernicus proposes to seek a solution on what is regarded as a certainty, that "the earth, enclosed between highest peaks [of its mountains] is bounded by a spherical surface."[105]

With there being no definitive answer on the finiteness or infinity of the universe, one cannot determine whether the heavens move or not (and consequently whether the earth moves or not). Therefore it is more appropriate, reasons Copernicus, to assign the motion to the earth than to the heavens. This holds all the more true as the earth's motion is naturally suited to its spherical form. Since the natural motion of a sphere is "rotation in a circle," the most natural consequence would be that the earth, too, moves on its own axis. Against this dialectically prepared and rhetorically alluring backdrop Copernicus introduces his "optical argument" illustrated by the sentence from Virgil about the sailors and the shoreline drifting away from them, and the rhetorical question: "Why then do we still hesitate to grant it the motion appropriate by nature to its form rather than attribute a movement to the entire universe, whose limit is unknown

De defectu oracolum theocentric conception of causality: there is no phenomenon or argument, mathematical or physical, that could determine whether the universe in which God manifests its omnipotent nature is finite or infinite. It seems to me that Rheticus confirms such reading, but he is quoting Pliny the Elder, not Plutarch. See also, M.-P. Lerner, *Tre saggi sulla cosmologia alla fine del Cinquecento*, p. 69, and n. 54.

105 *De revolutionibus* I, 8 (Rosen, p. 16; Zekl, p. 114). Compare with Aristotle, *Meteorologica* I, 3, 340b30–341a5 (Webster, p. 558): "However, it may well be that the formation of clouds in that upper region is also prevented by the circular motion. For the air round the earth is necessarily all of it in motion, except that which is cut off inside the circumference which makes the earth a complete sphere. In the case of winds it is actually observable that they originate in marshy districts of the earth; and they do not seem to blow above the level of the highest mountains. It is the revolution of the heaven which carries the air with it and causes its circular motion, fire being continuous with the upper element and air with fire. Thus its motion is a second reason why that air is not condensed into water." I agree with A. De Pace, *Niccolò Copernico e la fondazione del cosmo eliocentrico*, p. 349, who is translating "terra verticibus conclusa" as "che la Terra è delimitata da una superficie sferica conchiusa delle vette [delle sue montagne]." For the justification of such translation, see p. 186, and n. 555, and p. 316, n. 65. See also her translation. pp. 318–9, and n. 73 on p. 319, of *De revolutionibus* I, 2: "Vnde manifestum est terram quoque verticibus includi et propter hoc globosam esse." / " Di qui è manifesto che anche la Terra è delimitata dal [cerchio che congiunge le] cime delle sue montagne più alte e perciò è sferica."

and unknowable? Why should we not admit, with regard to the daily rotation, that the appearance is in the heavens and the truth [or reality] in the earth?"[106]

2. Copernicus, Ptolemy, and falling bodies (and clouds)

One is tempted to add an additional – albeit at first sight relatively circumstantial – piece of evidence for the claim that Copernicus' motivation in formulating Ptolemy's objections to the rotation of the earth is not only dialectical, but even contradictory and incoherent, by exposing paradoxes and contradictions that are implicit in Copernicus' presentation of the *next* Ptolemy's argument against the daily rotation of the earth when measured against Copernicus' responses to it and his theory of gravity and elemental motion. In contrast to the first argument, for which Copernicus must have felt, as Koyré puts it, that "he ought to make his objection in a stronger way," he presents this one in a *weaker* form.

Ptolemy's argumentation. Ptolemy discusses three possible observational tests to the earth's rotation in Chapter 7 of the Book I of the *Almagest* as well as three possible counter-arguments, the first two of which are of particular relevance to our discussion.

According to Ptolemy, "certain people" suppose that the heavens are motionless and the earth revolves from east to west about the same axis as the heavens, making roughly one revolution each day. The second possibility advocated by them is that both the earth and the heavens revolve about the same axis and in such a way that one moves faster than the other. But Ptolemy dismisses the earth's rotation as impossible, even ridiculous, because it violates several physical principles. To accept this thesis, we would have to concede either that the aethereal heavens remain motionless while the earth moves or (if both should move) they move in the same way as the earth, despite being of completely opposite nature. In either case, we would also have to concede that heavy bodies composed of the biggest parts have their own proper motion, even though such bodies are not readily moved even under the constraint of an external force:

> However, they do not realize that, although there is perhaps nothing in the celestial phenomena which would count against that hypothesis, at least from simple considerations (that this is so), nevertheless from what would occur here on earth and in the air, one can see that such a notion is quite

[106] *De revolutionibus* I, 8 (Rosen, p. 16; Zekl, p. 114).

> ridiculous. Let us concede to them [for the sake of argument] that such an unnatural thing could happen as that the most rare and light of matter should either not move at all or should move in a way no different from that of matter with the opposite nature (although things in the air, which are less rare [than the heavens] so obviously move with a more rapid motion than any earthy object); [let us concede that] the densest and heaviest objects have a proper motion of the quick and uniform kind which they suppose (although, again, as all agree, earthy objects are sometimes not readily moved even by an external force).[107]

Despite these "unnatural" things being implicitly present in the affirmation of the earth's rotation, Ptolemy is willing to address this assumption and reject it on the basis of what is seen around us "on earth and in the air." But what is actually seen around us "on earth and in the air"? Or better, what should be seen, but is not, if the earth rotates on its axis? It is a fundamental fact that the earth's rotation would be the most violent and the most unnatural motion for the element of earth, if the earth were to make a complete axial revolution in as short a time as twenty-four hours. The result of such rotation would be that all objects not fixed to the earth would appear to have the same motion as the earth, only in the opposite direction. Therefore, clouds or other objects that fly or are thrown upwards would not move in the same direction as the earth but rather the reverse, seemingly lagging behind the earth's motion towards the west.

> Nevertheless, they would have to admit that the revolving motion of the earth must be the most violent of all motions associated with it, seeing that it makes one revolution in such a short time; the result would be that all objects not actually standing on the earth would appear to have the same motion, opposite to that of the earth: neither clouds nor other flying or thrown objects [i.e. thrown up into the air] would ever be seen moving towards the east, since the earth's motion towards the east would always outrun and overtake them, so that all other objects would seem to move in the direction of the west and the rear.[108]

This particular objection to the earth's rotation is open to a counter-argument. Proponents of a rotating earth might contend that the air is moved in the same direction and at the same speed as the earth. But Ptolemy also objects to this

[107] Ptolemy, *Almagest* I, 7 (Toomer, p. 45; Heiberg, pars 1, pp. 24–5).

[108] Ibid. (Toomer, p. 45; Heiberg, pars 1, p. 25).

position. If it were so, he maintains, objects thrown up into the air and composed of other elements or simple bodies (e.g. water and earth) should be left behind the motion of the air moving with the earth. But his response, too, can be countered: these objects could be "fused" to the air in a way, "carried around" with it, thus having the same motion as the air and the earth. In this case, Ptolemy argues, the objects would always appear to be still, with their motion being neither forward nor backward, never wandering about, nor changing position – even though these are precisely the sorts of motion they would, in fact, be undergoing:

> But if they said that the air is carried around in the same direction and with the same speed as the earth, the compound objects in the air would none the less always seem to be left behind by the motion of both [earth and the air]; or if those objects too were carried around, fused as it were, to the air, then they would never appear to have any motion either in advance nor rearwards [i. e. with respect to the motion of the earth and the air]: they would always appear still, neither wandering about nor changing position, whether they were flying or thrown objects. Yet we quite plainly see that they do undergo all these kinds of motion, in such a way that they are [perhaps] not even slowed down or sped up at all by any motion of the earth.[109]

Copernicus' summary. Copernicus sums up Ptolemy's entire argument in Chapter 7 of Book I of *De revolutionibus* in two sentences: "Nor would objects falling (*cadentia*) in a straight line descend perpendicularly to their appointed place, which would meantime have been withdrawn by so rapid a movement. Moreover, clouds and anything else floating in the air (*alia in aere pendentia*) would be seen drifting always westward."[110] Copernicus' recapitulation can be paraphrased more extensively as follows: If the earth rotated about its axis, bodies falling in a straight line perpendicular to it would in no way reach their appointed place, for the place which the bodies falling in a straight and perpendicular line should reach, would in the meantime have been withdrawn by a

[109] Ibid. (Toomer, p. 45; Heiberg, pars 1, pp. 45-6).

[110] *De revolutionibus* I, 7 (Rosen, p. 15; Zekl, pp. 110–2). The first argument is originally Aristotle's – see *De caelo* II, 14, 296b21–25 (Leggatt, p. 163): "It is evident, therefore, that the earth must be at the centre and motionless, both for the reasons given, and because weights thrown straight upward by force return to the same point, even if the force flings them an unlimited distance."– but was repeated by many authors. Copernicus, however, ascribes it *explicitly* to Ptolemy. See the quotation from *De revolutionibus* I, 7, supra, p. 157.

rapid movement. But as we see, free falling bodies descend "in a straight line" at right angles to the earth's surface to reach their appointed place. Finally, if the earth rotated eastwards, clouds and all other objects floating in the air (as they are not fixed to the earth) should be seen to move westwards, which again is not the case. In short, the earth's motion should produce effects that are not seen.

It is immediately clear that Copernicus' presentation of Ptolemy's fairly extensive argument drawing from "appearances" (or experience or observational tests) is, again, quite liberal in that it almost completely disregards the subtle dialectic of Ptolemy's objections and counter-arguments to the earth's rotation. In addition to that, a comparison between Ptolemy's original and Copernicus' formulation of the argument reveals a small and at first glance completely insignificant difference. Whereas Ptolemy speaks of "thrown objects," that is, objects thrown up into the air, projectiles – *proiecta* in George of Trebizond's translation – Copernicus mentions "objects falling in a straight line [...] and perpendicularly (*cadentia in directum [...] ad perpendiculum*)." These objects fail to reach their appointed place, "which would meantime have been withdrawn by so rapid a movement [of the earth]." Since almost every falling body must have been thrown upwards into the air at some point, it seems that Ptolemy and Copernicus talk about one and the same thing. However, the slight difference in formulation, whether Copernicus is aware of it or not, signifies a greater quandary that he might face if he formulated the argument in a way as to preserve Ptolemy's *proiecta*. This quandary will become more obvious after we consider Buridan's interpretation of the same argument.

Buridan's argument against the rotation of the earth and Copernicus' falling bodies. Buridan refers to the argument in the 22nd question in the second book of his *Quaestiones in Aristotelis De caelo.*[111] He inquires whether it is possible to "save appearances" by assuming that the earth revolves around its centre and poles. He thus examines the question whether the celestial phenomena can be equally well accounted for by assuming that the earth rotates and the heavens are at rest as they are by the widely accepted thesis that the heavens move and the earth does not. In Buridan's view, many people have held as probable, and not contrary to the phenomena, for the earth to move circularly about its axis and the sphere of the fixed stars to remain at rest. By way of illustration, he gives the classic example of relative motion: a man on a moving ship. In so doing, Buridan contends that the problem with the assumed axial rotation of the earth

---

[111] See *Quaestiones in Aristotelis De caelo* II, 22 (Clagett, pp. 592–9; Patar, pp. 500–8). See also E. Grant's discussion of Buridan's arguments in *Planets, Stars & Orbs*, pp. 639–42.

is that of the perception of relative motion. Although the earth on which we stand appears to be at rest and the sun appears to revolve around us, the reverse might be true just as well, since the observed celestial phenomena will be the same.

Drawing from the example of a man on a ship, Buridan then considers potential objections to the rotation of the earth, including those based on appearances, as he puts it, like the example of an arrow projected vertically upwards and the question of where it will land. While other appearances testifying against the earth's rotation are fairly easy to rebut, this one, he declares the most demonstrative of all: it is *magis demonstrativa apparentia*.[112] An arrow projected from a bow vertically upwards always falls back to the spot from which it was projected. This would not happen if the earth moved at a speed necessary for it to complete its revolution in twenty-four hours. To the contrary, the part of the earth from which the arrow was projected would have been withdrawn for some distance. Buridan has an appropriate response to this appearance even under the assumption that the earth rotates: it is carried with the air, which is moved with the earth. Hence, the arrow is not moved only in an upward and downward rectilinear motion but also in a circular motion together with the earth, water and the lower region of air. Since we, too, share the same motion as the earth, water and air, we fail to perceive the movement by which the arrow is carried with the air: "But still the supporters would respond that it so happens because the air, moved with the earth, carries the arrow, although the arrow appears to us to be moved simply in

[112] See *Quaestiones in Aristotelis De caelo* II, 22 (Clagett, pp. 595–6; Patar, pp. 503–4). In response to the first objection to the earth's motion, which refers to the observed fact that the stars move from east to west, it can be said that according to the thesis of relative motion, the same stellar motion would also be observed if the stars were at rest and the earth rotated from west to east. The second appearance in support of the earth's motion draws on an analogy between the motion of the earth and the swift motion of a man riding a horse. If we move swiftly on horseback, we can feel the air resisting us. Therefore, by analogy, we should also feel air resistance with the swift whirling motion of the earth. Buridan's response is that earth, water and the lower region of the air are moved simultaneously with the earth's diurnal motion, so that there is consequently no air resistance. Another appearance that would be expected to occur as a result of the earth's diurnal rotation but does not, is its heating. The argument stems from the assumption that local motion produces heat. Therefore, if the earth rotated, it should become hot. In his response, Buridan explains that the motion does not produce heat, except by the friction, rubbing, or separation of bodies. This, however, does not occur as a result of the earth's rotation, as the earth, water and air would be moved together.

a straight line motion because it is being carried along with us. Therefore, we do not perceive the motion by which it is carried with the air."[113]

However, Buridan finds this response insufficient and grounds his criticism on the theory of impetus. As is well known, Buridan's theory of impetus responds to a problem facing Aristotle's theory concerning the violent motion of a moved body separated from the mover, i.e. a projectile separated from the projector. This is precisely the same motion as that imparted to a thrown stone or an arrow projected into the air.

According to Aristotle, everything that is moved has a mover. In the Middle Ages, this requirement was formulated in the famous phrase *omne quod movetur ab alio movetur* ("whatever is moved is moved by another").[114] For objects, things, and creatures that do not have a principle of motion in themselves, this requirement should translate to the following: the mover must be in direct contact with the moved body. Hence, everything that does not move by itself is moved by an external mover which must remain in contact with the moved body for the entire duration of motion. Aristotle's solution to the problem presented in Book VIII of *Physica* is that the first mover grants the medium (air or water) not only the ability to move but also the power or capacity to impart motion.[115] Therefore, although the medium itself stops moving once it is no longer moved by the mover, it still continues to cause motion. Air moves the air adjacent to it, which in turn moves the unit of air adjacent to it, etc. Thus, air is not only moved, but also moves the thrown object; hence, the first unit of air moves the stone and simultaneously activates the next unit of air to push the stone further and activate the next unit of air, etc., until the power of successive units of air gradually diminishes and the stone reaches a unit of air that can activate the next one to be moved but is no longer capable of granting it the power to function as a mover. At this point, the stone starts falling with its natural motion towards the centre of the universe.

The role of the medium set forth in the explanation of projectile motion, was criticized by Philoponus in the sixth century.[116] His view of the problem was adopted by Arabic philosophers and no later than the thirteenth century

---

[113] *Quaestiones in Aristotelis De caelo* II, 22 (Clagett, p. 596; Patar, pp. 500–8).

[114] See *Physica* VII, 1, 241b34.

[115] See ibid. VIII, 10, 266b27–267a20.

[116] On Philoponus, see R. Sorabji, ed., *Philoponus and the Rejection of Aristotelian Science*, in particular the contributions of M. Wolff, "Philoponus and the Rise of Preclassical Dynamics," and F. Zimmermann, "Philoponus' Impetus Theory in the Arabic Tradition."

also transmitted to the Latin West as the theory of impetus, where it was very strongly associated with the name of John Buridan.[117] His solution to the problem, as formulated in his *Quaestiones in Aristotelis De caelo*, is that the motion of a projectile no longer in contact with the initial mover is the result of an impetus, a motive power or force impressed directly on the body in motion. Impetus is a motive power transferred directly from the first mover to the moved object. It is a *vis impressa*, "force [or power] impressed" on the moved object. Once the projectile is separated from the initial mover, it is no longer moved by the external mover but by an internal force or power granted to it by the initial mover.[118] According to Buridan, the more swiftly the initial mover moves the projectile, the stronger its impetus will be, and it would remain in the body infinitely, if the moving body did not encounter any resistance. The initial amount of impetus, Buridan maintains, would not diminish unless opposed by any of the following two forms of resistance acting on the body: external resistance, such as that of air, or internal resistance manifested as the natural tendency of the body to move to its natural place. In the absence of any resistance, under ideal circumstances, impetus would keep the body in constant motion. If unimpeded, the moving body would continue to move in a straight line and at uniform speed to infinity. Moreover, it would neither have to change direction nor deviate from its initial speed, because its natural tendency to move towards its natural place would likewise be inoperative.

Let us return to the problem of the earth's motion. What does theory of impetus mean for the earth's motion in connection with the "appearance" of

---

117 See A. Maier, *Zwei grundprobleme der scholastischen Naturphilosophie*, pp. 111–314; M. Wolff, *Geschichte der Impetus Theorie;* J. Sarnowsky, "Concepts of Impetus and the History of Mechanics"; A. Funkenstein, "Some Remarks on the Concept of Impetus and the Determination of Simple Motion." For more extensive treatment of Buridan's theory of impetus, see also E. Grant, *Physical Science in the Middle Ages*, pp. 43–53.

118 Buridan, *Quaestiones in Aristotelis De caelo* III, 2 (Patar, p. 523): "[…] I am more inclined to believe that a moving body acquires not only motion onto itself from its mover but also an impetus, or a force, or a quality (*movens imprimit moto non solummodo motum sed consequenter aliquem impetum vel aliquam vim aut aliquam qualitatem*) – as it may also justly be called – which is naturally predisposed to move a body on which it is impressed (*impetus habet naturam movendi illud cui est impressus*), just as a force impressed on iron by a magnet moves the iron to the magnet. And the swifter the motion is, the stronger the impetus will be; the impetus of the projectile or the arrow is continually decreased by contrary forces resisting it, until it can no longer move the projectile." See also his formulation of the theory of impetus from his *ultima lectura* of *Quaestiones in Physicam* VIII, 12 (Clagett, *The Science of Mechanics*, pp. 534–5).

an arrow projected vertically into the air? Why is Buridan not satisfied with the response that an arrow projected vertically upwards falls back to the same spot from which it was projected, having been moved with the air, which moves together with the earth? In his opinion, the "impetus of the violently moved arrow (*impetus violentiae sagittae*)" in its upward movement would resist the motion of the air created by a rotating earth, and therefore would not travel the same distance as the air. Buridan reinforces his conclusion by invoking a situation in which the air is moved by high wind: even on such an occasion the arrow projected upwards is not moved as much as the wind itself, although it is moved somewhat.[119] If the earth revolved about its axis, an arrow projected into the air at a right angle to the surface should, owing to the impetus impressed on it, resist the motion of the air, which moves together with the earth, so that it should nevertheless fall behind the spot from which it was projected. The impetus should cause the arrow to continue traveling in the same direction for a certain period of time and to a certain measure. Therefore, if in the meantime the earth moved forward, the arrow should fall a certain distance behind the spot from which it was projected. But this does not happen, since an arrow or a stone is always seen to fall back to the spot from which it was projected or thrown. Therefore, the earth does not move; it does not rotate.

Now we can answer why Copernicus only weakly formulates the argument of the body thrown vertically into the air and why he uses the term *cadentia* instead of Ptolemy's term *proiecta*. One is tempted to say that it is because Copernicus knows Buridan's argument concerning *magis demonstrativa apparentia* against the movement of the earth, because he accepts some of the basic premises of the impetus theory, and because he obviously wants the reader of his book to avoid the conclusion reached by Buridan. But even if he did not know Buridan's argument, and even if did not use *proiecta* instead of *cadentia* in order to avoid Buridan's conclusion, he should have known better: since he accepts the basic tenets of the impetus theory, he should come to the same conclusion as Buridan, that is, the earth does not rotate.

There is only circumstantial evidence to support Copernicus' knowledge of Buridan. However, in his discussion of the earth's daily rotation and elsewhere in Book I of *De revolutionibus*, Copernicus uses too many notions, ideas and reasons that can also be found in Buridan's treatment of the possible rotation of the earth in Question 22 of Book II of *Quaestiones in Aristotelis De caelo* for it to be pure coincidence. Let me enumerate just the most obvious ones, though I

---

119 See Buridan, *Quaestiones in Aristotelis De caelo* II, 22 (Clagett, p. 596; Patar, p. 504).

will reveal more later on:[120] the earth is of a spherical figure and a spherical figure means aptitude to move in a circle;[121] saving the appearances with the rotation of the earth;[122] relativity of motion;[123] distance–period relationship;[124] rest is a nobler state than movement;[125] economy or simplicity of explanation;[126] earth, water and the lower region of the air are moved simultaneously with the earth's diurnal motion,[127] comparison of the rectilinear motion of a piece of the earth towards the natural place with the process of healing of the sick.[128] This certainly testifies to

---

[120] Compare with M. Kokowski, *Copernicus's Originality*, pp. 82–3, and E. Grant, *Planets, Stars, & Orbs*, pp. 647–8, who lists only several correspondences between Buridan and Copernicus: "optical relativity of motion, as illustrated by the movement of ships; that it is better for the earth to complete a daily rotation with a very much slower velocity than would be required for the vast heavens; that the air shares the daily rotation of the earth; that the motion of bodies rising and falling with respect to the rotating earth results from a motion compounded of rectilinear and circular elements; and finally, that since the state of rest is nobler than motion, it is more appropriate for the ignoble earth to rotate than it is for the noble heavens."

[121] See Buridan, *Quaestiones in Aristotelis De caelo* II, 22 (not in Clagett's translation; Patar, p. 500): "The earth is of a spherical figure and the spherical figure is an aptitude for the spherical or circular movement (*Terra est figurae sphericae, et figura sphaerica est aptitudo quaedam ad motum sphaericum sive circularem*)."

[122] See ibid. (Clagett, p. 594; Patar, p. 501): "[…] whether, in positing that the earth is moved circularly around its own centre and about its own poles, all the phenomena that are apparent to us can be saved."

[123] See ibid. (Clagett, pp. 594–5; Patar, p. 501): "If anyone is moved in a ship and he imagines that he is at rest, then, should he see another ship which is truly at rest, it will appear to him that the other ship is moved."

[124] See ibid. (Clagett, p. 595; Patar, p. 501): "And, therefore, this opinion imagines that any of the spheres of the planets moved evidently like the earth from west to east, but since the earth has a lesser circle, hence it makes its rotation in less time. Consequently, the moon makes its rotation in less time than the sun. And *this is universally true*, so that the earth completes its rotation in a natural day, the moon in a month, and the sun in a year, etc." Emphasis added.

[125] See ibid. (Clagett, p. 595; Patar, p. 502): "The celestial bodies ought to be attributed the nobler conditions, and to the highest sphere, the noblest. But it is nobler and more perfect to be at rest than to be moved. Therefore, the highest sphere ought to be at rest."

[126] See ibid. (Clagett, p. 595; Patar, p. 503): "Just as it is better to save the appearances through fewer causes then through many, if this is possible, so it is better to save them by an easier way than by one more difficult."

[127] See ibid. (Clagett, p. 595; Patar, p. 504). See supra, pp. 174–5.

[128] See ibid. IV, 2 (Patar, p. 530): "[…] sicut se habet sanitas ad sanabile et sanationem, et perfecta magnitudo ad augmentabile et augmentationem, ita locus deorsum se habet

Copernicus' familiarity with some of the key elements of Buridan's discourse, but it does not prove that he was drawing directly or exclusively from Buridan or that he was familiar with the argument against the rotation of the earth based on his impetus theory. The majority of these ideas were also used by Oresme in his *Livre du ciel et du monde*, and Copernicus might have become familiar with them while studying in Padua, where Oresme's work was well known,[129] or through Albert of Saxony's derivative and repetitive accounts of the views of John Buridan and Nicole Oresme, which were well known in Cracow.[130] Some of notions cited by Copernicus, however, are found exclusively in Buridan, for example, the notion that rest is a nobler state than movement and should accordingly be attributed to the heavens and not to the earth, or that the spherical figure has an aptitude to rotate; this notion was mentioned neither by Nicole Oresme nor by Albert of Saxony. But still, there is no evidence that Copernicus derived his arguments

---

ad grave et ad motus eius. Sed sanitas est finis ad quem tendit sanatio, et est perfectio formalis ipsius sanabilis cum acquisita fuerit; et sic etiam est de perfecta magnitudine ad augmentationem at ad augmentabile. Ergo similiter debet concedi de loco deorsum ad ipsum grave et ad totum eius." Compare with Aristotle, *De caelo* IV, 3, 310b15 et seq., and *Physica* V, 229a7–229b14. See also, R. Hooykaas, "The Aristotelian Background to Copernicus's Cosmology," p. 115.

129 See A. De Pace, *Niccolò Copernico e la fondazione del cosmo eliocentrico,* pp. 178, 523 and 524, and infra, p. 286. She lists, p. 184, the following elements that are according to her common to Oresme and Copernicus: optical relativity of motion; simplicity achieved if all heavenly motions move in one direction from west to east with a progressively smaller velocity from the centre to circumference; abolition of the excessive velocity of the last sphere, which is also in violation of the distance - period relationship. Perhaps there is also an echo of Oresme' explanation of the earth's circular, rotational movement and to its proper place when gets outside it, by its nature or form, which is similar to the nature which draws iron to the magnet in Giese's magnetic analogy in *Narratio prima* (see supra, p. 80). See *Le livre du ciel et du monde* II, 25 (Denomy and Menut, pp. 528–9): "To the fourth argument, we can say that the force causing this lower region of the world to move in a circle is its nature or form; and this same force – similar in nature to that which draws iron to the magnet – moves the earth to its proper place when it gets outside."

130 On Albert of Saxony's questions on *De caelo*, see A. Goddu, *Copernicus and the Aristotelian Tradition*, pp. 128–123, and more extensive treatment in "Sources of Natural Philosophy at Kraków in the Fifteenth Century." Unfortunately, he does not discuss Albert's reasoning on the earth's rotational movement. See also G. Federici-Vescovini, "Note sur circulation en Italie du commentaire d'Albert de Saxe sur le *De caelo*"; E. Grant, "The Unusual Structure and Organisation of Albert of Saxony's Questions on *De caelo*"; J. Sarnowsky, "Albert von Sachsen und die Astronomie des 14. Jahrhundert."

from medieval sources,[131] and even if he had, he would have had to have access to Buridan's *Questiones* or to some report of it, since neither Oresme nor Albert took into account Buridan's impetus argument against the rotation of the earth.

There is only scarce but clear textual evidence for Copernicus' acceptance of the impetus theory.[132] As we have seen, Copernicus claims at the beginning of Chapter 8 of Book I of *De revolutionibus* that *vis vel impetus* causes things to which it is applied to disintegrate and to perish. This indicates that he accepts the basic tenet of the impetus theory, or transmission of force directly to a moving object.[133] But he also qualifies it; he uses the term *impetus* to denote some kind of violence and un-naturalness of motion produced by it on two (or three) other occasions. The next occurrence is in his objection to "Ptolemy's" centrifugal force argument: "For the higher they [i. e. the heavens] are driven by the impetus of their motion (*ipse motus impetu rapiatur*), the faster that motion will be […]."[134] He uses it next in the context of his explanation how both air and objects suspended in the air share in the earth's rotational motion: "[t]he air closest to the earth will accordingly seem to be still. And so will the things suspended in it, unless they are tossed to and fro, as indeed they are, by the wind or some other impetus (*vel alio quovis impetus*). For what else is the wind in the air but the wave in the sea?"[135] The fourth and last occurrence of the term *impetus* is only partially associated with violence – Copernicus speaks of *impetus ponderis*, his reason for why the rectilinear motion of falling bodies is not uniform.

Thus Copernicus uses impetus to explain the difference between natural and violent motions in order to reject "Ptolemy's" "centrifugal force" or "projection"

---

[131] See E. Grant, *Planets, Stars, & Orbs*, p. 648.

[132] See also A. Goddu, *Copernicus and the Aristotelian Tradition*, p. 353, who also claims that Copernicus does not adhere to the traditional theory of impetus. Copernican scholars only recently acknowledged that Copernicus adheres to some variant of *impetus* theory. For a brief overview of this issue, see M. Kokowski, *Copernicus's Originality*, pp. 218–20. A. De Pace almost completely avoids the issue.

[133] See also M. Kokowski, *Copernicus's Originality*, pp. 221–31, and especially A. Goddu, *Copernicus and the Aristotelian Tradition*, pp. 336–45. There is a question whether Copernicus is referring to *impetus* also in Chapter 4 of Book I of *De revolutionibus*, where he speaks of *virtus* (virtus = vis = impetus): "[Q]uoniam fieri nequit, vt caeleste corpus simplex vno orbe inaequaliter moueatur. Id enim euenire oporteret vel propter virtutis mouentis inconstantiam, siue asciticia sit siue intima natura, vel propter reuoluti corporis disparitatem." While Kokowski, pp. 229–30, believes that he does, I agree with A. Goddu, pp. 336, and 339–40, who is of the opinion that he does not.

[134] *De revolutionibus* I, 8 (Rosen, p. 15; Zekl, p. 112).

[135] Ibid. (Rosen, p. 16; Zekl, p. 116).

argument. This means that he differs from Buridan with regard to the *effects* of impetus on the moving body. Contrary to Buridan, who does not think impetus causes any disruption on the part of the moving object, Copernicus believes that motion produced by *impetus* is unnatural and violent and therefore has disastrous effects on the things to which it is applied.[136] However, this does not change the fact that he accepts the transfer of the force from the motor to the moved object. But if this is the case and if his argumentation in Chapter 8 is thorough, Copernicus should also have been aware that impetus serves as a strong tool for rejecting the earth's rotation: an object thrown or an arrow projected vertically into the air on which the mover has impressed an impetus should not fall to the same spot from which it was thrown, but at least some distance behind it. Since this does not happen, the earth does not move. Copernicus, of course, wants to avoid such a conclusion. And this might be, one is inclined to think, the reason why he formulates Ptolemy's argument so weakly, using *cadentia* instead of *proiecta*. The former expression bears – at least at first sight – no implication that the thrower has impressed an impetus on the projectile. Rather, the objects appear to move by themselves, without any visible external force to impart *impetus* to them.

Copernicus, impetus and the acceleration of falling bodies. But this is not the end of the story; Copernicus' reasoning is still contradictory, whether he omitted Buridan's impetus argument against rotation of the earth out of ignorance (very likely) or by design (less likely). Either way, he should arrive at the same conclusion as Buridan. His acceptance of some sort of impetus theory should lead him to the conclusion that the earth does not rotate, especially since he believes, like Buridan and some other partisans of the impetus theory, that impetus is the cause of acceleration in falling bodies.

As is well known, the theory of impetus is used not only to explain projectile motion but also to explain some other phenomena as well. Buridan considers the possibility that the celestial spheres move under the influence of impetus (God imparted an impetus to celestial spheres at the creation and they have revolved

---

136 This is an additional indication that Copernicus is not drawing directly or exclusively on Buridan. Buridan and Albert of Saxony were an exception to the generally held view that *impetus* produces violent and self-dissipating effects in sublunary bodies. See D. Knox, "Copernicus's Doctrine of Gravity and the Natural Circular Motion of the Elements," p. 176. Knox, n. 93, refers to Alessandro Achillini, *De elementis*, and Fridericus Sunzell, *Collecta et exercitata [...] in octo libros Physicorom Aristotelis*, for arguments *pro et contra* the idea that impetus in projectiles is violent. See also M. Clagett, *The Science of Mechanics in the Middle Ages*, pp. 538–40.

thereafter).[137] He also – more pertinent to our consideration – employs impetus to account for the accelerated rectilinear motion of falling bodies. The accelerated motion of a falling body, Buridan maintains, is not caused by its weight, which is the natural tendency of the body to remain in its natural place or move towards it. If a body is not in its natural place and encounters no resistance, its weight will only cause it to move in uniform motion at minimum speed. However, uniform motion can only be maintained for a short period of time, as the motion imparts to the moved body an impetus which adds a motion of a certain minimum speed for the subsequent minimum period of time. The accelerated motion, in turn, produces impetus accelerating the motion of the body, and this accelerated motion, again, imparts impetus to the body in motion, and so forth, so that during the course of motion, a continual increase in impetus and consequently continual acceleration takes place.

> From this theory also appears the cause of why the natural motion of a heavy body downward is continually accelerated. For from the beginning only gravity was moving it. Therefore, it moved more slowly, but in moving it impressed in the heavy body an impetus. This impetus now [acting] together with its gravity moves it. Therefore, the motion becomes faster; and by the amount it is faster, so the impetus becomes more intense. Therefore, the movement evidently becomes continually faster.[138]

Copernicus, like Buridan, uses impetus in order to explain the acceleration of falling bodies. He mentions impetus once again in Chapter 8 of Book I of *De revolutionibus* in his explanation of the accelerated rectilinear motion of heavy objects.

Copernicus broaches this topic in the first of his four arguments – which I cover below – in favor of the earth's rotation, intended to devise his own, alternative theory of natural motion of the elements and of heaviness and lightness and thus to refute the first two arguments against its rotation, based on Aristotelian theory of "heaviness and lightness" and "nature of motion." Copernicus wants to show that the Aristotelian concept of simple, rectilinear and natural motion is internally inconsistent and that the earth's rotation, albeit circular, is nonetheless natural and simple. This means that the rectilinear motion is not simple motion. According to Copernicus, objects moving rectilinearly also move with the earthly sphere (or any other sphere for that matter) in a circular motion and are therefore

---

137 See Buridan, *Questiones super octo Phisicorum libros Aristotelis* VIII, 12 (Clagett, p. 536). See also M. Clagett, *The Science of Mechanics in the Middle Ages*, p. 523 et seq.

138 Buridan, *Questiones super octo Phisicorum libros Aristotelis* VIII, 12 (Clagett, pp. 535–6).

composed of circular and rectilinear motions. But even if we disregard that fact, there is yet another reason why rectilinear motions are not simple (and hence not natural or only partially natural); rectilinear motions themselves "do not execute a simple, constant and uniform motion." Falling bodies move faster and faster as they fall; Copernicus explains this non-uniformity and inequality by employing a concept of impetus of their weight (*impetus ponderis*).[139]

Copernicus obviously adopts at least two fundamental theses of the theory of impetus: firstly, that violent motion is produced by an impetus which exhausts itself in the course of motion; and secondly, that accelerated free fall is a compound motion, and that the acceleration of free fall is due to *impetus ponderis*.[140] Copernicus believes that weight impresses an impetus on a moving heavy body and that this impetus, together with the weight of the object, moves it, thus increasing its speed. Now, if Copernicus believes in this second application of the impetus, this means that even if he phrases Ptolemy's objection in terms of "falling bodies (*cadentia*)" instead of "projectiles (*proiecta*)," and even if he did not know Buridan's argument, he cannot avoid the conclusion that the earth does not rotate.

Ptolemy's problem with the rotation of the earth, in Copernicus' words in Chapter 7 of Book I of *De revolutionibus*, is that the "objects falling in a straight line" would "not descend perpendicularly to their appointed place, which would meantime have been withdrawn by so rapid movement" of the earth and that "clouds and everything floating in the air would be seen drifting always westward." Or, as he puts it in Chapter 8: "Then what about the clouds and the other things that hang in the air in any manner whatsoever, or the bodies that fall down, and conversely those that rise aloft?"[141]

---

[139] D. Knox, "Copernicus's Doctrine of Gravity and the Natural Circular Motion of the Elements," pp. 174–5, points out that this unusual expression derives from George of Trebizond's translation of the *Almagest* I, 7, where it denotes the impact caused by heavy bodies as they hit the earth (*minimi ponderis impetu*). But Copernicus uses the expression in a different sense as Ptolemy. See also A. Goddu, *Copernicus and the Aristotelian Tradition*, p. 343, who also believes that this passage suggests that Copernicus adopted the notion of an accidental impetus produced by the weight of a body and that this impetus causes a falling body to accelerate.

[140] According to M. Wolff, "Impetus Mechanics as a Physical Argument for Copernicanism: Copernicus, Benedetti, Galileo," pp. 223–4, and 230–1, Copernicus borrows three elements from the traditional impetus theory, the third one being that all motions require an immediately moving cause which, however, is not the final cause in the case of natural motions. According to Wolff, Copernicus also adopts several other elements of the classic impetus theory and transforms them in part.

[141] *De revolutionibus* I, 8 (Rosen, p. 16; Zekl, p. 116).

Copernicus' reply is simple and more or less classic. Without delving into a detailed analysis of Ptolemy's arguments, and without taking into account the implications of the impetus theory applied to falling bodies, he responds that the air moves together with the earth and with the air all the things that are somehow linked to the earth:

> We would only say that not merely the earth and the watery element joined with it have this motion, but also no small part of the air and whatever is linked (*cognationem habent*) in the same way to the earth. The reason may be either that the nearby air, mingling with earthy or watery matter, follows the same nature as the earth, or that the air's motion, acquired from the earth by contiguity, shares without resistance in its perpetual rotation.[142]

Copernicus' first explanation for the circular motion of the "nearby air," that it "follows the same nature as the earth," may derive, as Dilwyn Knox suggests, "from the scholastic view, based on Aristotle and Averroes, that water and earth cooled the neighboring air, thickening and endowing it with 'preternatural' heaviness."[143] But in light of what follows, this interpretation is inadequate. Copernicus is saying that air mixed with the earthy and watery matter conforms to the same nature as earth, that is, its circular motion.[144] The nature in question is that of natural circular motion and not of the "'preternatural' heaviness." As for the second option, Knox believes that Copernicus "adapted the scholastic idea that fire and air, though naturally inclined to rest in their spheres, did not resist the circular motion imposed by a superior agency."[145] Copernicus carries this classic response even further and links the question concerning the movement of the lower region of air to the problem of the circular movement of comets in the upper region of air. [146]

---

142 Ibid.

143 D. Knox, "Copernicus's Theory of Gravity," p. 167.

144 See A. De Pace, *Niccolò Copernico e la fondazione del cosmo eliocentrico*, p. 189, and n. 559.

145 D. Knox, "Copernicus's Theory of Gravity," p. 167. This presupposes that air retains its authentic nature and its propensity to move or remain still and participates without resistance in the earth's diurnal rotation.

146 There is also a similar passage in M. Ficino's *Compendium in Timaeum*, 25 (Farndell, pp. 40–1; *Opera omnia*, p. 1449): [Chapter title] "Circular motion is the property of every sphere in constant movement; and light is the principal property of fire // But let us not be troubled by those who, on the grounds that the motion of the heavens is circular, while the motion of fire is in a straight line, doubt that the heavens are fiery.

Copernicus draws on Aristotle's division of the cosmos in *Meteorologica,* nicely described in Pierre d'Ailly's *Ymago mundi*, into the superlunary and sublunary realms.[147] According to Aristotle the former contains the heavens with the

---

For if any part of the heavens happens to be placed in the middle realm, it will at once rise up in a straight line, seeking its native home by the shortest, and yet it remains true that the proper motion of the heavens is circular; and if the earth is moved away from the centre, it will move downwards, although its true property is to stay motionless. In fact, every moving sphere, whether elemental or celestial, moves naturally in a circuit in accordance with its own form, so that it may take fuller advantage of its own position. But wherever there is movement in a straight line, this is not so much a natural movement as a return to nature. If circular motion suits any body, it suits fire most of all; for since it is by nature furthest from earth, it is ever in movement, just as the earth is ever motionless. If it is always in movement, then it is in circular motion (*in orbem*); for whatever moves in a straight line stops moving when it has traversed space, which is finite. But the circle made by comets shows that both fire and air move in a circle. Unending inflow and outflow demonstrate that water, too, has its cycle. But motionlessness was required of earth, that it might resemble an indivisible unmoving centre and offer its benefits with the greater convenience." Compare also with the passage from *Theologia Platonica* IV, 1 (Allen and Hankins, pp. 284–5): "Again, the natural movement of the elements is what does not trespass beyond the bounds of nature and of natural place; in other words, it is an everlasting circular movement in its place and sphere perfectly resembling the sphere's natural shape. Fire and air move in a circle like the moon, as the revolution of the comets shows. Water moves in a circle, ceaselessly flowing back. If earth were moved, as Hegesias claimed, it would be moved in a circle. It stays still, most people believe, on the surface. [But] its parts in a way make a circle around the centre of the world insofar as they pack themselves together on all sides around the centre continuously pressing it. But ascent and descent are not strictly speaking natural motion, but a sudden restoration by way of a straight line to a natural location and motion, a restoration which, because it is ended by another and is always directed along a singular path, clearly shows us that the parts of the elements are not moved of their own accord." See also D. Knox, "Ficino, Copernicus and Bruno," p. 337, and, "Copernicus' Doctrine of Gravity," pp. 200–1. Compare also with Giambattista Capuano da Manfredonia's *Commentary on the Sphere of Sacrobosco*, 30[va], in the M. Shank, "Setting up Copernicus? Astronomy and Natural Philosophy in Giambattista Capuano da Manfredonia's *Expositio* and the *Sphere*," p. 298. Shank also points out, pp. 298–9, and n. 22, that the association of comets with the rotation of the spheres of fire and air also appears in *Commentary on the Sphere* by Robert Anglicus. See L. Thorndike, *The Sphere of Sacrobosco and its Commentaries*, p. 50. Robert does not discuss the rotation of the earth in this context. Compare also with the Arabic astronomer Nasīr al-Dīn al-Tūsī in his book on theoretical astronomy, *Tadhkira*. On this, see F. J. Ragep, "Tūsī and Copernicus: The Earth's Motion in Context," especially pp. 146–8.

147 See *Meteorologica* I, 3, 340b10–14; 340b32–341a5; I, 7, 344a11–13; II, 4, 361a23–25. For d'Ailly, see supra, pp. 121–2.

sphere of fire and the upper region of air, and moves with diurnal rotation. The latter realm contains the middle and lower regions of air, together with water and earth, as well as mixed bodies; it remains still and has no diurnal motion. Copernicus links this division to Aristotle's theory that comets are formed in the upper region of air and moved in it with the diurnal motion of the superlunary spheres.[148] If, according to the peripatetic views, comets, which belong to the sublunary part of the universe, can conform to the superlunary spheres and share their diurnal motion, then, Copernicus concludes, the lower spheres of air can conform to the earth and share its motion. But the supreme sphere of air is too far away from the earth to be affected by its motion.

> No less astonishingly, on the other hand, is the celestial movement declared to be accompanied by the uppermost region of air. This is indicated by those bodies that appear suddenly, I mean, those that the Greeks called "comets" and "bearded stars." Like the other heavenly bodies, they rise and set. They are thought to be generated in that region. That part of the air, we can maintain, is unaffected by the earth's motion on account of its great distance from the earth.[149]

In sum: If we accept that no small part of air is "linked with the earth," it also follows that "[t]he air closest to the earth will accordingly seem to be still. And so will the things suspended in it, unless they are tossed to and fro, as indeed they are, by the wind or some other impetus. For what else is the wind in the air but the wave in the sea?"[150] The *propinquus aer* moves together with *terraquea sphaera*, and we, participating in the same movement, cannot notice any movement in the air, except those caused by wind.

Obviously, Copernicus is not the first to deal with the problem of observational tests (appearances, experiences), which, along with the commonly accepted physics, only further convinced Ptolemy of the earth's immobility. Besides Ptolemy, Buridan and Oresme, there were other scholars in both the Arabic and Latin traditions who provided different answers to this question. One basic distinction between them, however, was whether or not they considered observational tests sufficient to determine the question of the earth's motion or rest. In the western tradition of the high Middle Ages, there were two diametrically opposed positions

---

148 On Aristotle's theory of comets, see, for instance, T. Heidarzadeh, *A History of Physical Theories of Comets, From Aristotle to Whipple*, pp. 1–19.

149 *De revolutionibus* I, 8 (Rosen, p. 16; Zekl, p. 116).

150 Ibid.

on this question. One was held by Oresme, who maintained that no such observation could have ever been made.[151] The other one, as we already know, was advocated by Buridan, who argued that at least one observational test, one *apparentia*, tipped the scales in favor of the earth's rest. Buridan, as we have seen, rejected it on the basis of the theory of impetus. Copernicus, too, holds that observations of what occurs around us "on earth and in the air," as Ptolemy would put it, cannot lead to a definite conclusion whether it is the earth that moves or the heavens, as there is a plausible answer to every relevant observational test or sensory experience that speaks against the earth's rotation, that is, in favor of its rest. But, as already said, he should not![152] He should, if he would thoroughly consider his own theory

---

151 See *Le Livre du ciel et du monde* II, 25 (Denomy and Menut, pp. 520–1): "At the outset, I wish to state that it is impossible to demonstrate from any experience at all that the contrary is true; second, that no argument is conclusive; and third, I shall demonstrate why this is so." Oresme finally rejects the earth's rotation for theological reasons.

152 And neither, then, should Nicole Oresme. Oresme's solution to this "experience," which he calls *plus forte*, is the following: "Concerning the third experience, which seems more complicated and which deals with the case of an arrow or stone thrown up into the air, etc., one might say that the arrow shot upward is moved toward the east very rapidly with the air through which it passes, along with all the lower portion of the world which we have already defined and which moves with daily motion; for this reason the arrow falls back to the place from which it was shot into the air. Such thing could be possible in this way, for, if a man were in a ship moving rapidly eastward without being aware of the movement and if he drew his hand in a straight line down along the ship's mast, it would seem to him that his hand were moving with a rectilinear motion; so, according to this theory it seems to us that the same thing happens with the arrow which is shot straight down or straight up. Inside the boat moved rapidly eastward, there can be all kinds of movements – horizontal, criss-cross, upward, downward, in all directions – and they seem to be exactly the same as those when the ship is at rest. Thus, is a man in this boat walked toward the west less rapidly than the boat was moving toward the east, it would seem to the man that he was approaching the west when actually he was going east; and similarly as in the preceding case, all the motions here below seem to be the same as though the earth rested." M. Clagett, *The Science of Mechanics in the Middle Ages*, p. 587, calls Oresme's solution a "concept of a closed mechanical system," – "[…] Buridan hinted at, and Oresme rather specifically outlined, the concept of a closed mechanical system, wherein, due to the relativity of the perception of motion, the observer describes all movements as if they were a part of his system only. This concept is used by Oresme to explain what he believes to be the apparent rectilinear movement of objects falling to the earth. The analogy is, again, to the man on the ship. As Oresme showed, it would appear to the observer on the ship that his hand was descending in rectilinear motion if he made it slide down the mast. […] Oresme also seems to suggest that the basic reason for the phenomena

of impetus in connection to observational tests, come to the conclusion that the earth does not move.

According to him, falling bodies, too, receive in their fall an impetus which is due to their weight and which accelerates their movement. If this is the case, then the same reasoning that is valid for *proiecta* should also apply to *cadentia*. The weight of falling bodies impresses an impetus on them which should, just like in the case of projectiles, function as an inhibitor of the movement of whatever falling body (a falling stone, a descending arrow). If the earth revolved about its axis, a body, falling perpendicularly to the surface of the earth should, owing to an impetus impressed on it by its weight, resist the motion of the air, which moves together with the earth, so that it should nevertheless fall behind the "appointed place." The impetus should cause the falling body to continue traveling in the direction of the fall for at least a certain period of time and to a certain measure. Therefore, if in the meantime the earth moved "forward," the body should fall a certain distance "behind" the "appointed place." But this does not happen, since an arrow or a stone falling down is always seen to fall back to the "appointed place." Therefore, the earth does not move. This line of thought is, however, something that does not occur to Copernicus. He uses this classic counter-argument to reject all consequences that should, as Ptolemy maintained, affect objects in the air: continual motion in a contrary direction to that of the earth; a stone thrown perpendicular into the air should not fall back to the same spot from which it was thrown due to the earth's rapid rotation; the movement of clouds, etc.

Copernicus' formulation of Ptolemy's objections to the rotation of the earth, paired with the implicit contradictions in his solution to the observational tests against the earth's rotation, only confirm my argument that in Chapter 7, which is dedicated to the formulation of the objection against the earth's rotation,

---

of the closed mechanical system is that all objects therein share the same horizontal velocity (as, per example, all objects on the board of the ship)." – but this does not solve the problem. To move the hand along the mast is not the same thing as to shoot an arrow in the air or its fall down. To move a hand one does not need an impetus. And since Oresme is also a partisan of impetus (for *loci* where Oresme speaks of it, see Menut and Denomy, *Introduction*, pp. 18–9), his response is contradictory. See also M. Clavelin, *La philosophie naturelle de Galilée*, pp. 124–5, n. 159: "L'explication est en contradiction ouverte avec la théorie de l'*impetus*, acceptée ailleurs par Oresme, et revient à donner un rôle moteur au milieu. Buridan, plus cohérent, la rejetait." See also A. De Pace, *Niccolò Copernico e la fondazione del cosmo eliocentrico*, p. 182. The contradiction may have been the reason why Oresme did not mention Buridan's impetus argument, although he must have been familiar with his discussion.

Copernicus is more concerned with preparing a dialectically alluring terrain for the refutation of the objections in Chapter 8 than with all the intricacies of the arguments. Likewise, he should provide consistent refutation of objections to the earth's rotation in his Chapter 8, but he instead settles for answers which, though classical, are not very consistent. Copernicus fails to offer a thorough reflection on the implications the theory of impetus should have for the observational tests, which indicates that Chapters 7 and 8 are – at least so far – more about dialectical refutation than about demonstration and certainty.

3. Copernicus' positive arguments in favor of the earth's rotation

This feeling is further strengthened by Copernicus' four positive arguments in favor of the earth's motion in the final part of Chapter 8 of Book I of *De revolutionibus.* They are his answer to the first question posed at the beginning of Chapter 5, regarding the earth's rotation as logically following from its spherical shape. The second question, regarding its place in the universe, is answered in Chapters 9 and 10.[153]

The most important of the four arguments is the first one dealing with Copernicus' alternative theory of gravity and nature of the motion of the elements. His second and third arguments fall, at least partly, under the philosophy of nature as well. In the second he claims that immobility, since it is deemed nobler and more divine than change and instability, should belong to the heavens and movement to the earth. In the third he again asserts that motion should not be attributed to the outermost sphere of the universe, constituting the space or place of the bodies, but to the bodies in it, that is to the earth. Since these arguments weigh certain explicit and implicit difficulties of traditional philosophy of nature against each other, all three may be regarded as dialectical rather than strictly physical. The last argument underlines the fact that the distance between the planets and the earth varies, which means that their circuits are not homocentric, and that there are several centres of the celestial movement, which has implications for the traditional theory of motion around the universe's centre. Let us first deal with the second, third, and fourth arguments.

---

[153] On the four arguments, see also M. Wolff's analysis in "Impetus Mechanics as a Physical Argument for Copernicanism. Copernicus, Benedetti, Galileo," pp. 217–8; J.-J. Szczeciniarz, *Copernic et la révolution copernicienne*, pp. 141–53.

Copernicus' last three arguments. Copernicus' second argument in support of the earth's motion is that "immobility is deemed nobler and more divine than change and instability, which are therefore better suited to the earth than to the universe."[154] As already noted,[155] Copernicus could find it in Buridan or some other source common to them both – it is worth noting again that this one is not to be found in Oresme – who used it as one of *persuasiones* for the daily rotation of the earth in the 22nd question in Book II of his *Questiones in Aristotelis De caelo:*

> The third persuasion is this: To celestial bodies ought to be attributed the nobler conditions, and to the highest sphere, the noblest. But it is nobler and more perfect to be at rest than to be moved. Therefore, the highest sphere ought to be at rest. The proof of the minor premise is as follows: if a heavy body moves downwards, it does so not with the goal to move but to proceed to its natural place and remain at rest there. Therefore, rest is the goal of this motion, and the goal is nobler. This has also been confirmed: as the Commentator explains in Fourth Book of the *Physics*, there is nothing unnatural about the natural rest of a heavy body in a lower place. But there is always something unnatural about the downward motion of a heavy body, owing to the presence of a part of the upper place, for the elimination of which the heavy body moves (*quia est aliquid de loco sursum ad quod removendum grave movetur*). Therefore, to put it in simple terms, it is more perfect for a heavy body to rest in a lower place than to move downwards. And this stands as proof that the state of rest is indeed nobler than motion.[156]

Obviously, Copernicus does not bother to explain this argument in any detail and he, of course, does not take into account Buridan's response to this "persuasion." The state of rest is nobler than motion only "for bodies which move in order to arrive at their natural place. By contrast, for objects which are always at rest in their natural place, and which have no other goal than to move and thus acquire their ultimate perfection, I say that motion is nobler than rest. This holds for celestial bodies."[157] The fact that there is a consistent objection to this persuasion does not deter Copernicus from citing the argument in favor of the earth's motion.

---

154 *De revolutionibus* I, 8 (Rosen, p. 17; Zekl, p. 120).

155 See supra, p. 178, n. 125.

156 Buridan, *Quaestiones in Aristotelis De caelo* II, 22 (Clagett, p. 595, translated only the beginning of this *persuasio;* Patar, pp. 502–3).

157 Ibid. (not in Clagett's translation; Patar, p. 505).

Copernicus' third argument in Chapter 8 basically reiterates what has already been said in Chapter 5: it would seem quite absurd to attribute motion to "what contains and locates," and not, more appropriately, to "what is contained and located, namely, the earth.[158] Copernicus, who has already in the *Commentariolus* related the immobility of the firmament and its function of container and furnisher of place – "The highest is the immovable sphere of the fixed stars which encloses and fixes the positions of all [the others] (*immobilis et omnia continens et locans*)." –[159] seems to be reasoning along the following lines. According to the Aristotelian doctrine, the sphere of the fixed stars is the *locus*, the place of the universe which cannot move. It follows that the *locatum*, that which is in place, moves while the *locans* remains immobile. But, as has been noted by several scholars, this is hardly an argument for an Aristotelian. It would be valid if the universe were considered at least immense if not actually infinite.[160] But the Aristotelian cosmos is nothing like that; it is finite, and therefore measurable. Besides, there is a qualitative difference between the heavens and the earth. While the heavens are made of aether and endowed with natural circular movement, the natural state of the earth is rest in the middle of the universe.

But Copernicus may also be reasoning along the Platonic lines, disregarding Aristotelianism. In Chapter 5 he says that the motion should rather be attributed to the enclosed (*contentum*) than to the enclosing (*continens*) or the located thing (*locatum*) rather than to the framework of place (*locans*), and provides a special justification for it: the heavens are something "that contains and *adorns* everything, being the common place of all things (*caelat omnia communis universorum locus*).[161] According to Plato's *Timaeus*, the demiurge has given to the heavens the spherical form and the function of containing everything. It is also part of the universe which provides beauty, order and unity and is as such an ontological reference frame which gives orientation to the human intellect and should, as such, rest immobile.[162]

---

[158] See *De revolutionibus* I, 8 (Rosen, p. 17; Zekl, p. 100).

[159] *Commentariolus* (Swerdlow, p. 440 ; Zekl, p. 8).

[160] See A. Koyré, *The Astronomical Revolution*, pp. 55–6. See also J.-J., Szczeciniarz, *Copernic et la révolution copernicienne*, pp. 144–7, for additional contradictions in Copernicus' reasoning in the Aristotelian context of theory of *locus*.

[161] Emphasis added. For the full quotation of the argument, see supra, p. 153, n. 73.

[162] See also A. De Pace, *Niccolò* Copernico *e la fondazione del cosmo eliocentrico*, p. 336, n. 119.

In the fourth argument Copernicus repeats the known fact that according to the traditional Ptolemaic astronomy planets in approaching the earth and then receding farther away from it are moving with two different movements. They move around the middle of the universe (i.e. the centre of the earth), as well as towards and away from it. So, concludes Copernicus, the motion around the middle should be understood in the sense that every single circular motion rests on its own centre:

> Last of all, the planets obviously approach closer to the earth and recede farther from it. Then the motion of a single body around the middle, which is thought to be the centre of the earth, will be both away from the middle and also toward it. Motion around the middle, consequently, must be interpreted in a more general way, the sufficient condition being that each such motion encircle its own centre.[163]

This argument relies on the first one, in which Copernicus devises a theory of motion in an effort to respond to theoretical challenges of the peripatetic tradition to his affirmation of the earth's motion.

Copernicus' first argument: theory of gravity and elemental motion. The first and most important of Copernicus' positive arguments in favor of the earth's rotation is his theory of elemental motion and related theory of gravity, which represent his answer to Aristotle's arguments against what Giese calls "the opinions of the *Timaeus* and the Pytagoreans" or, more directly, to Giese's own questions and suggestions concerning the direction in which the loose fragments of the earth would move, if the latter were raised to the lunar sphere (would they move to the centre of the universe or to the earth's sphere?), violent or non-violent nature of the daily rotation attributed to the earth and Aristotle's division of motion away from the centre, towards the centre, and about the centre (can they really be separated?).[164]

Copernicus has already suggested at the beginning of Chapter 8 that rotation is the earth's natural motion and that its circular motion is "appropriate by nature to its [spherical] form." Now his task is to further persuade the reader that the Aristotelian theory of the motion of a single simple body which has a single natural and simple motion – the elements of earth, water, air and fire move with rectilinear (downward or upward) motion and aether with circular

---

[163] *De revolutionibus* I, 8 (Rosen, p. 17; Zekl, p. 120).

[164] For the full quotation, see supra, p. 80. For Aristotle's arguments, see supra, pp. 86–91, and for Copernicus' formulation of them, see pp. 155–7.

motion around the middle – is inconsistent, and he has to present an alternative to it. He has to show how it is possible that the earth's rotation, albeit circular, is nonetheless a natural and simple motion, which also means that its rectilinear motion is not simple and natural, and, consequently, that Aristotle's division of all simple motions of simple elements into rectilinear and circular is wrong. Or, in other words, Copernicus has to demonstrate that Aristotle's criticism of Plato's *Timaeus* (and the Pythagoreans) was wrong and that Plato was, at least in some fundamental aspects, correct.[165]

Copernicus does so by defending two claims, each of which has already been briefly mentioned. Firstly, he asserts that "the motion of falling and rising bodies in the framework of the universe is twofold, being in every case a compound of straight and circular,"[166] and secondly, he maintains that "bodies that are carried upward and downward, even when deprived of circular motion, do not execute a simple, uniform and equal motion."[167]

(i) Compound, circular and rectilinear, motion of falling and rising bodies. Let us first take a look at his claim that the motion of downward and upward moving bodies in the universe is actually composed of circular and rectilinear motions.

Rectilinear downward motion of heavy bodies arises from their weight, which, in turn, arises from their predominantly earthy composition. Their circular component, their circular motion, which we cannot perceive, can be explained by their retaining the same nature as their whole: "For, things that sink of their own weight, being predominantly earthy, undoubtedly retain the same nature as the whole of which they are parts."[168] Since *eadem est ratio totius et partium*, the parts, fragments of earth, or bodies predominantly composed of earth, retain the same circular motion (= rotation) which is natural to the spherical earth as a whole. And since it is in the nature of the spherical earth to move about its own axis, the falling bodies, composed predominantly of earthy matter, move with two

---

[165] On Copernicus' theory of elemental motion and gravity, see also M. Wolff, "Impetus Mechanics as a Physical Argument for Copernicanism. Copernicus, Benedetti, Galileo"; D. Knox, "Copernicus's Doctrine of Gravity and the Natural Circular Motion of the Elements"; A. Goddu, *Copernicus and the Aristotelian Tradition*, pp. 345–55; and especially A. De Pace, *Niccolò Copernico e la fondazione del cosmo eliocentrico*, pp. 186–208.

[166] *De revolutionibus* I, 8 (Rosen, p. 16; Zekl, p. 116).

[167] Ibid.

[168] Ibid. (Rosen, p. 17; Zekl, p. 118).

movements: the natural, circular movement of their whole, that is, the earth, and with forced, partially unnatural, rectilinear movement with which they strive to reach the whole. Parts of the earth, forced from their natural place in one way or another, move in rectilinear motion towards their natural place and, retaining the earthy nature of their whole, also move in a circular motion (rotation), which is their natural motion.

With respect to upward motion, Copernicus only mentions the motion of fire, which he apparently does not consider a special element, but only a result of violence applied to earthy matter.[169] It is a property of "earthy fire," which "feeds mainly on earthy matter," "to expand what it enters" and moves "from the centre to the circumference":

> Nor is the explanation different in the case of those things, which, being fiery, are driven by the fiery violence upward. For also earthly fire feeds mainly on earthy matter, and flame is defined as nothing but blazing smoke. Now it is a property of fire to expand what it enters. It does this with such great force that it cannot be prevented in any way by any device from bursting through restraints and completing its work. But the motion of expansion is directed from the centre to the circumference. Therefore, if any part of the earth is set afire, it is carried from the middle upwards."[170]

---

[169] On that, see A. Goddu, *Copernicus and the Aristotelian Tradition*, pp. 345–6; D. Knox, "Copernicus' Doctrine of Gravity and the Natural Circular Motion of the Elements," pp. 167–8, and 192–3; M. Kokowski, *Copernicus's Originality*, p. 55; A. De Pace, *Niccolò Copernico e la fondazione del cosmo eliocentrico*, p. 190, n. 570.

[170] Ibid. (Rosen, p. 17; Zekl, pp. 116–18). For the fire being defined as a "burning smoke (*fumus ardens*)," see Aristotle, *De generatione et corruptione* II, 4, 331b 25–26; *Meteorologica* IV, 9, 388a2. See also D. Knox, "Copernicus' Doctrine of Gravity and the Natural Circular Motion of the Elements," p. 168, n. 52. Traditional interpretation, E. Rosen, *Commentary*, p. 353, of the last sentence of this passage is that Copernicus refers to "the recently introduced firearms and cannon" which were "personally familiar to Copernicus, who served as a military commander in his Chapter's war against the Knights of the Teutonic Order." See also M. Wolff, "Impetus Mechanics as a Physical Argument for Copernicanism," p. 221: "Copernicus probably alludes to the recent invention of firearms"; D. Knox, "Copernicus' Doctrine of Gravity and the Natural Circular Motion of the Elements," p. 167: "Copernicus was thinking, perhaps, of exploding gunpowder in particular." It seems to me that A. De Pace, *Niccolò Copernico e la fondazione del cosmo eliocentrico*, p. 188, might be right, again, in suggesting that Copernicus' inspiration here is very probably Plato, *Timaeus* 56d et seq. See Ficino, *Compendium in Timaeum*, 5 (Farndell, p. 7; *Opera omnia*, p. 1439): "[…] fire is the most effective of all the elements […]. fire has the power to divide and penetrate." A similar view is expressed by Cicero, *De natura*

In sum, Copernicus seems to be saying that the *terraquea sphaera*, including the nearby air, moves naturally with circular motion. All bodies predominantly composed of earth (whether falling bodies and rising bodies) are parts of this earthy sphere and therefore retain its nature. Since parts retain the nature of their whole, bodies composed of earth move with not only rectilinear motion, which is perceptible to us, but also with circular motion, which is not. This, then, means that falling bodies are not moving towards the earth because it occupies the centre of the universe, but because they seek their whole. Consequently earthy rising bodies retain the nature of the whole, that is, of the earthy sphere and move with both circular and rectilinear motion.

Thus Copernicus develops his own point by "explaining" and transforming Aristotle, stressing that the only natural and simple motion of a simple body is circular motion, that is, rotation. This is his major point: circular motion is the only natural motion of a simple body, but there is one condition. A simple body must remain "in its natural place and with its whole." Here is the difference between Copernicus' and Aristotle's conceptions of the natural place of a simple body. While Aristotle believes that natural places of earth, water, air and fire are cosmologically determined by their absolute or relative heaviness (*gravitas*) or lightness (*levitas*), Copernicus maintains that the natural place of a simple body is to be with its whole, no matter where in the universe this whole is situated, and regardless of what matter composes it. As long as the parts of a simple body remain with their whole, they retain its simple, natural, circular movement. Circular motion is the only natural motion of simple bodies: "Hence the statement that the motion of a simple body is simple holds true primarily for circular motion, as long as the simple body abides in its natural place and with its whole [or in its

deorum II, 15, 41 (Rackham, p. 161), who ascribes it to Cleanthes: "Now our ordinary fire that serves the needs of daily life is a destructive agency, consuming everything, and also wherever it spreads it routs and scatters everything." Copernicus later on, *De revolutionibus* I, 10 (Rosen, p. 19; Zekl, p. 128), explains that space between earth and moon contains "air and, if you please, also what is called 'fiery element'." The holograph includes *aethera* between *etiam* and *quod*. And "aether" reappears a few lines down. Venus' epicycle, according to conventional astronomy, encompasses "terram, aerem, aethera, Lunam, atque Mercurium." See D. Knox, "Copernicus's Theory of Gravity," p. 168 and especially p. 193. He points in the direction of Cicero, *De natura deorum*, Pliny the Elder, *Historia naturalium*, and Plutarch, *Placita philosophorum*, as possible sources. In *Placita philosophorum* is a record that Plato identified five elements, earth, water, air, sublunary aether, superlunary fire, adding 'he sometimes blends aether with fire."

unity]. For when it is in place, it has none but circular motion, which remains wholly within itself like a body at rest."[171]

It seems that the simple body in question here is a sphere, or more precisely, a *terraquea sphera* together with the nearby air, which is said to adhere to the earthy nature. On the other hand, this body could be any other sphere of indefinite matter whose natural movement is circular. Therefore, contrary to Aristotle, Copernicus does not conceive of rectilinear motion (*rectus*; up and down relative to the centre of the universe, which is the centre of the earth) as natural but – at least partly – as unnatural, violent or forced motion. This motion affects things which leave their natural place or are thrust out of it or are in any manner whatsoever out of it. The reason why this rectilinear motion happens is – being out of their place – the parts of the whole contravene the order of the whole and the structure of the world:

> Yet nothing is so incompatible with the orderly arrangement of the whole and the structure of the universe (*ordinationi totius et formae mundi*) as something that is out of its place. Therefore rectilinear motion occurs only to things that are not in proper condition [or not well] (*rebus non recte se habentibus*) and are not in complete accord with their nature, when they are separated from their whole and forsake its unity (*separantur a suo toto, et eius deserunt unitatem*).[172]

As noted by Hooykaas, this passage contains a play on words which must have given pleasure to the physician in Copernicus.[173] *Rectus* has two meanings: when speaking about motion it means "rectilinear" and when speaking about the state of health it means "well"; *recte se hebere* means to feel well, to be in good condition. Rectilinear, upward and downward motion, which the peripatetic tradition holds as completely natural to earth, water, air, and fire, when not in their natural regions, which are cosmologically determined, is according to Copernicus unnatural or violent, because it occurs only to things which are outside their natural place, separated from their whole and have forsaken its unity. Circular motion belongs

---

171 *De revolutionibus* I, 8 (Rosen, p. 17; Zekl, p. 118). Compare with Aristotle, *Physica* VIII, 9, 265b1–8.

172 *De revolutionibus* I, 8 (Rosen, p. 17; Zekl, p. 118).

173 See "The Aristotelian Background to Copernicus's Cosmology," p. 113. Compare with the beginning of Chapter 8 (Rosen, p. 15; Zekl, p. 112): "Quibus enim vis vel impetus infertur, dissolui necesse est, et diu subsistere nequeunt. Quae vero a natura fiunt, *recte se habent* et conseruantur in optima sua compositione." Emphasis added.

to wholes, whereas rectilinear motion belongs to their parts which are in contradiction to the order of the universe. And Copernicus continues with the medical metaphor. Rectilinear motion is in comparison to the circular motion a deviation, a sickness: "since circular motion belongs to wholes, but parts have rectilinear motion in addition, we can say that 'circular' subsists with 'rectilinear' as 'animal/being alive' with 'being sick' (*Cum ergo motus circularis sit universorum, partium vero etiam rectus, dicere possumus manere cum recto circularem, sicut cum aegro animal*)."[174] Parts of the earth and all bodies of earthy matter that fall to the earth seek their whole, returning to their natural place or their centre. By virtue of this unnatural, non-circular motion, they in fact re-establish the order of the whole, the structure of the universe. Though themselves a sickness, in performing their action they heal the universe, so to speak.[175]

It is important to stress here that the "healing" rectilinear motions of elements are in addition to their natural circular motions. The earth's rotation implies that rectilinear motions are composed of rectilinear and circular motions. Therefore, Aristotle's division of simple motion into motion away from the middle (upward motion), towards the middle (downwards) and around the middle (circular) cannot be but an *actus rationis*, because rectilinear motion cannot exist without circular motion, just as a point, a line and a surface cannot exist without one another, and none of them can exist without a real body: "Surely Aristotle's division of simple motion into three types, away from the middle, toward the middle, and around the middle, will be construed merely as an act of the reason. In like manner we distinguish line, point, and surface, even though one cannot subsist without another, and none of them without body (*unum sine alio subsistere nequeat, et nullum eorum sine corpore*)."[176]

Copernicus' account on the nature of motion finds its supplement in his account on gravity in Chapter 9, where he explains:

> For my part I believe that gravity is nothing but a certain natural desire, which the divine providence of the Creator of all things has implanted in parts, to gather as a unity and a whole by combining in the form of a globe.

---

174 *De revolutionibus* I, 8 (Rosen, p. 17; Zekl, p. 120).

175 Compare with *De caelo* IV, 3, 310b16–20 (Stocks, p. 507): "Thus to ask why fire moves upward and earth downward is the same as to ask why the healable, when moved and changed *qua* healable, attains health and not whiteness." See also *Physica* V, 229a7–229b14, and Buridan, *Quaestiones in De caelo* IV, 2; see supra, pp. 178–9, n. 128. See also D. Knox, "Copernicus's Theory of Gravity," p. 173.

176 *De revolutionibus* I, 8 (Rosen, p. 17; Zekl, p. 120).

> This impulse (*affectio*) is present, we may suppose, also in the sun, the moon, and the other brilliant planets, so that through its operation they remain in that spherical shape which they display. Nevertheless, they swing round their circuits in diverse ways.[177]

As has been pointed out by Knox, there is, however, a tension between the two accounts.[178] While in the first, physical account, rectilinear movement is implicitly considered *not natural* (rectilinear movement is like "sickness" to circular movement of the wholes; it occurs only in things *non recte se habentibus*), the second, teleological account, explicitly states that gravity is a *natural* appetite implanted in parts in order to seek and find their wholes and to gather together in the form of a globe. Since Copernicus accepts the axiom that there is only one natural movement of simple bodies (circular), these two accounts are evidently contradictory: the rectilinear movement of parts to their wholes is at the same time considered to be non-natural and natural. Gravity should, as a natural appetite, produce natural motion. If that is true, than Copernicus must also believe that simple bodies have two natural motions: circular in their natural places and rectilinear when returning to them. But this contradicts his claim that

---

177 Ibid. I, 9 (Rosen, p. 18; Zekl, p. 122). For possible sources of Copernicus' doctrine of gravity and natural circular motion, see D. Knox's article "Copernicus' Doctrine of Gravity and the Natural Circular Motion of the Elements," who – after examining a number of possible origins of Copernicus' first (physical) and second (teleological) account (Nicole Oresme, *Le livre du ciel et du monde* II, 25; Plutarch of Chaeronea, *De facie quae in orbae lunar apparet*, *De defectu oraculorum*, and *De Stoicorum repugnantiis*; Pliny the Elder, *Historia naturalium*; Cicero, *De natura deorum*; Xenarchus, Ptolemy, Plotinus, Proclus, Philoponus, Ficino and others), – finally, p. 202, limits the list to two Plotinus' treatises in Ficino's Latin translation, Thomas Aquinas' *De caelo commentary*, Ficino's *Theologia Platonica*, his *Compendium in Timaeum* and his commentary on Plotinus, and Valla's *De expetendis et fugiendis rebus* of which "the most promising [...] possibilities are Valla's *De expetendis* and Ficino's *Timaeus* commentary." Although Copernicus' doctrine itself points in the direction of Ficino's *Compendium in Timaeum* (for the quotation, see supra, pp. 184–5, n. 146), Knox finally argues in favor of the Byzantine dictionary, the so-called *Suda*. Compare also with A. Goddu's translation from Latin *Suidae lexicon*, *Copernicus and the Aristotelian Tradition*, p. 349, and his discussion on pp. 347–53. A. De Pace, *Niccolò Copernico e la fondazione del cosmo eliocentrico*, pp. 166–76, on the other hand, argues for Plutarch, *De facie in orbae lunae*, and *De defectu oraculorum*. See also her "Plutarco e la rivoluzione copernicana," and F. Krafft, "Copernicus Retroversus II: Gravitation und Kohäsionstheorie."

178 See "Copernicus' Doctrine of Gravity and the Natural Circular Motion of the Elements," pp. 172–9.

things in a natural state produce opposite effects to those that are not in natural state. According to Knox, Copernicus recognized the problem and solved it with "characteristic ingenuity" by claiming that the rectilinear motion even in itself, without the circular motion, is not simple, is not uniform but accelerated, and therefore not natural.[179]

(ii) Rectilinear motion is not a simple motion in itself: acceleration of rectilinear motion. Because of its actual twofold, circular and rectilinear nature, Copernicus, unlike Aristotle, obviously does not regard rectilinear motion "away from the middle" and "toward the middle" as a simple motion of a simple body. However, there is yet another reason why rectilinear motions are not simple and hence not natural. Even if we disregard the fact that objects moving rectilinearly also move with the earthy sphere (or any other sphere that constitutes their totality, their whole) in a circular motion and are therefore composed of circular and rectilinear motion, they in themselves "do not execute a simple, uniform, and equal motion." Contrary to Aristotle, Copernicus argues that accelerated rectilinear motion is not simple and provides two reasons for it. Firstly, he holds that the circular motion of the sphere(s) is the only simple motion, while rectilinear motions are essentially deviations from circular motions. The second reason that rectilinear motions are not simple is their acceleration or deceleration:

> Furthermore, bodies that are carried upward and downward, even when deprived of circular motion, do not execute a simple uniform and equal motion. For they cannot be regulated by the lightness or by the impetus of their weight. Whatever falls moves slowly at first but increases its speed as it drops. On the other hand, we see this earthly fire (for we behold no other), after it has been lifted up high, slacken all at once, thereby revealing the reason to be the violence applied to the earthy matter (*Vbi rursum vicissim ignem hunc terrenum (neque enim alium videmus) raptum in sublime statim languescere cernimus, tamquam confessa causa violentiae terrestris materiae*).[180]

While circular motion "always rolls along uniformly," rectilinear motions are non-uniform and unequal. Falling bodies, Copernicus holds, move more slowly at the beginning and accelerate their speed during the course of their downward

---

[179] See ibid., p. 174.

[180] *De revolutionibus* I, 8 (Rosen, p. 17; Zekl, p. 118).

motion due to their *impetus ponderis*. On the other hand, "earthy fire," being the only one we can see, slackens "after it has been lifted up high."

Neither kind of rectilinear motion (downward or upward) can be temperate. The reason for this is the fact that the cause of the motion of falling and rising bodies, that is, the "impetus of their weight" or their "lightness," does not work for a long time; the cause of the motion of falling and rising bodies "quickly stops functioning." Bodies brought to their designated places by their heaviness or gravity (*gravitas*) and lightness (*levitas*) stop being heavy and light. Both causes of rectilinear motion, *impetus ponderis* and *levitas*, are "failing." Circular motion, on the other hand, has an "unfailing" cause which enables it to be always simple, equal and uniform: "Circular motion always rolls along equally [or uniformly], since it has an unfailing cause. But rectilinear motion has a cause that quickly stops functioning. For when rectilinear motion brings bodies to their own place, they cease to be heavy or light, and their motion ends."[181]

With this theory of impetus Copernicus is able to develop his theory of gravity as a natural appetite imposed by God on parts to reach their wholes, that parts returning rectilinearly to their wholes do not observe simple, uniform motion, and that this motion is meant to restore a natural state of affairs rather than being itself natural. Gravity is a natural appetite restoring parts to their wholes, but it also produces impetus and acceleration. The free fall of a body is a composite of natural and non-natural motion.[182]

While it is relatively clear what Copernicus means by *gravitas*, the exact functioning of *levitas* is less discernible. According to Knox, Providence, while implanting gravity, does not impose the complementary principle of levity. Although some things are obviously lighter than others, lightness is relative. If air is trapped under water, it will rise upwards, because it is less heavy than the body in which it is held. But as Copernicus maintains, its raising has nothing to do with the natural tendency to seek air sphere occupying a fixed cosmic position. The propinquous air adheres to the terraqueous globe because it mixes or moves preternaturally with it, and there is, according to Copernicus, no (sublunary) fire sphere.[183] Rectilinear motion indicates that things are out of their place, separated

---

181 Ibid. I, 8 (Rosen, p. 17; Zekl, pp. 118–20).

182 See also D. Knox, "Copernicus' Doctrine of Gravity and the Natural Circular Motion of the Elements," p. 176.

183 See ibid., p. 177. This is, *grosso modo*, compatible with Plato's explanation of "heavy" and "light" in *Timaeus* 62c–63e. Compare with A. De Pace, *Niccolò Copernico e la fondazione del cosmo eliocentrico*, pp. 168–70.

from their whole. But there is a distinction between downward and upward motion. While the former serves to bring things to their natural place, that is, to their whole, the latter takes them out of it.

***

What, then, does the analysis of Chapter 7 and 8 of Book I of *De revolutionibus* reveal about Copernicus' refutation of the ancients' thoughts on the earth as the unmoving centre of the universe? First: Copernicus' presentation of the arguments against the rotation of the earth is relatively meager. As we have seen, Aristotle, Ptolemy and others, especially Buridan and Oresme, developed a host of very subtle, philosophical and experiential reasons (called also appearances or observational tests) against the rotation of the earth.[184] Of all of these, Copernicus lists just two philosophical arguments, which he ascribes to Aristotle, and three more or less experiential reasons (all concerning the centrifugal force argument), which he ascribes to Ptolemy.

What is more: Copernicus is not very faithful to these authors when he summarizes their arguments. He simplifies Aristotle's philosophical reasons from the nature of the heaviness and lightness and from the nature of motion, and he seriously distorts Ptolemy. Ptolemy does not discuss the earth's rotation and it's supposed centrifugal effects at all, but Copernicus constructs, from various elements of Ptolemy's discourse, a stronger argument than Prolemy's original. Instead of things projected perpendicularly in the air, that is, projectiles, which should according to Ptolemy (and Aristotle, for that matter) in case of the earth's movement not reach the spot of their projection, Copernicus talks about bodies falling down.

This reveals the dialectical nature of Copernicus' discourse. He distorts Ptolemy's argumentation and formulates his so called centrifugal force objection to the rotation of the earth for dialectical reasons. Thus he can bring out the contradictions in "Ptolemy's" reasoning and prepare his argumentation (that takes him to the question whether the universe is finite or infinite) which leads him to the "natural" affirmation of the earth's rotation. This might also be the case with the formulation of the next reason concerning projectiles in Ptolemy and falling bodies in Copernicus. By such formulation he prepares terrain so he can more easily overcome the objection.

This strategy, however, does not help him much, as we have seen. Copernicus actually contradicts his own reasoning, and not Ptolemy's, in his answer to "Ptolemy's" objections concerning what would happen to falling bodies if the

[184] See supra, pp. 86–94.

earth rotated. Copernicus responds to Ptolemy's reasons in a traditional fashion, claiming that not only earth rotates but also the adjacent air, which carries the falling bodies along, and saying that this is the cause that the falling bodies do not fall behind but reach the appointed place. This would be a coherent theory if Copernicus did not adhere to at least one tenet of the impetus theory. Since he believes that *impetus ponderis* is the cause of the acceleration of falling bodies, and since the impetus imparted on the falling bodies causes them to resist the movement of the air, they should, as has been shown by Buridan, fall behind the appointed place. But they do not fall behind, but reach the appointed spot, which in this case means that the earth does not move.

This indicates that his responses to the experiential objections to the rotation of the earth and his positive physical doctrine arising thereof are not very thoroughly elaborated. This is confirmed also by analysis of his four positive arguments, including his theory of gravity and elemental motion. Copernicus is striving to provide some positive alternative to the Aristotelian explanation of gravity, levity and the nature of motion. But on the whole, his alternative is very rudimentary and relatively inconsistent.

Copernicus seems to believe that the whole of the earth is composed of the terraqueous sphere and adjacent air. There is no place for the sublunary element of fire (Copernicus speaks only of *earthly* fire), while there seems to be a place for celestial fire, which he considers to be the same substance as aether. The rotation of the earth is, according to him, natural and follows from its form or shape, but the shape is not the cause of the movement. Why does, then, the circular motion have an unfailing cause (*causam indeficientem*) which enables it to always roll along equally or uniformly? What is the "unfailing cause" of rotation? If it is not the spherical form (or shape or figure) of the whole, is it impetus implanted in the spheres by God at the creation?[185] There is no textual evidence to support this claim, and since Copernicus conceives of impetus as some kind of force that has disastrous effects when applied to objects – see the beginning of Chapter 8 of Book I of *De revolutionibus* – this is also impossible.[186]

---

185 So believe M. Kokowski, *Copernicus's Originality*, pp. 229–30, and M. Wolff, "Impetus Mechanics as a Physical Argument for Copernicanism," pp. 218–31.

186 See also A. De Pace, *Niccolò Copernico e la fondazione del cosmo eliocentrico*, p. 192, n. 575. Compare with Nicolaus Cusanus (1401–1464) who believed that the outermost celestial sphere moves because of the impetus imparted on it by God. He did not, however, conceive of the impetus as having disastrous effects when applied. See *De ludo globi* I, 21–3 (Hopkins, pp. 1191–2).

The only thing Copernicus claims is that the earth's rotation is natural within the order of the universe and this is, in my opinion, *causa indeficiens* of the rotation. This explains, to a certain extent, why not all of the spherical wholes in the universe exhibit circular movement. Spherical shape is by nature the most appropriate for circular movement and circular movement is the one natural, simple movement. When in its natural place the spherical earth moves as a whole with simple, uniform circular movement. Uniform circular motion of the spherical whole, when they all things are in their natural places, is the only natural movement. The earthly parts retain the nature of their whole, which means that even when we see them moving rectilinearly up or down they are actually moving with two movements: circular and rectilinear. This also means that their rectilinear motion is not natural. But Copernicus also explains gravity as a natural appetite of the parts to reach their spherical whole(s), by which they return to their natural place and restore orderly arrangement of the whole and the order of the universe. The parts of the earth thus do not return to the centre the universe, as in Aristotelian physics, but to their natural whole, wherever in the universe it is situated. But this also means that the rectilinear motion of the parts of the whole is at the same time non-natural and natural. To amend this, that is, in order to show that rectilinear movement of falling bodies (explanation for bodies moving upward is different) even when considered in abstraction of the circular motion is not simple or uniform but accelerated and therefore not really natural, Copernicus adopted the impetus of the weight theory. But *impetus ponderis* theory is, as we have seen, in contradiction to his solution of the Ptolemy's experiential reasons against the rotation of the earth.

This is not to say that his physical argumentation is not, in a way, original and above all non-Aristotelian, although it is presented as if it would naturally follow from Aristotelian premises.[187] Although his explanations are sometimes difficult to follow, and although he uses some traditional, Aristotelian arguments (he probably also knew Buridan's arguments in one way or another), it seems that his positive views are very much influenced by Platonist ideas adopted to suit the rotation of the earth.

This is especially true for his conception of gravity. Rheticus describes this very clearly in his *Holy Scripture and the Motion of the earth* where he defends the movements of the earth against those who would consider them to be monstrous. Rheticus believes that one should not tie God's power to the disputations of the Peripatetics about heavy and light. Is God "unable to give the earth natural

[187] Compare with R. Hooykaas, "The Aristotelian Background to Copernicus's Cosmology."

movements in accord with its shape?"[188] It is not necessary to agree with the Aristotle who teaches that there are three distinct movements, from the centre, to the centre and around the centre. One, circular movement of the earth around its axis is enough, since this motion is also present in all its parts:

> Nor therefore is it necessary for us to agree with Aristotle, when he teaches that movements from the centre, to the centre, and around the centre are distinct from each other. We may indeed rightly insist, especially as mathematical reasoning compels us, that the whole earth moves in a circle, and that therefore this motion is present in all parts. Furthermore, that motions from the centre and to the centre appear by accident in the elements, as when air, confined by water or earth, tends upwards, and breaks out towards the place, which nature has assigned to it, and earth that has been thrown up into the air, keeps falling downward, until it stops on the earth, or occupies the centre of the earth, that being its naturally allotted place. Wherefore we would say that to bring something into its place is nothing else than to bring it into its proper 'Form' (*in suum locum nihil aliud esse quam in suam formam fierri*), and that the ancients rightly held – in opposition to Aristotle – that like is attracted by like. For this is the reason and divine ordinance, on account of which the sun, the moon, the stars, and the earth are spherical, and it is not proved that heavy objects tend to reach the centre of the universe, but only towards the centre of their own globe.[189]

Copernicus' theory of gravity, regardless of which author was his immediate source – Ficino, Plotinus, Plutarch or somebody else – is evidently Platonist. While Aristotle considers heaviness to be an absolute quality possessing an onto-cosmological function (to carry heavy objects towards the centre of the universe), Copernicus, like Plato and the Platonists, regards it as a natural tendency of parts to gather in their whole and thus form a unity in the shape of a sphere. Like is attracted to like.[190]

---

188 Rheticus, *Holy Scripture and the Motion of the earth* (Hooykaas, Latin, p. 57; English, p. 92).

189 Ibid. (Hooykaas, Latin, pp. 57–8; English, pp. 91–2).

190 See Plato, *Timaeus* 63c (Cornford, p. 264): "Standing on the earth, when we are trying to distinguish between earthly substances or sometimes pure earth, we are dragging the two things into the alien air by violence and against their nature; both cling to their own kind, but the smaller yields more readily to our constraint than the larger and follows it more quickly into the alien element." See also ibid. 63e (Cornford, p. 264): "The one thing to be observed in all cases, however, is that it is the travelling of

Against this backdrop, it is even more evident that Chapters 7 and 8 of Book I of *De revolutionibus*, in which Copernicus considers the first motion of the earth, its rotation, possess a strong dialectical component. In short, they are but preparation for Chapters 9 and especially 10, in which he presents the harmony argument for the earth's second motion, that is, its motion around the sun. Here his principal intention is not so much to provide physical arguments for his system but to show that the earth's motion is physically possible and not completely incompatible with reasonable views about nature. In Michael Wolff's words, Copernicus "justifies his system, not by showing that it can successfully explain physical facts, but rather by attempting to make the current and more or less traditional ways of explaining motion compatible with the earth's motion."[191] His physical arguments in favor of the earth's rotation serve only one purpose: to bring the theory of motion in conformity with his cosmological-astronomical principles, that is, with the *harmonia* and *symmetria* of the world. Copernicus' essential concern is not the rotation of the earth but its movement around the sun. Or, if we look at it from another angle: when addressing the question whether the earth moves or not, it is not physical arguments "against" or "in favor" that are crucial for Copernicus, but the mathematical cosmological reason, that is, the harmony of the universe. His central argument for earth's motion is, therefore, the firm *symmetria* of the universe, that is commensurability of its parts, which can be achieved by taking into account the various motions of the earth.[192] No wonder then, that at the end of the argumentation of Chapter 8 he claims that "that all these arguments make it more probable that the earth moves than that it is at rest."[193] Likewise, he opens Chapter 9 by suggesting that "since nothing prevents the earth from moving [...] we should now consider also whether several motions suit it, so that it can be regarded as one of the planets."[194]

each kind towards its kindred that makes the moving thing 'heavy' and the region to which it moves 'below'." Compare also with *Timaeus* 81a. On this, see Cornford's commentary, pp. 264–6, and A. Gregory, *Ancient Greek Cosmogony*, pp. 140–1.

191 M. Wolff, "Impetus Mechanics as a Physical Argument for Copernicanism," p. 218.

192 See also D. Knox, "Copernicus' Doctrine of Gravity and the Natural Circular Motion of the Elements," pp. 208–10.

193 *De revolutionibus* I, 8 (Rosen, p. 17; Zekl, p. 120). Emphasis added.

194 Ibid. I, 9 (Rosen, p. 17; Zekl, p. 122).

CHAPTER TEN

The Second Motion of the earth and its Position in the "World Machine"

In Chapter 4 of Book I of *De revolutionibus*, Copernicus says that the apparent irregular celestial motions can be accounted for by two types of models, both presupposing uniform motions of spheres: either circles in the models "have different poles," or "that the earth is not at the centre of the circles on which they [i. e. the objects in the heaven] revolve."[195] Both account for some apparent "inequality" of planetary motion, but, as can be inferred from his reasoning, the eccentric models explain the approaching of the planets to the earth and their receding from it better. The explanation for that phenomenon is the observer's "off-centred" position. The approaching of the planets to the earth and their receding from it proves that the earth is not located at the centre of circles on which planets move embedded in spheres. It is not located at the centre of epicycles moving on a deferent.

Copernicus continues this line of reasoning in Chapter 5. If we accept the plausibility of the earth's axial motion, and its possibility of a second motion and displacement from the centre of the universe, Copernicus is able to push further the existing *Ptolemaic* explanatory logic pertaining to the irregular motions of the planets, that is, their movements towards the earth and away from it. But this

[195] See supra, pp. 135–7.

continuation also requires what in the perspective of this logic seems a minor shift: instead of moving the planets towards the earth and away from it, one can now assign this motion to the earth in relation to the planets:

> If we assume its daily rotation, another and no less important question follows concerning the earth's position. To be sure, heretofore there has been virtually unanimous acceptance of the belief that the middle of the universe is the earth. Anyone who denies that the earth occupies the middle or centre of the universe may nevertheless assert that its distance [from the centre] is insignificant in comparison with (the distance of) the sphere of the fixed stars, but perceptible and noteworthy in relation to the orbs of the sun and the other planets. He may deem this to be the reason why their motions appear nonuniform, as conforming to a centre other than the centre of the earth. Perhaps he can [thereby] produce a not inept explanation of the apparent nonuniform motion. For the fact that the same planets are observed nearer to the earth and farther away necessarily proves that the centre of the earth is not the centre of their circles. It is less clear whether the approach and withdrawal are executed by the earth or the planets. It will occasion no surprise if, in addition to the daily rotation, some other motion is assigned to the earth.[196]

The displacement of the earth from the universe's centre is possible only under certain conditions. Its "off-centredness" should not be comparable to its distance from the sphere of the fixed stars. The displacement should be imperceptible to the earth observer, which Copernicus achieves by expanding the dimensions of the universe. Because no parallax is detected from the earth, the universe must be vast enough to allow the perceptible motion of the fixed stars to be the same, irrespective of whether they are observed from the centre of the universe or from a point in its relative proximity.[197] On the other hand, the displacement of the

---

196 *De revolutionibus* I, 5 (Rosen, p. 12; Zekl, p. 100).

197 See ibid. I, 6 (Rosen, p. 13; Zekl, p. 102), where Copernicus demonstrates that "the massive bulk of the earth does indeed shrink to insignificance in comparison with the size of the heavens." See also ibid. I, 10 (Rosen, p. 20; Zekl, p. 132): "In comparison with any other spheres of the planets, the distance from the earth to the sun has a magnitude which is quite appreciable in proportion to those dimensions. But the size of the universe is so great that the distance earth-sun is imperceptible in relation to the sphere of the fixed stars.." And ibid. (Rosen, p. 22; Zekl, p. 138): "Yet none of these phenomena appears in the fixed stars. This proves their immense height, which makes even the sphere of the annual motion, or its reflection, vanish from before our eyes. For, every visible object

earth from the centre should be sufficiently large or perceptible to bring it into relation with the sun and the planets. Once these two conditions have been satisfied, the earth's displacement from the centre and its motion around some other centre, that is, the sun, can be understood as the continuation of the *Ptolemaic* "geocentric logic" of accounting for irregular celestial motions.

It was already established in Ptolemaic geocentric astronomy that the earth is not the exact centre of planets circuits on account of the varying distances of the planets or their "unequal" motions. If in addition we consider the "universalized" principle of relativity of motion – the "optical argument" – the apparently varying distances of the planets can be explained by their actual approach and withdrawal from the earth and the observer on the earth or vice versa: by the approach and withdrawal of the earth and the observer on the earth from the planets. Consequently, "[i]t is less clear whether the approach and withdrawal are executed by the earth or the planets."[198] The general acceptability of the earth's motion enables the "off-centred" earth to be transposed from the Ptolemaic into the Copernican system. The earth is "off-centred"; in addition to its rotation, it also performs "some other motion" around the centre of the universe, upon which the planets are "centred." And, again, Copernicus seeks the support of some time-honored authorities. That "[i]t will occasion no surprise if, in addition to the daily rotation, some other motion is assigned to the earth," was also maintained by some remarkable ancient philosophers: "That the earth rotates, that it also travels with several motions, and that it is one of the heavenly bodies are said to have been the opinions of Philolaus the Pythagorean. He was no ordinary astronomer, inasmuch as Plato did not delay going to Italy for the sake of visiting him, as Plato's biographers report."[199]

Copernicus then offers two kinds of demonstrations to support the earth's rest at the centre of the universe: geometrical and physical. The physical reasons

---

has some measure of distance beyond which it is no longer seen, as is demonstrated in optics. From Saturn, the highest of the planets, to the sphere of the fixed stars there is an additional gap of the largest size. This is shown by the twinkling lights of the stars. By this token in particular they are distinguished from the planets, for there had to be a very great difference between what moves and what does not move. So vast, without any question, is the divine handiwork of the most excellent Almighty."

198 Ibid. I, 5 (Rosen, p. 12; Zekl, p. 100).

199 Ibid. For the analysis of Copernicus' argumentation, see, for instance. A. De Pace, *Niccolò Copernico e la fondazione del cosmo eliocentrico*, pp. 117–20. As already said, Copernicus presents the physical reasons for the motionless earth in Chapter 7 of *De revolutionibus*; his solution to them is presented in Chapter 8

in Chapter 7 and his response in Chapter 8 have already been analyzed. The geometrical reason in Chapter 5 is as follows:

> But many have thought it possible to prove by geometrical reasoning that the earth is in the middle of the universe; that being like a point in relation to the immense heavens, it serves as their centre; and that it is motionless because, when the universe moves, the centre remains unmoved, and the things nearest to the centre are carried most slowly.[200]

The answer to that demonstration is given in Chapter 6, in which Copernicus' investigation of these proofs shows that they are inconclusive and that at that moment of the exposition we still do not know anything about the earth's motion or rest and its position in the universe. In Chapter 9 Copernicus employs the same reasoning, the same dialectical strategy as in Chapter 5, except that here he already pretends to consider the earth's rotation almost as a "fact" – or at least something that is more probable than its rest – demonstrated by the arguments set forth in Chapter 8. Since nothing prevents the earth from rotating, it should now be considered whether several motions could be assigned to it, "so that it can be regarded as one of the planets."[201]

If we take into account also the homocentric astronomy mentioned in the *Preface* to *De revolutionibus*, we can summarize Copernicus' argument that leads him to the question of the earth's second motion "as conforming to a centre other than the centre of the earth" as follows. There are irregular motions of the celestial bodies in the heavens that need to be brought in conformity with the principle of uniform circular motion. This can be accomplished by either one of the combinations of spheres, which are the real causes of celestial motions: the theory of homocentric spheres or the theory of eccentrics and epicycles. However the homocentric astronomy, which places the earth at the centre of the universe and the motion of every single planet, is unable to explain why the planets move towards the earth and away from it, because it considers them permanently fixed at the same distance from it. This oscillation can be explained by the system of eccentrics and epicycles, which places the earth in an "off-centred" position relative to the planetary motion. Since the earth is not the centre of all revolutions, having already been replaced by other centres in the Ptolemaic astronomy, there is every reason to consider whether the earth is at the centre of the universe or, as Copernicus puts it, "whether the centre of the universe is identical with the centre of terrestrial gravity":

---

[200] Ibid. I, 5 (Rosen, p. 12; Zekl, p. 102).

[201] Ibid. I, 9 (Rosen, p. 17; Zekl, p. 122).

> For, it is not the centre of all the revolutions. This is indicated by the planets, apparent nonuniform motion and their varying distances from the earth. These phenomena cannot be explained by circles concentric with the earth. Therefore, since there are many centres, it will not be by accident that the further question arises whether the centre of the universe is identical with the centre of terrestrial gravity or with some other point.[202]

And just as the principle of relative motion perception or phenomenological equivalence applies to the earth's rotation about its axis and the motionless heavens *versus* the motionless earth and the moving heavens, it also applies to a possible earth's revolution around the motionless sun in the centre of the universe *versus* the sun's motion around the motionless earth in the centre of the universe. Whatever the case, the observed phenomena are the same: "For if this is transformed from a solar to a terrestrial movement, with the sun acknowledged to be at rest, the risings and settings which bring the zodiacal signs and fixed stars into view morning and evening will appear in the same way."[203]

Copernicus uses the "optical argument" and phenomenal equivalence once again. The same appearances as we see under the assumption of a motionless earth in the middle of the universe we also see under the assumption of its yearly motion around some other centre: "If, then, the earth too moves in other ways, for example, about a centre, its additional motions must likewise be reflected in many bodies outside it."[204] Therefore, "appearances" in many celestial bodies are also produced by the earth's motion around *some other centre* and its motion must also in this case be reflected in the apparent motion of other bodies. And another such motion is the earth's "yearly revolution" around the sun, which is the centre of its second movement.

While not taking anything away from the celestial phenomena, the transformation from a solar to a terrestrial movement brings Copernicus an additional bonus. The retrograde motion of the planets proves to be an illusion, a reflection of the earth's yearly revolution: "The stations of the planets, moreover, as well as their retrogradations and [resumptions of] forward motion will be recognized as being, not movements of the planets, but a motion of the earth, which the planets borrow for their own appearances."[205]

---

[202] Ibid.

[203] Ibid. (Rosen, p. 18; Zekl, p. 122).

[204] Ibid.

[205] Ibid. (Rosen, p. 18; Zekl, pp. 122–4).

Thus, the earth's motion is made the explanatory principle of several phenomena, but with different consequences for them. The daily "risings and settings" of the fixed stars will appear unchanged to the eye, and the earth's second, yearly motion around the sun will not alter the impact of its first, diurnal motion. The lack of impact of the yearly motion of the earth on the celestial phenomena, that is, the absence of a parallax, can be explained by the "immensity" of Copernicus' universe. On the other hand, the earth's second motion brings about an important epistemological shift: its yearly revolution around the sun makes it possible for the "stations of the planets [...] as well as their retrogradations and [resumptions of] forward motion" to be recognized as a consequence of the earth's motion rather than their own (see *Figure 11*). The planets "borrow" the motion of the earth "for their own appearances." A planet comes to an apparent stop when in the same line with the earth, it retrogrades when being overtaken by the earth, and resumes forward motion when moving faster than the earth.

The power of the principle of descriptive relativity is made manifest once again. *We* are the ones who project the actual effects of the earth's motion around the sun on the apparent motions of the celestial bodies. Nonetheless, this is still not the critical moment that convinces Copernicus that the earth actually moves.

Dialectical argumentation for the second movement of the earth: Chapter 10 of Book I of *De revolutionibus*. Copernicus' momentous realization that the earth moves – not only with the first motion around its axis but also and especially with the second one around the sun – largely rests on the order of the planets and the *mundi totius harmonia*, which follows as a consequence of the earth's motion around the sun. As we have seen, Copernicus' fundamental objection to the Ptolemaic astronomy refers not to its violation of the principle of uniform circular motion but to its inability to deduce the essential: *formam mundi* and *certam symmetriam* of its parts.[206]

In Chapter 10 of Book I of *De revolutionibus* Copernicus examines the order of the universe, that is, the order of the planets with respect to the "duration of the [planetary] revolution,"[207] or, as Goldstein calls it, "distance–period relationship."[208]

---

206 See *Preface* (Rosen, p. 4; Zekl, p. 70), and supra, p. 69.

207 Compare with *De revolutionibus* I, 6 (Rosen, p. 14; Zekl, p. 106): "But things separated by the diversity of their situations are subject to a very different relation: those enclosed in a smaller circular path revolve faster than those traversing a bigger circle."

208 See his "Copernicus and the Origin of his Heliocentric System." He analyses sources, commentators and translations of this particular concept on pp. 223–6. We shall deal

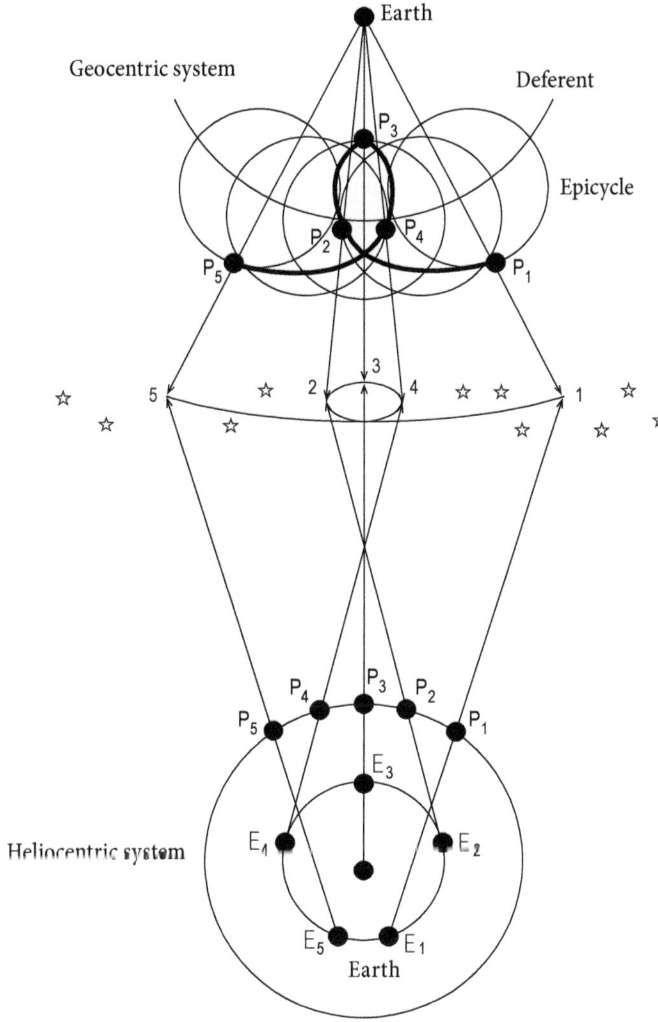

Figure 11.
Retrograde motion of the planets in the geocentric system and explanation of these motions in the heliocentric system.

---

with the question of Copernicus' sources infra, pp. 300–6. For Copernicus' account in Chapter 10, see also N. Swerdlow, "Copernicus and Astrology, with an Appendix

Copernicus gives this criterion as a given, indisputable principle, one assumed by ancient philosophers, not astronomers. He is, he seems to suggest, addressing a philosophical question not purely astronomical. Later in the chapter he will explain that "nobody will propound a more suitable principle (*ratio*) than that the size of the orbs is measured by the length of the time,"[209] which implies that there were and still are other criteria of ordering the celestial spheres but that they are not satisfactory. Another thing that should also be kept in mind is that Copernicus continues his dialectical argumentation, not doctrinal exposition. On one hand he argues against the reasoning behind the known orderings of the planets; on the other hand he demonstrates that the one already established and accepted principle of ordering, which is partly but not fully already at work in existing geocentric arrangements, that is, the distance–period relationship, works perfectly in heliocentric cosmos. This means, in other words, that he is not explaining *how* he arrived to the solution nor is he presenting *an isolated* argument for his solution. He already has a solution. All he wants to do at this particular moment is to present it as a completely coherent, logical, natural consequence of already existing principles.

Chapter 10 of Book I of *De revolutionibus* can be divided in five logical parts. In the first part Copernicus presents the existing arrangements of planetary spheres and reasons for particular arrangement. The second part is dedicated to the refusal of these arrangements. He then presents a dilemma: either the earth is not at the centre of the world or there is no order in the universe. Since he wants to preserve order he sacrifices the central position of the earth. But what is, then, at the centre of the universe? Copernicus then presents his solution to the vexing question of the order of the planets: a heliocentric cosmos in which the sun takes place at the centre of the universe and assumes a role of the reference point for the ordering of the planets according to distance–period relationship. The chapter ends with the explanation of the absence of any detectable parallax in such cosmos.

(i) The existing arrangements of planetary spheres. Pursuant to the old principle of ancient philosophers, a planet's distance from the earth is determined by its visible angular speed: the slower its motion, the greater its distance from the observer. In other words: the lower its angular speed, the greater its distance from the centre of the universe, that is, the stationary earth on which the

of Translations of Primary Sources," pp. 364–6; M. di Bono, *Le sfere omocentriche di Giovan Battista Amico nell' astronomia del Cinquecento*, pp. 31–6.

209 *De revolutionibus* I, 10 (Rosen, p. 21; Zekl, pp. 132–4).

observer is located. If a planet maintains equal linear speed, the duration of its passage is proportionate with the size of its sphere. Therefore, different sphere sizes can be determined on the basis of various angular speeds:

> Of all things visible, the highest is the heaven of the fixed stars. This, I see, is doubted by nobody. But the ancient philosophers wanted to arrange the planets in accordance with the duration of the revolutions. Their principle assumes that of objects moving equally fast, those farther away seem to travel more slowly, as is proved in Euclid's *Optics*.[210]

Nevertheless, failing to ensure an unproblematic order of *all* the planets in the geocentric universe, this principle is not universally valid.

When arranging the celestial spheres between the heaven of the fixed stars and the motionless earth in the middle of the universe on the basis of the traditional principle of the "duration of the [planetary] revolution," the *prisci philosophi* had the least difficulty with the moon and Saturn. The former runs on the smallest circle as the closest to the earth, while the latter, being the farthest from the earth, "completes the biggest circuit in the longest time." Below Saturn are Jupiter and Mars: "The moon revolves in the shortest period of time because, in their opinion, it runs on the smallest circle as the nearest to the earth. The highest planet, on the other hand, is Saturn, which completes the biggest circuit in the longest time. Below it is Jupiter, followed by Mars."[211] However, the general consensus on the traditional arrangement of the celestial spheres ends with the Mercury and Venus, to which different authors attribute different positions in relation to the sun: "With regard to Venus and Mercury, however, differences of opinion are found. For, these planets do not pass through every elongation from the sun, as the other planets do."[212]

Copernicus now changes the criterion for establishing the planetary order by reintroducing Ptolemy's principle from the *Almagest* based on the elongation from the sun. Some planets do have "every elongation," that is, they reach 180° from the sun, and some do not. As seen earlier,[213] Ptolemy briefly addresses the arrangement of the planetary spheres in the beginning of Book IX of the *Almagest*. He refers to the agreement among astronomers that the sphere of the fixed stars is the most distant from the earth, the sphere of the moon the nearest, and the

---

[210] *De revolutionibus* I, 10 (Rosen, p. 18; Zekl, p. 124).

[211] Ibid.

[212] Ibid.

[213] See supra, pp. 70–2.

sphere of Saturn the largest (and hence the farthest away from the earth), followed by the spheres of Jupiter and Mars. He, too, finds the arrangement of the sun, Venus and Mercury problematic. Some place Venus and Mercury below the sun, others above it, because neither planet has ever been observed to transit across the solar disk, "eclipsing" or "occulting" the sun in proportion to their sizes.

Copernicus then continues by presenting the opinions and reasons of "Plato's followers," Ptolemy, and al-Bitrūjī (Alpetragius; fl. ca. 1190). Each of them furnished a different arrangement of Mercury and Venus in regard to the sun and their own reasons for it. Plato placed Mercury and Venus above the sun, Ptolemy placed them below it, whereas al-Bitrūjī placed Venus above and Mercury below the sun: "Hence Venus and Mercury are located above the sun by some authorities, like Plato's *Timaeus*,[214] but below the sun by others, like Ptolemy and many of the moderns. Al-Bitrūjī places Venus above the sun, and Mercury below it."[215]

Copernicus maintains that "Plato's followers" advocated placing Venus and Mercury above the sun for two reasons. According to them, all planets are opaque, dark bodies having no light of their own. The planets shine by light they receive from the sun. Situated near the sun – for they only have a limited elongation – Venus and Mercury would, if positioned below it, only be seen halved or at least not fully round. That is, they would exhibit phases like the moon:

> According to Plato's followers, all the planets, being dark bodies otherwise, shine because they receive sunlight. If they were below the sun, therefore, they would undergo no great elongation from it, and hence they would be seen halved or at any rate less than fully round. For, the light which they receive would be reflected mostly upward, that is, toward the sun, as we see in the new or dying moon.[216]

Another reason why those who follow Plato disagreed with the placement of Venus and Mercury below the sun and proposed to position them above it comes from an observed fact or, rather, the absence of one. As already pointed out by

---

[214] See *Timaeus* 38c–d (Cornford, p. 105): "Having made a body for each of them, the god set them in the circuits in which the revolution of the Different was mowing – in seven circuits seven bodies: the moon in the circle nearest the earth; the sun in the second above the earth; the Morning Star (Venus) and the one called sacred to Hermes (Mercury) in circles revolving so as, in point of speed, to run their race with the sun, but possessing the power contrary to his; whereby the sun and the star of Hermes and the Morning Star alike overtake and are overtaken by one another."

[215] *De revolutionibus* I, 10 (Rosen, p. 18; Zekl, p. 124).

[216] Ibid.

Ptolemy, the sun should be eclipsed by Venus and Mercury passing between it and the earth and its "light cut off" in proportion to their sizes. "Since this is never observed, these planets do not pass beneath the sun at all, according to those who follow Plato."[217]

According to Copernicus, their opponents, the partisans of Ptolemy, proposed the placing of Venus and Mercury below the sun by proceeding from a completely different starting point. Their first argument in favor of placing Venus and Mercury below the sun, that is, between the sun and the earth, is of cosmological or, better, philosophical nature: "On the other hand, those who locate Venus and Mercury below the sun base their reasoning on the wide space which they notice between the sun and the moon."[218] Pursuant to the first "Ptolemaic" argument, Venus and Mercury must be below the sun to satisfy the "postulate of plenitude" (if I may call it so), which takes into account the hypothesis of nesting or contiguous spheres. According to the hypothesis of contiguous spheres the greatest distance of one planet is equal to the least distance of the planet above it.[219] Although Regiomontanus presented the same argument in *Epytoma Almagesti*,[220] according to majority of scholars Copernicus here draws on a Proclus' *Hypotyposis astronomicarum positionum*, translated to Latin in Giorgio Valla's *De expetendis et fugiendis rebus*. In *Hypotyposis* Proclus determines the order of the planets by relying on Ptolemy's *Planetary Hypotheses*, without mentioning him.[221] The computations of unknown origin are, however, based upon the relative least and greatest distances established in the *Almagest*. In *Hypotyposis* it is stated that Venus and Mercury must be placed between the sun and the moon to avoid having so vast a space remain empty as though it were forgotten by nature. Calculations show that there is a vast empty space between the spheres of the moon and the sun which should be filled.[222] Therefore it seems the most appropriate to do so by placing there Venus and Mercury (see *Figure 12*):

---

[217] Ibid.

[218] *De revolutionibus* I, 10 (Rosen, p. 18; Zekl, p. 124–6).

[219] See supra, p. 54.

[220] See *Epytoma Almagesti* IX, 1.

[221] See supra, pp. 53–5.

[222] Copernicus' figures only partially agree with those provided by *Proclus*. Could it be that he did not draw on Giorgio Valla's translation but had Proclus' manuscript with different calculations? See Swerdlow's explanation of the figures in "Copernicus and Astrology," p. 365.

Consequently, to avoid having so vast a space remain empty, they announce

> that the same numbers almost exactly fill up the apsidal distances, by which they compute the thickness of those orbs. Thus the moon's apogee is followed by Mercury's perigee. Mercury's apogee is succeeded by the perigee of Venus, whose apogee, finally, almost reaches the sun's perigee. For between the apsides of Mercury they calculate about 177 ½ earth-radii. Then the remaining space is very nearly filled by Venus' interval of 910 earth-radii.[223]

The "plenitude argument" is then followed by the mathematical-astronomical argument. Compared to the theory of "Plato's followers," this viewpoint also implies a different opinion on whether the planets – in general, which is more likely, or just these two in question – shine with the light absorbed from the sun or some other light source.[224] Therefore "the Ptolemaists" cannot argue that Venus and Mercury would imitate the moon and its visible phases, because the moon shines only by reflecting sunlight. That no transit of Venus and Mercury across the sun's disc has been observed they explain with the assumption that Venus, Mercury and the sun usually deviate in latitude or, in other words, "because it rarely happens that they interfere with our view of the sun." Moreover, their imperceptible passage across the sun's face is also due to the fact that they are tiny bodies in comparison to the sun. And there is even a possibility that somebody actually spotted the transit of Mercury. Averroes reports that he saw some dark spot traveling across the sun's disc and his calculations made him believe that it was Mercury:

> Therefore they do not admit that these heavenly bodies have any opacity like the moon's. On the contrary, these shine either with their own light or with the sunlight absorbed throughout their bodies. Moreover, they do not eclipse the sun, because it rarely happens that they interfere with our view of the sun, since they generally deviate in latitude. Besides, they are tiny bodies in comparison with the sun. Venus, although bigger than Mercury, can occult barely a hundredth of the sun. So says Al-Battani of Raqqa, who thinks that the sun's diameter is ten times larger (than Venus'), and therefore so minute a speck is not easily descried in the most brilliant light. Yet in his Paraphrase of Ptolemy, Ibn Rushd [i.e. Averroes] reports having seen something blackish when he found a conjunction of the sun and Mercury

---

223 *De revolutionibus* I, 10 (Rosen, p. 19; Zekl, p. 126).

224 Copernicus relies on Pico della Mirandola, *Disputationes contra astrologiam divinatricem* X, 4.

indicated in the tables. And thus these two planets are judged to be moving below the sun's circle.[225]

(ii) Copernicus' criticism. Copernicus disagrees with these positions, claiming that their reasoning is "weak and unreliable." Here are his reasons. Copernicus first addresses the "principle of plenitude." His argument is, again, dialectical. He wants to show that those who argue for such order of the spheres on the basis of their principles are in contradiction with themselves. If they establish a "principle of plenitude," the latter should apply universally, to the entire universe. But this is not so.

There are two facts that prove it. On the one hand there is a vast space between the earth and the moon – even vaster according to Copernicus' more accurate calculation – that "contains nothing but air" and "what is called 'the element of fire'." Homogeneity implied by the "principle of plenitude" should also apply to the sublunary region – or vice versa: if there is an empty space between the earth and the moon, why should not there also be an empty space between the moon and the sun? The demonstration in favor of placing Mercury and Venus between the moon and the sun according to the "principle of plenitude" is inconsistent and contradictory if the same cannot apply also for the space between the earth and the moon. He writes: "But this reasoning also is weak and unreliable. This is obvious from the fact that there are 38 earth-radii to the moon's perigee, according to Ptolemy, but more than 49 according to a more accurate determination, as will be made clear below. Yet so great a space contains, as we know, nothing but air and, if you please, also what is called 'the element of fire'."[226]

On the other hand their solution – the spheres of Mercury and Venus occupy what would otherwise be just a large empty space between the spheres of the moon and sun – produces another large empty space. Copernicus now challenges the diameter of Venus' epicycle. As mentioned earlier,[227] in order to preserve the distance between Venus and the sun, which changes only within certain limits (= maximum elongation), Ptolemy must fix the epicycle inside the sun's orb, so that it would correspond to Venus' assumed position in the Ptolemaic system (see *Figure 12*). The diameter of the epicycle of Venus that account for its elongation from the sun of about 45°, is six times greater than the distance from the earth

[225] *De revolutionibus* I, 10 (Rosen, p. 19; Zekl, 128). Copernicus' source is Pico della Mirandola, *Disputationes contra astrologiam divinatricem* X, 4.

[226] Ibid. Reference to Ptolemy is to *Almagest* V, 13 (Toomer, pp. 247–51; Heiberg, pars 1, pp. 408–16).

[227] See supra, p. 72.

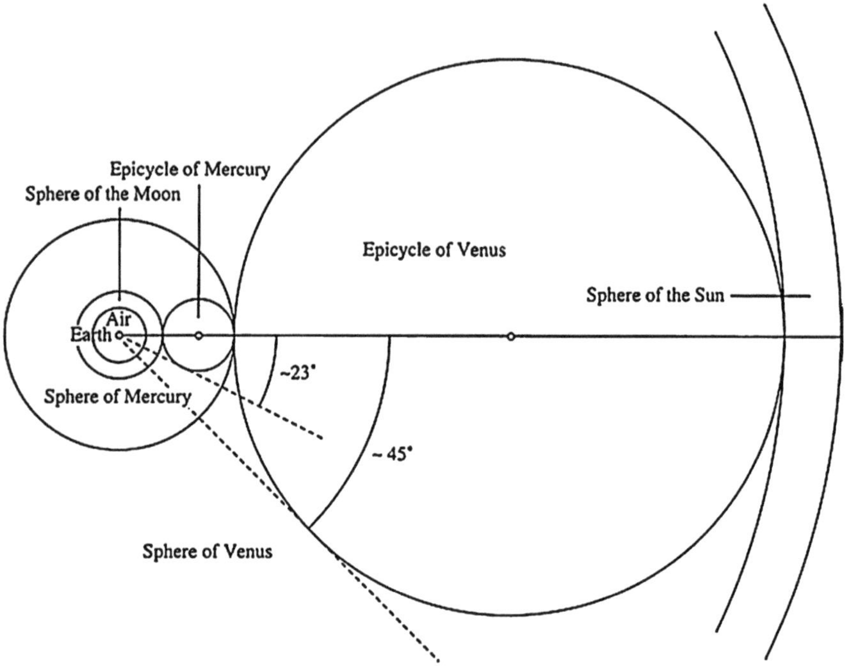

Figure 12.
The relative sizes of the sphere of the moon, the spheres of Mercury and Venus, and the sphere of the sun in nesting spheres model. Reconstruction by N. Swerdlow, "Copernicus and Astrology," p. 366.

to its lower apsis.[228] The resulting epicycle would have a vast diameter with its radius being almost equal to that of the deferent or, more accurately, covering two thirds of the deferent: "Moreover, the diameter of Venus' epicycle which carries it 45° more or less to either side of the sun, must be six times longer than the line drawn from the earth's centre to Venus' perigee, as will be demonstrated in the proper place."[229]

Venus' epicycle would therefore be immense. In observance of the principle of plenitude, "the Ptolemaists" would have to fill this space, just as they filled the space between the earth and the moon: "In this entire space which would be taken

[228] For calculations, see, again, N. Swerdlow, "Copernicus and Astrology," p. 365.
[229] *De revolutionibus* I, 10 (Rosen, p. 19; Zekl, p. 126).

up by that huge epicycle of Venus and which, moreover, is so much bigger than what would accommodate the earth, air, aether, moon, and Mercury, what will they say is contained if Venus revolved around a motionless earth?"[230]

Copernicus finally challenges the very foundation of Ptolemaic arrangement of the planets, that is, the criterion of full and limited elongation. The arbitrariness of Ptolemy's arrangement of the planets (Saturn, Jupiter and Mars above the sun and Mercury and Venus below the sun) on the basis of their elongation becomes even more evident to Copernicus from the fact that the moon, which Ptolemy places the closest to the earth, that is, below the Mercury, does not behave in the same way as Mercury and Venus, but as if it were one of the superior planets (Saturn, Jupiter and Mars), by digressing from the sun as they do: "Ptolemy argues also that the sun must move in the middle between the planets which show every elongation from it and those which do not. This argument carries no conviction because its error is revealed by the fact that the moon too shows every elongation from the sun."[231]

There is one final argument against all who "locate Venus and Mercury below the sun" or "separate these planets [from the sun] in some other order." Copernicus obviously takes all the different arrangements of Mercury and Venus in relation to the sun as a target *en bloc*: "Ptolemy and many of the moderns," who located Venus and Mercury below the sun, al-Bitrūjī, who placed Venus above and Mercury below the sun, and Plato and his followers, who placed Mercury and Venus above the sun. His objection is the following: how do they explain the fact that Mercury and Venus do not make similar elongations from the sun as the other planets, that is, Saturn, Jupiter, and Mars, if they do not violate the order of the planetary spheres, that is, if they are ordered correctly, and at the same time observe the distance–period relationship: "What cause will they adduce to explain why Venus and Mercury do not likewise traverse separate circuits divergent from the sun, like the other planets, without violating the arrangement [of the planets] in accordance with their [relative] swiftness and slowness?"[232]

To put it differently: if any one of these arrangements is correct and if in this way the principle of "duration of the [planetary] revolution" is observed, then they should also have maximum elongation. Put in another way, if the Mercury and Venus are arranged in accordance with their relative swiftness and slowness (slower movement = larger orb = the more distant orb from the earth) they should

[230] Ibid.

[231] Ibid.

[232] Ibid.

also have maximum elongation, or 180° from the sun. At first sight it seems that Copernicus is being unfair, since neither of them (nor Ptolemy and many of the moderns, nor Alpetragius, nor Plato and his followers), have not been presented as explicitly relying on the distance–period principle for arrangement of the planets. As we have seen they all had different criteria for their arrangements. But this is not completely so. The arrangement of Saturn, Jupiter, Mars, and the moon in all the above mentioned arrangements follows from principle of "duration of the planetary revolution." They all respect this principle to certain extent. Because of the lack of parallax even Ptolemy had to respect it for the superior planets and for the moon, which all have "every elongation." Now, if the principle is respected also for Venus and Mercury, regardless of which arrangement we take into account, that is, regardless of their differences in arrangement and the reasons for the arrangements, then they should also have the maximum elongation as do the Saturn, Jupiter, Mars, and the moon. But they do not. They have limited elongation.

Copernicus targets the internal conflict in the existing arrangements of the universe. There are basically two criteria: distance–period relationship and arrangement according to limited or full elongation. Existing arrangements use both principles and produce inherently inconsistent solution which has as a result non-homogeneous universe. If they apply the "principle of limited and full elongation" as criterion of differentiating among planets and arranging them in geocentric order, there is the moon that proves them inconsistent. The moon has full elongation but is still considered to be nearest to the earth because it has the shortest period. If they apply the principle of "duration of the [planetary] revolution" (= period) which works so well for Saturn, Jupiter, Mars and the moon (which all have full elongation) then Mercury and Venus should also have full elongation. But they do not.

Now one can see why Copernicus, in the *Preface,* accuses the "eccentric astronomy": (1) to be incapable of deducing the structure of the universe; (2) of being uncertain in regard of the organization of its parts; (3) of allowing the rearrangements of the order of the planets without causing any consequence to the whole; and (4) of being monstrous. There are, first, different arrangements of the parts (= planetary orbs) of the universe, which means that there is no certainty in regard of the organization of the universe. These arrangements, apart from being different, are also all inherently inconsistent and therefore not necessary, not certain. One can shift some below or above the sun without causing any problem for the arrangement of the whole, which proves that these arrangements are not necessary but arbitrary. And finally, parts of the universe when put together do not respond to each other, they do not belong to each other. There is no common, unique principle of organization and as a consequence, the parts of the universe, when put together, create a monster.

It is therefore obvious that Copernicus does not have an astronomical problem with the ordering of the planetary spheres but a philosophical one. As can be inferred from his rebuttal of the arguments in favor of the existing arrangements of the Mercury and Venus in regard to the sun (Ptolemy, Plato, al-Bitrūjī), Copernicus seeks a principle of arrangement that can be universalized, a principle that applies to the entire universe evenly and that would not produce inherently inconsistent and contradictory results. This requirement is satisfied neither by the "principle of plenitude" nor by the principle of full and limited elongation. Instead of many principles of arrangement, which are all inconsistent and contradict each other, there must be *one* principle, one common measure that can be unambiguously applied to all the planets, or, in other words, to *All* (= universe). Cosmos must be governed by one principle that would arrange its parts in an inherently coherent and *homogenous* way.

(iii) Copernicus' dilemma. Copernicus dismisses the principle of plenitude and the principle of elongation because neither can be applied universally to all the planets. There is always something that breaks the logic and produces a heterogeneous universe. Since these two criteria of arrangement are ruled out the only principle that remains is the distance–period relationship – but not as it has been applied so far. This ancient philosophical principle while arranging the Saturn, Jupiter, Mars and the moon perfectly, does not work for Mercury, Venus and the sun because in the geocentric arrangements they all have a period of one year. But it does not mean that principle itself is wrong; its application is. Therefore, Copernicus announces a major dilemma: either we change the centre of the universe and stop explaining the motions of the planetary spheres by referring to the earth as their centre, or we retain the same centre at the expense of a planetary order that explains the dependency of planetary motions on the sun in all its forms. If we insist that the earth is the (immovable) centre, we obtain an implausible and arbitrary order of the universe: Therefore "[…] one of two alternatives will have to be true. Either the earth is not the centre to which the order of the planets and spheres is referred, or there really is no principle of arrangement nor any apparent reason why the highest place belongs to Saturn rather than to Jupiter or any other planet."[233]

233 *De revolutionibus* I, 10 (Rosen, p. 20; Zekl, p. 130).

(iv) Copernicus' solution: *symmetria* of the universe and firm, harmonious linkage between the period and distance of the orbs. So far we have followed negative part of Chapter 10, that is, Copernicus' criticism of existing arrangements of the planets because of the contradictions and inconsistencies. What follows is his solution of the problem, and, as is usual with him, he presents it as a continuation of already existing Ptolemaic logic. The starting point is the distance–period relationship. To that foundation special relationship of Venus and Mercury to the sun is adduced. This special relationship is an indisputable fact. But this special link between them and the sun is in Ptolemaic arrangement geometrically described by limited elongation, without any consequence for the rest of the universe. In line with this logic of special relationship of Mercury and Venus with the sun Copernicus finds a special relationship of Saturn, Jupiter and Mars with the sun as well. While conceding the fact that Venus and Mercury are by necessity related to the sun, he also maintains that there is also a connection between the sun and Mars, Jupiter and Saturn (see *Figures 13a and 13b*), albeit a different one.

Since Venus and Mercury are never in opposition to the sun, they always remain in its proximity when moving in their assumed paths around the earth. Venus' greatest elongation is never more than 47° from the position of the sun and Mercury' greatest elongation is never more than 28°. Their special relation to the sun should be taken into account whatever the planetary order. However, there is also a special relation between the sun and the superior planets Mars, Jupiter, and Saturn. They are the closest to the earth when in opposition to the sun and the farthest away from it when in conjunction. Copernicus sizes this opportunity to introduce an arrangement that can account for the already existing double dependency of planetary motions on the sun.

The first step for Copernicus is an explanation of Mercury's and Venus' limited elongation by the transformation of their reference point. Their special relationship with the sun is better explained if we assume that they do not circle the earth but the sun. The sun, and not the earth, is the centre of Mercury's and Venus' motion, which subsequently also explains their constant connection to the sun (see *Figure 14*).

There were ancient authorities that changed their reference point. Martianus Capella (fifth century) and "certain other Latin writers" had already aranged that:

> Venus and Mercury revolve around the sun as their centre. This is the reason, in their opinion, why these planets diverge no farther from the sun than is permitted by the convexity of their orbs. For they do not encircle the

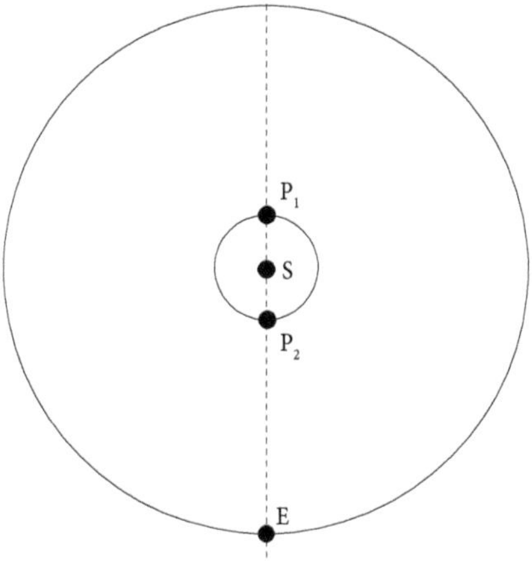

Figure 13a.
The inferior planets passing between the earth and the sun can be only in conjunction and never in opposition. An inferior planet is in conjunction when forming the earth–planet–sun or the earth–sun–planet alignment. The former alignment is called inferior conjunction (P2), and the latter is called superior conjunction of an inferior planet (P1).

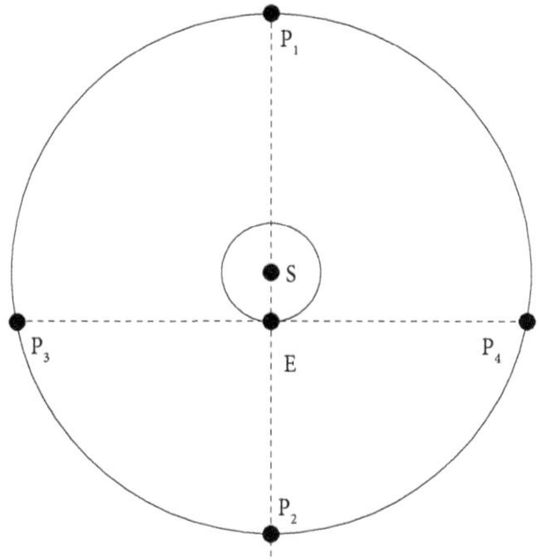

Figure 13b.
A superior planet is in conjunction when in line with the sun and the earth, describing the earth–sun–planet (P1) alignment. This configuration occurs when the planet is the farthest away from the earth, facing it with its non-illuminated side. It has the smallest radius and is visible in the sky during daytime. A superior planet is in opposition when in line with the sun and the earth, describing the sun–earth–planet (P2). This configuration occurs when the planet faces the earth with its illuminated side and comes the nearest to it. For this reason it also has apparently the largest diameter and is visible all night.

earth, like the other planets, but 'have opposite circles'. Then what else do these authors mean but that the centre of their orbs is near the sun? Thus Mercury's orb will surely be enclosed within Venus', which by common consent is more than twice as big, and inside that wide region it will occupy a space adequate for itself.[234]

---

234 Ibid. Reference is to *De nuptiis Philologiae et Mercurii* VIII, 857. For fuller account on Martianus Capella and "certain other Latin writers" see infra, pp. 358–60.

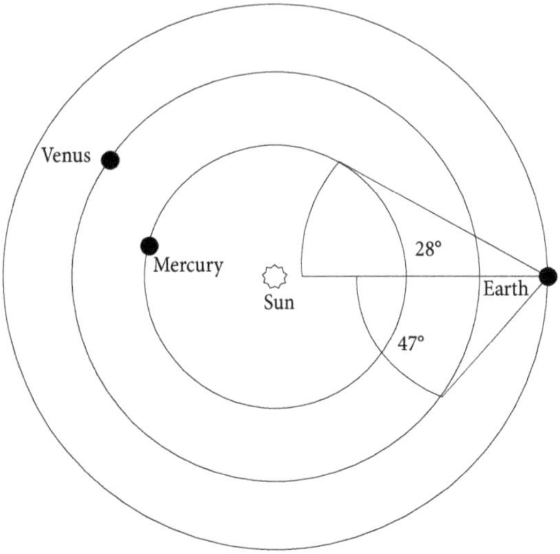

Figure 14.
Copernicus' explanation of the inferior planets' limited elongation. Maximum elongation is the point at which a planet reaches its greatest angular separation from the sun. The value of the greatest elongation cannot exceed the value of the angle at which the radius of the planet's path can be seen from the earth.

The limited elongation is, if we accept this arrangement, a consequence of the geocentric perspective. And it is not only Mercury and Venus that have a special relationship to the sun; Copernicus finds that Saturn, Jupiter and Mars are also more related to the sun than to the earth:

> For [these outer planets] are always closest to the earth, as is well known, about the time of their evening rising, that is, when they are in opposition to the sun, with the earth between them and the sun. On the other hand, they are at their farthest from the earth at the time of their evening setting, when they become invisible in the vicinity of the sun, namely, when we have the sun between them and the earth. These facts are enough to show that their centre belongs more to the sun, and is identical with the centre around which Venus and Mercury likewise execute their revolutions.[235]

There is a connection, then, between the sun and the superior planets Mars, Jupiter and Saturn. The superior planets are the closest to the earth when in opposition to the sun and the farthest from it when in conjunction. They make the retrograde cycle when they are at their brightest, that is, when they are the closest to the earth. To accommodate this, Ptolemy forced the superior planets to move on the epicycle in such a way as to make its diameter always parallel

235 *De revolutionibus* I, 10 (Rosen, p. 20; Zekl, p. 130).

to the earth–sun line and place the planets in perigee of the epicycle when in opposition. On the other hand, a planet is in conjunction when it reaches the apogee of the epicycle.

With this special connection in mind, it remained for Copernicus only to relate the motion of Saturn, Jupiter and Mars to the sun as well. The resulting arrangement of the planets is inherently necessary and justified by its uniformity and homogeneity: "If anyone seizes this opportunity to link Saturn, Jupiter, and Mars also to that centre, provided he understands their orbs to be so large that together with Venus and Mercury the earth too is enclosed inside and encircled, he will not be mistaken, as is shown by the regular pattern of their motions."[236] The first step, the centering of Venus and Mercury on the sun, which explains their specific relationship with it, prepares the ground for the second step, in which the sun also becomes the centre of the motions of the superior planets. As the centre of Venus' and Mercury's motion, the sun accounts for their peculiar motion and relation to it. Moreover, the sun, being likewise the centre of the motions of the superior planets, also accounts for their specific motion.

In contrast to that of Ptolemy, Copernicus' arrangement of the planetary spheres around the sun enables him to *explain* the observed facts pertaining to the motion of the superior planets within a necessary, inherently coherent system of mutually interdependent elements. The appearance, or the observed fact of a superior planet being the closest to the earth when in opposition to the sun, can, just like that of a planet being the farthest away from the earth when in conjunction, be explained by the newly arranged planetary system, in which the earth and the superior planets (much as the inferior ones) revolve around the sun. The observed facts arise from combining the earth's motion around the sun with that of each respective superior planet.

Mars, for instance, is the closest to the earth (and shines its brightest) when it arrives at P_4 (see *Figures 15a and 15b*). At that time it is in opposition. When reaching P_1 or P_7, it is the farthest away from the earth and at its least bright. At that time, Mars is in conjunction. To account for it, Ptolemy has to make Mars move on an immense epicycle (see *Figure 8,* supra, p. 71). Copernicus, on the other hand, has no need to do so. He considers the change in Mars' brightness and consequently its distance from the earth as the outcome of combining Mars's movement with the earth's motion around the sun between the spheres of Mars and Venus. Hence, what Ptolemy only geometrically describes by the planet's motion on an epicycle carrying it closer to the earth and farther away from it,

---

236 Ibid.

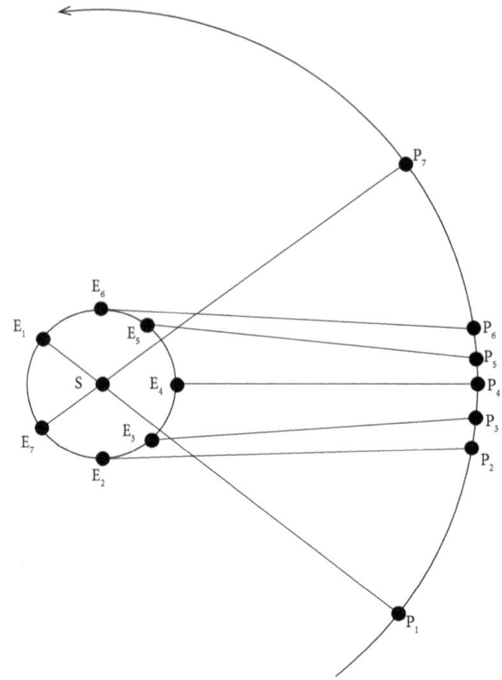

Figure 15a.
The retrogradation of the superior planets. A superior planet passes from E1P1 conjunction to E7P7. During this time the earth and the planet pass through all intermediary positions between E1 and E7, and P1 and P7. The planet retrogresses when it is in opposition to the sun (at P4). This happens because the earth passes by on the inside track. See also Figure 11, supra, p. 213.

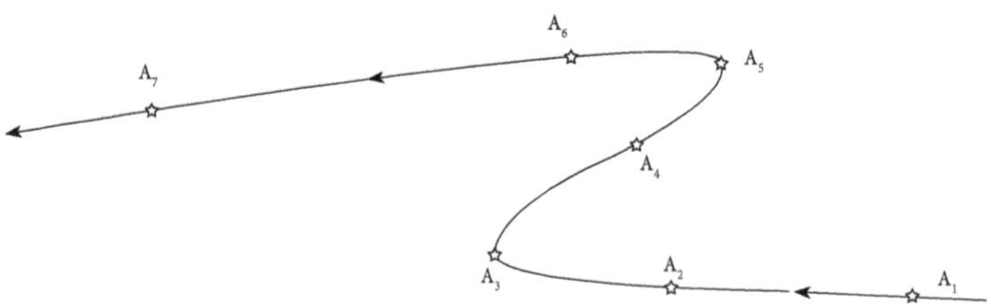

Figure 15b.
The apparent motion of a superior planet against the sphere of the fixed stars. Changes in latitudinal motion combine with the changes in longitudinal motion. Letters A_1, A_2, etc. correspond to letters P_1, P_2, etc.

Copernicus explains as causal necessity by combining the motion of the earth and the planet around the sun.

On the basis of the arguments discussed above, Copernicus concludes that the earth revolves around the motionless sun situated near the centre of the cosmos and that it can consequently also rotate about its axis. These arguments follow from the premise that the planetary motion is, one way or another, relative to the sun or dependent on it:

> But since all these planets are related to a single centre, the space between Venus' convex orb and Mars' concave orb must be set apart as also an orb or a sphere, both of whose surfaces are homocentric with those [orbs]. This [intercalated orb or sphere] receives the earth together with its attendant and whatever is contained below the globe of the moon.[237]

Copernicus' insistence on the distance–period relationship in its heliocentric variant has additional bonuses. Heliocentric arrangement explains the strange connection of the motion of the sun and the motions of the planets. First, heliocentrism *explains* why Mercury and Venus have limited elongations. It also *explains* why superior planets retrogress when in opposition to the sun. But there could be one problem for Copernicus' logic and his universalization of the distance–period principle: the moon. Where is the place of the moon in this scheme? Copernicus has nothing much to say about it, but since – one would infer – the moon does not exhibit any connection to the motion of the sun, and since, as Copernicus explicitly states ,"[w]e can by no means detach the moon from the earth," because it is "incontrovertibly nearest to the earth, mainly for the reason that in this space we find quite a convenient and abundant place,"[238] he makes it one unit with the earth, and thus avoids the break-up of his logic.

The earth and the moon are therefore located in the former place of the sun and the sun takes place of the earth, that is, near the centre of the universe. He writes: "I feel no shame in asserting that this whole region engirdled by the moon, and the centre of the earth, traverse this great orb amid the rest of the planets in an annual revolution around the sun. Near the sun is the centre of the universe."[239] Then Copernicus again underlines the principle of descriptive relativity – "Moreover, since the sun remains stationary, whatever appears as a motion of the sun is really due rather to the motion of the earth." – and explains that the distance from the earth to the sun is of large enough magnitude to be noticeable within these dimensions, but imperceptible in relation to the size of the universe. That is why there is no detectable parallax.[240] And he concludes

---

[237] Ibid. (Rosen, p. 20; Zekl, p. 132).

[238] Ibid.

[239] Ibid.

[240] See ibid. Quoted supra, p. 208, n. 197.

by stressing again that importance of there being *one explanatory principle* for multitude of phenomena. One explanation – the distance–period principle, which translates into movement of the earth – accounts for so many phenomena, which have been so far explained by multitude of spheres:

> This should be admitted, I believe, in preference to perplexing the mind with an almost infinite multitude of spheres, as must be done by those who kept the earth in the middle of the universe. On the contrary we should heed the wisdom of nature. Just as it especially avoids producing anything superfluous or useless, so it frequently prefers to endow a single thing with many effects.[241]

The concept of the earth's motion around the sun (and around its axis) should be accepted because it enables the arrangement of all motions on the basis of *one single* manner of explanation, which is uniform, homogeneous, causal and links the universe into inherently linked whole that can not be rearranged without falling apart. Copernicus had found what he sought: order of the planets that can be accounted for by a single principle, which does not produce heterogeneity in the universe and has additional bonus of explaining many other celestial phenomena.

Here is also the answer to our initial question about the true reason for Copernicus' belief in the earth's motion and solution to the question what entitles him – against the backdrop of the optical argument (or relativity of motion) – to assert the movement of the earth with certainty. The grand finale, the critical moment for Copernicus' realization that the earth moves and that it must move – primarily – around the sun situated at the centre of the universe, rests above all on the arrangement of the planets in relation to the sun and in the harmony of the universe as a whole. "All these facts," he argued already in Chapter 9, "are disclosed to us by the principle governing the order in which the planets follow one another, and by the harmony of the entire universe (*mundi totius harmonia*), if only we look at the matter, as the saying goes, with both eyes."[242] By taking into

---

[241] Ibid.

[242] *De revolutionibus* I, 9 (Rosen, p. 18; Zekl, p. 124). In his *Commentary*, p. 355, Rosen explains that Copernicus here expands the meaning of the *ambobus oculis*, taken from Giorgio Valla's *De expetendis et fugiendis rebus*, by adding *ut aiunt*, "to signify unimpeded intellectual insight." According to A. De Pace, *Niccolò Copernico e la fondazione del cosmo eliocentrico*, p. 202, and n. 595, Copernicus here evokes Books VI and VII of Plato's *Republic*. Compare also with Rheticus' quote in *Narratio prima* (Rosen, p. 132; Roche and Segonds, p. 53) of pseudo-Aristotle, *De mundo* I, 391a15: "with the divine eye of the soul."

account the motion of the earth around the sun, Copernicus can arrange the celestial spheres in the heliocentric order, which reveals a "marvelous symmetry of the universe, and *certain* harmonious linkage between the motion of the orbs and their size, such as can be found in no other way (*admirandam mundi symmetriam ac certum harmoniae nexum motus et magnitudinis orbium, qualis alio modo reperiri non potest*)."[243] This *certain* harmonious order and *symmetria* of the world is derived from the principle of ancient philosophers that the size of the orb is determined by its period. This principle does not work fully in geocentric arrangement but reveals its full potential in heliocentric order. It is, as Copernicus showed in the first part of the Chapter 10, the only principle that can organize the universe without internal contradictions and inconsistencies (see *Figures 16a and 16b*):

> Consequently, with the first principle remaining intact, for nobody will propound a more suitable principle than that the size of the orbs is measured by the length of the time, the order of the spheres is the following, beginning with the highest. // The first and the highest of all is the sphere of the fixed stars, which contains itself and everything, and is therefore immovable. It is unquestionably the place of the universe, to which the motion and position of all the other heavenly bodies are compared. [...] [The sphere of the fixed stars] is followed by the first of the planets, Saturn, which completes its circuit in 30 years. After Saturn, Jupiter accomplishes its revolution in 12 years. Then Mars revolves in [around the sun] 2 years. The annual revolution takes the series' fourth place, which contains the earth, as I said, together with the lunar orb as an epicycle. In the fifth place Venus returns in 9 months. Lastly, the sixth place is held by Mercury, which revolves in a period of 80 days.[244]

This is, then, the heliocentric solution to the question of celestial order, which respects the principle of the ancient philosophers according to which the distance should be accounted for by the period. This order observes that principle, does not introduce any heterogeneity into universe, explains a host of celestial phenomena with one stroke, and has also the advantage to fully respect the admiration of the sun as "the lamp," "the ruler," "the lantern," "the mind," "the visible god," "the all-seeing," and the governor of the universe by different ancient authorities. Copernicus mentions only two by name, Hermes Trismegistus and Sophocles, but

[243] *De revolutionibus* I, 10 (Rosen, p.22; Zekl, p. 136). Emphasis added.
[244] Ibid. (Rosen, p.21; Zekl, pp. 132–6).

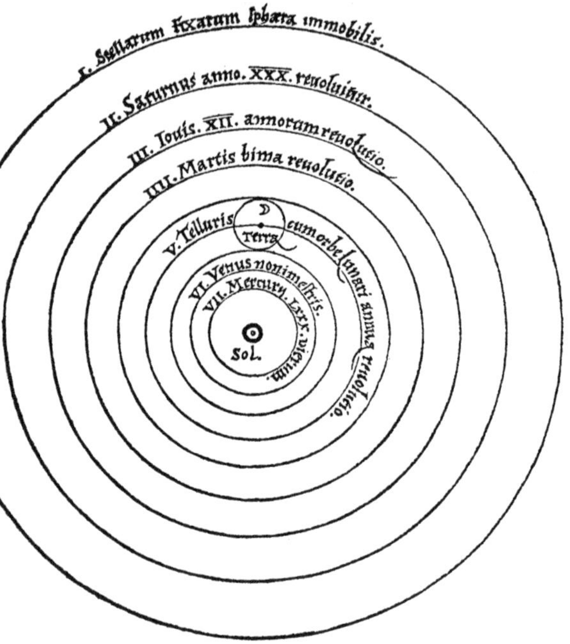

Figure 16a. Copernican system from De revolutionibus. *Image courtesy of the History of Science Collections, University of Oklahoma Libraries.*

metaphysical background of this position is, as has been already demonstrated by many scholars, Platonist or Neo-Platonist.[245]

> At rest, however, in the middle of everything is the sun. For in this most beautiful temple, who would place this lamp in another or better place than that from which it can light up the whole at the same time? For, the sun is not inappropriately called by some people the lantern of the universe, its mind by others, and its ruler by still others. [Hermes] the Thrice Greatest labels it a visible god, and Sophocles' Electra, the all-seeing. Thus indeed, as though seated on a royal throne, the sun governs the family of planets revolving around it."[246]

[245] See, for example, E. Garin, "La rivoluzione copernicana e il mito solare"; T. Kuhn, *The Copernican Revolution*, p. 130; A. Koyré, *The Astronomical Revolution*, p. 66; F. A. Yates, *Giordano Bruno and the Hermetic Tradition*, pp. 151–5; B. Biliński, *Il Pitagorismo di Niccolò Copernico*, pp. 39–43, 71–3, 96–103, 127–128; F. Schmeidler, *Kommentar zu De revolutionibus*, p. 88.

[246] *De revolutionibus* I, 10 (Rosen, p. 22; Zekl, p. 136).

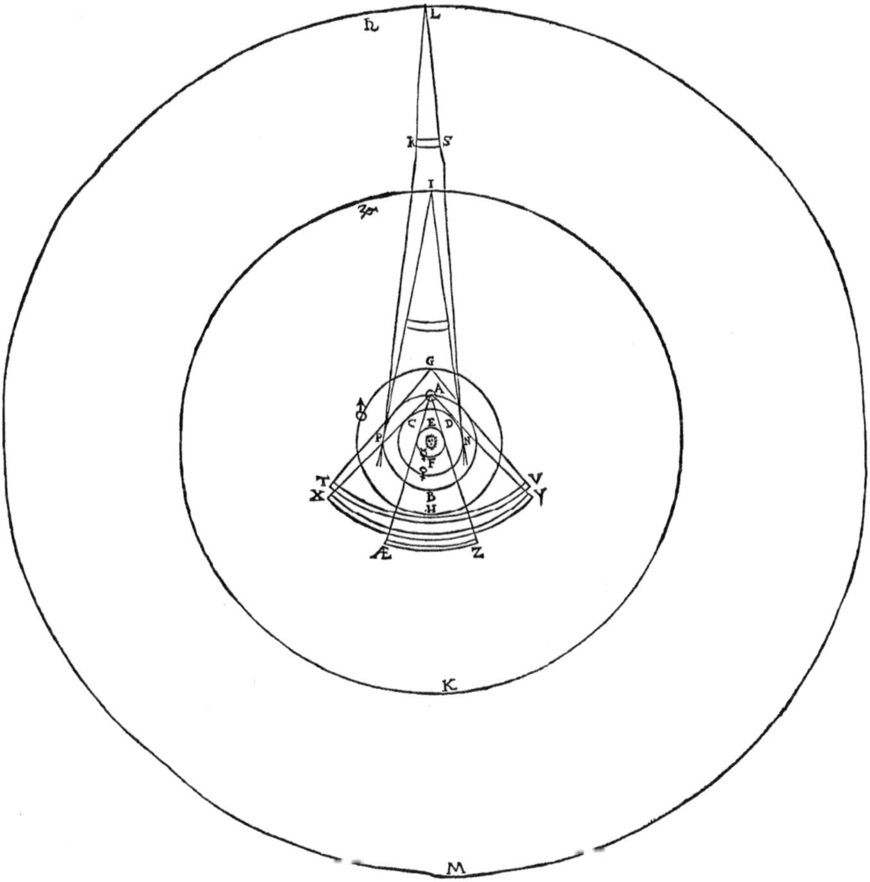

Figure 16b.
The Copernican system according to Johannes Kepler, Mysterium cosmographicum.
Image courtesy of the History of Science Collections, University of Oklahoma Libraries.

And a few lines below Copernicus concludes:

> In this arrangement, therefore, we discover a marvelous *symmetria* of the universe, and certain harmonious linkage between the motion of the orbs and their size (*mundi symmetriam, ac certum harmoniae nexum motus et magnitudinis orbium*), such as can be found in no other way. For this permits a not inattentive student to perceive why the forward and backward arcs appear greater in Jupiter than in Saturn and smaller than in Mars, and on the other hand greater in Venus than in Mercury. This reversal in direction appears more frequently in Saturn than in Jupiter, and also more rarely in Mars and Venus than in Mercury. Moreover, when Saturn, Jupiter, and Mars rise at sunset, they are nearer to the earth than when they set in the evening or appear at a later hour. But Mars in particular, when it shines all night, seems to equal Jupiter in size, being distinguished only by its reddish color. Yet in the other configurations it is found barely among the stars of the second magnitude, being recognized by those who track it with assiduous observations. All these phenomena proceed from the same cause, which is in the earth's motion.[247]

To sum it up: The strength of Copernicus' astronomical *system* is that taking into account the motion of the earth – the principle of uniform circular motion being presupposed – he is able to create not just a harmonious universe but a universe where everything is connected by a firm, certain bond, which does not permit to arrange and rearrange the order of the parts without any consequence for the whole. His system, neither omitting anything essential nor admitting anything extraneous, can fulfill the following *two* basic duties of astronomy as Copernicus understands it: provide the calculations for the positions of celestial bodies *and* explain the certain order and the inner orderly coherence of the universe. The linkage between motions of the planets and various motions of the earth, especially its revolution around the sun, not only enables calculations to concur with the observations but also accounts for the *necessary* order of all spheres and planets. Unlike the Ptolemaic astronomical tradition, the new astronomical system, which Copernicus nevertheless presents in a way as a logical derivation from commonly accepted Ptolemaic principles, establishes the "harmony of the entire universe," in which all its "parts" are so perfectly fit together that no individual "part" can be arbitrarily transposed from one place

[247] Ibid. (Rosen, p. 22; Zekl, pp. 136–8). Emphasis added.

to another without causing disruption in other parts of the universe and the universe as a whole. The earth's motion around the sun ensures the necessary rather than arbitrary order of the "parts" of the universe, so that they all can be, as Horace would put it, "assigned to a single shape."[248] The harmonious order of the universe which results out of the respect for single bond that connects all the planets into firm, internally coherent system also entails movement of the earth around the stationary sun. This is the ultimate reason why Copernicus believes that earth moves and can move also around its axis.

[248] *De arte poetica*, 8–9 (Fairclough, pp. 450–1). On the explanatory advantages of heliocentric system, see also J. Evans, *The History and Practice of Ancient Astronomy*, pp. 411–3.

PART THREE

COPERNICUS IN CONTEXT

Analysis of Copernicus' argumentation in Book I of *De revolutionibus* led us through the "optical argument" and refutation of Aristotle's and Ptolemy's reasons for the stationary earth to Chapter 10, where the motion of the earth around the sun was made a key element for the establishment of the harmonious universe. We have seen that Copernicus' main concern was not the earth's daily rotation but its circular movement around the sun. With the heliocentric arrangement of planetary spheres Copernicus solves his principal objection to Ptolemaic astronomy: its incapability to establish the form of the world and certain *symmetria* of its parts. A heliocentric universe enables Copernicus not only to "save the appearances" but also to determine the order of the planetary spheres in such a manner that heaven is, as he maintains in the *Preface*, so "linked together that in no portion of it can anything be shifted without disrupting the remaining parts and the universe as a whole."[1] Thus Copernicus' system fulfills two tasks: provides models which account for the celestial appearances *and* establishes the harmony of the entire universe.

Now it is time to consider why Copernicus expanded the task of astronomy to include not only "saving the appearances" but also the philosophical question of the order of the universe. What was the question he was trying to answer? What were the motives for such reform of astronomy? Was his project part of a broader astronomical or philosophical movement or was he a lone ranger? Which philosophical and epistemological changes enabled his project? Did he – apart from the movement of the earth – bring anything new to astronomy? Was he already part of the Copernican and Scientific Revolution or did he merely start one?

---

1 *Preface* (Rosen, p. 5; Zekl, p. 74).

CHAPTER ELEVEN

Astronomy before Copernicus

In order to answer to all these questions, we must reverse the familiar convention. Instead of looking at Copernicus through Galileo's *occhiale* and the prism of his own Copernicanism, interests and struggles, we shall examine him through the eyes of his contemporaries, against the backdrop of late medieval and Renaissance astronomy and philosophy. An excellent starting point is the work of two outstanding astronomers of the fifteenth century, Georg Peurbach, and his student and colleague, Johannes Müller, known as Regiomontanus. A cross-section of their works reveals what astronomy was like when Copernicus was a student, as well as four major elements that help illuminate his project of the restoration of astronomy:

(1) high quality of mathematical astronomy, coupled with some (sometimes rather severe) criticism of Ptolemy and awareness of the achievements of the Muslim astronomers;

(2) ultimate understanding of the motions of the celestial objects as a result of the motions of real three-dimensional orbs;

(3) constant awareness of the Averroist, philosophical critique of the Ptolemaic astronomy of eccentrics and epicycles;

(4) humanism and the restoration of Greek learning.

The *Epytoma Almagesti* of Peurbach and Regiomontanus elevated technical mathematical astronomy to a level that no medieval or renaissance scholar had been able to achieve. Peurbach undertook the preparations for the *Epytoma Almagesti* in 1460 at the behest of Basilios Bessarion known as Johannes Bessarion (1403–1472), but only finished the first six volumes before his untimely death. The work was then completed by his student and collaborator Regiomontanus. The *Epytoma Almagesti* provided a fairly comprehensible rendering of Ptolemy's complex work, but also supplemented his findings with contemporary observations and revised calculations and critical considerations, particularly in regard to his moon theory. It was printed in Venice as late as 1496, twenty years after the death of Regiomontanus and at the time when Copernicus was delving into astronomy. He later used it as one of his fundamental sources.

Another work of astronomy that is of equal significance for our understanding of Copernicus' astronomy is Peurbach's textbook *Theoricae novae planetarum*.[2] The work embodied a series of Peurbach's lectures given in 1454 at the Collegio Civium (Bürgerschule) in Vienna, but was published only between 1472 and 1474. It was written with a view to replacing the traditional *Theorica planetarum*, often attributed to Gerard of Cremona (ca. 1114–1187), a text which had, along with Sacrobosco's *De sphaera*, remained the standard introduction to the study of astronomy for over two hundred years. After Regiomontanus committed it to print, *Theoricae novae* became the favourite astronomical textbook of the sixteenth century. Between 1472 and 1653, it appeared in more than fifty editions and prompted a massive production of commentaries. The first was that of Albert of Brudzewo (ca. 1445–1495), the most important astronomer at Cracow University. Peurbach's textbook corrected the errors in the traditional *Theorica planetarum*, explained its terminology more fully, and introduced the motions of the planets by employing sets of three-dimensional spheres or orbs to replace two-dimensional geometric models composed of circles. Though several other scholars had introduced orbs by the end of the thirteenth and in the fourteenth centuries, Peurbach's work, in light of its popularity and the profusion of commentaries it inspired, definitively established the models of real spheres to account for the celestial motions.

Several early modern authors, such as Francesco (Giambattista) Capuano da Manfredonia (fl. 1494–1505), Gregor Reisch (1467–1525), Erasmus Reinhold

---

[2] For the English translation, see E. J. Aiton, "Peurbach's *Theoricae novae planetarum*: A Translation with Commentary." For the summary of *Theoricae novae*, see P. Barker, "The Reality of Peurbach's Orbs: Cosmological Continuity in Fifteenth and Sixteenth Centuries Astronomy," pp. 10–4, and M.-P. Lerner, *Le monde des sphères*, vol. 1, pp. 121–30.

(1511–1553), and Pierre Gassendi (1592–1655), suggested that Peurbach wrote his *Theoricae novae* in order to eliminate certain objections to the Ptolemaic astronomy that were usually raised by advocates of homocentric astronomy.[3] The account sheds new light on another and perhaps more important feature of astronomy of the time: the constant desire of Ptolemaic astronomers to devise an astronomical system that would concur with the principles of Aristotelian natural philosophy and thus be immune to the attacks of supporters of homocentric arrangement of celestial spheres. Peurbach and – as it has been firmly established only recently – Regiomontanus, each responded to the present criticism of the Averroist, Aristotelian purists who censured Ptolemaic astronomy for violating the accepted principles of Aristotelian physics.

1. Three approaches to studying the heavens in the Middle Ages

The astronomical science revealed by the works of Peurbach and Regiomontanus was not so much characteristic of the fifteenth century as quantitative and qualitative elaborations of the developments that had taken place over the preceding two centuries. Three elements of this science – the Ptolemaic tradition of mathematical astronomy, the Ptolemaic practice of employing combinations of real spheres, and the Averroist critique of Ptolemy, coupled with the attempts to introduce homocentric astronomy – had already been presented in the late thirteenth century by Roger Bacon (ca. 1214 or 1220–ca. 1292). He classified them in his *Opus tertium*, finished ca. 1267, as three approaches to the "study of the stars."[4] According to Bacon, the main protagonists of astronomical investigations were, first and foremost, Ptolemy and his *Almagest*, which accounted for

[3] See, Capuano, *Theorice nove planetarum Georgii Purbachii astronomi celebratissimi. Ac in eas [...] Domini Francisci Capuani de Manfredonia [...] sublimis expositio*; Reisch, *Margarita philosophica*, liber VII, tract. 1, cap. 28 (p. 261); P. Gassendi, *Georgii Purbachii et Ioannis Mulleri Regiomontani vita* (*Opera Omnia*, vol. 5, p. 521, col. 2). See also M.-P. Lerner, *Le monde des sphères*, vol. 1, p. 129, and n. 81 and 82, on pp. 318–9.

[4] See P. Duhem, *Un fragment inédit de l'Opus tertium de Roger Bâcon*, pp. 99 et seq., 108 et seq.; P. Duhem, *Le système du monde*, vol. 3, pp. 428–42. See also E. Grant, *Planets, Stars & Orbs*, p. 278–81, and "Celestial Motions in the Late Middle Ages," pp. 135–7. Bacon systematized the chaotic situation resulting from the profusion of translations of philosophical, astronomical, and astrological discussions from Arabic and Greek. In a matter of a few decades, between the end of the twelfth century and the early thirteenth century, the Western Latin world simultaneously confronted works of various provenances, from Aristotle's *De caelo* to Ptolemy's *Almagest*, through the textbooks of al-Farghānī and al-Battānī, as well as al-Bitrūjī's *Book on Astronomy*. On this, see

the celestial motions by means of eccentrics and epicycles; al-Bitrūjī, known as Alpetragius, with his attempt to introduce homocentric astronomy; and finally, an *ymaginatio modernorum* attempting to strike a compromise between the ability of the Ptolemaic astronomy to provide accurate computations and the requirements of the Aristotelian philosophy of nature. [5]

Bacon, like many of his contemporaries, preferred al-Bitrūjī's homocentric astronomy, with which he became familiar through Michael Scot's 1217 Latin translation titled *De motibus celorum circularibus*, in the original, the *Kitāb fī l-hay'a* (*Book on Astronomy* or *On the Principles of Astronomy*).[6] Al-Bitrūjī introduced an astronomical system that would provide calculations for celestial positions as accurate as Ptolemy's in *Almagest*, only on the basis of the Aristotelian physics as he understood it, with all celestial motion in the same direction and by using combinations of homocentric physical spheres moving in a uniform circular motion around the central earth. His endeavors were part of a broader criticism against Ptolemaic astronomy that emerged in twelfth-century Andalusia.[7] Al-Bitrūjī most probably drew inspiration from Ibn Tufayl, the architect of the "Andalusian revolt," who had reportedly arranged the celestial spheres so as to allow for the reproduction of the planetary motions without eccentrics and epicycles. A contemporary Andalusian who criticized Ptolemy's eccentrics and epicycles for violating Aristotelian physics was the philosopher Ibn Rushd, known as Averroes (1126–1198). However, unlike their counterparts from Marāgha (as well as Copernicus several centuries later), the Andalusians did not impugn Ptolemaic astronomy because of the concept of the equant but because they believed that

---

for example, C. D. Lindberg, "The Transmission of Greek and Arabic Learning to the West."

5 One must still agree, unfortunately, with C. D. Lindberg, *The Beginnings of Western Science*, p. 395, n. 41, that a more detailed historical analysis of late medieval astronomy is yet to be produced. For a short overview, see O. Pedersen, "Astronomy." See also M.-P. Lerner, *Le monde des sphères*, vol. 1, pp. 85–138. M. Shank, "Mechanical Thinking in European Astronomy (13th–15th Centuries)," and other studies quoted in notes. What I am presenting here is only a sketch of astronomy in this period, as a context for understanding Copernicus.

6 F. J. Carmody prepared a critical edition of *De motibus celorum, Sermo in radicibus naturalium et positionibus Tholomei*, and B. Goldstein critical edition of the Arabic text with the English translation, and the Hebrew translation by Moses ibn Tibbon (fl. between 1240 and 1283), in al-Bitrūjī, *On the Principles of Astronomy.*

7 See A. I. Sabra, "The Andalusian Revolt against Ptolemaic Astronomy: Averroes and al-Bitrūjī."

Ptolemy's eccentric-epicycle models contravened the requirements of Aristotelian physics and could not exist *in naturalibus.*

According to Averroes' *In Metaphysicam*, Ptolemy even failed to devise an astronomical system that would concur with his own principles, much less one that would be without natural impossibilities, since eccentrics and epicycles "are impossible in nature." Epicycles are completely contrary to nature because their centres do not coincide with the centre of the world. Admitting to their existence would inevitably lead to the presumption that there are several centres of circular motion in the universe; this, in turn, would imply the existence of several earths, which is impossible:

> The theory of eccentric spheres or of epicycles is contrary to nature. The epicycle is altogether impossible because the body moving in a circle moves about the centre of the universe, not outside it, since that which moves in a circle defines the centre. If there were a circular motion out of this centre, there would be another centre beside the former and another earth beside this earth. The impossibility of all this has been shown in natural philosophy.[8]

In like manner, eccentrics imply several centres of motion and thereby indicate the existence of a significant number of heavy bodies outside the earth. This leads to the presumption that the centre of the world is not unique and that it would have been extended and divided (between the centre of the world and the centre of the eccentric); all this is also impossible. Furthermore, the existence of realistically understood eccentrics would suggest the existence of superfluous bodies in the heavens whose only function is to fill empty spaces:

> This is what the situation seems to be with regard to the eccentric sphere postulated by Ptolemy: if there were several centres, there would be heavy bodies somewhere other than the earth; the middle would not be one but would have breadth and could be divided; all this is not correct. Also, if there were eccentric spheres, there would be superfluous bodies in heaven, with no purpose but as filling, as is thought to be the case in animal bodies. But there is nothing in the apparent motion of these stars that compels us to postulate the existence of epicycles or eccentric spheres.[9]

These philosophical or physical absurdities of the Ptolemaic astronomy led Averroes to announce in his *Middle Commentary on De Caelo,* composed in 1171,

---

8 *In Metaphysicam* XII, 8, comm. 45 (Genequand, p. 178).

9 Ibid.

that he would construct an astronomical system such as had been used in the "days of Aristotle," that is, a system in agreement with the Aristotelian physics.[10] But in the subsequent *Commentary on the Metaphysics,* written in 1186, he admitted in resignation that he was too old to fulfill this plan.[11]

Al-Bitrūjī met with greater success, at least as to the composition of a homocentric astronomy. In the *Kitāb fī l-hay'a*, probably written around 1200 and known to the Latin world as *De motibus celorum*, he stressed in particular the fact that all of Ptolemy's celestial spheres did not move in the same direction, a violation of Aristotle's physics, where it was stipulated that all spheres moved from east to west: "It is explained in the *Physics* that the mover of the heavens is unique and the natural motion of the heavens is one. Thus the heavens do not have two opposite motions because they have only one natural mover. Therefore, the heavens have no more than one motion, which results from one mover in one direction." [12] He also reiterated Averroes's *horror vacui* objection. In his opinion, the concept of eccentrics as real spheres would imply the existence of a vacuum in the heavens and the condensation of the immutable and indivisible celestial substance, which was impossible according to Aristotelian physics:

> That those assumptions are defective in arrangement and far from perfection was not hidden from him, for it follows from the principles which he assumed that a vacuum would result from the motions of eccentric spheres. Even if the spheres which encompass these spheres were filled by another body – one not moving in circles, as do the bodies moving in it – (the spheres) would still fill and vacate place after place. All this is repugnant and far from truth and contradicts the true facts of the heavens.[13]

---

10 See *De celo et mundo* II, 6, comm. 35 (Carmody, p. 332).

11 See *In Metaphysicam* XII, 8, comm. 45 (Genequand, p. 179): "We must examine this ancient astronomy from the beginning. It is the true astronomical scheme which is in accordance with the natural principles. It is based, I think, on the motion of one single sphere about one single centre and two different poles and more, according to what appears to the senses, because motions like these can make a star go faster and slower, forwards and backwards and have all the motions for which Ptolemy was unable to find a model. On account of that, a star can appear to get nearer or remoter as the moon appears to do. In my youth, I hoped to make a complete study of this, but now that I have grown old, I have given up this idea because of the obstacles I found in my way before."

12 *On the Principles of Astronomy*, vol. 1, Introduction, 10 (Goldstein, p. 57): See also Scot's Latin translation, *De motibus celorum* II, 1–6 (Carmody, p. 74).

13 Ibid., 17 (Goldstein, p. 61). This means that eccentrics ought to be understood as real

Al-Bitrūjī's homocentric system, which differed from those of Aristotle and Averroes, proved to be an utter failure in its ambition to reproduce celestial motions, and al-Bitrūjī was well aware of it. Nevertheless, his work inspired a series of new attempts in the Latin West to introduce homocentric astronomy that continued more or less intensely until Copernicus' day.

The Aristotelian line of astronomy with its three tenets – a single immovable physical centre for all celestial spheres, the one-way circular motion of all the stars, and the impossibility of a vacuum and condensation of celestial substance – found its medieval supporters in Bacon and even in some earlier scholars. Michael Scot (before 1200–ca. 1235) advocated such astronomy in *Theorice planetarum* and *Commentary on Sacrobosco's Sphera*; a similar view was also propagated by William of Auvergne (1180/90–1249) and Robert Grosseteste (ca. 1175–1253) in *De motu supercaelestium*. William the Englishman (first half of the thirteenth century), also called William of Marseille for his adoptive city, cited it as well. The Averroist attacks on Ptolemaic astronomy in defense of the Aristotelian worldview were directed not so much against the Ptolemaic geometric models as they were against the eccentric and epicycle models as representations of celestial reality. Alpetragius' criticism of eccentrics, that they leave behind a vacuum and cause condensation of celestial substance, makes sense only insofar as Ptolemy's eccentrics and epicycles are not considered in terms of geometrical models presented in the *Almagest* but as bodies that have physical existence resulting from the combinations of real three-dimensional orbs presented in the *Hypotheses planetarum*. This fact, however, throws a new light on the common assertion made by some historians of medieval astronomy that medieval astronomy was a predominantly mathematical enterprise.

The tradition of Ptolemaic mathematical astronomy was, however, undoubtedly the predominant trend in thirteenth-century university astronomy. Although based on the *Almagest*, which was translated to Latin from Greek around 1160 and from Arabic in 1175, Ptolemaic astronomy only became well-known through the widespread and popular *Tractatus de sphaera* of Johannes de Sacrobosco (end of the twelfth century–1256), written between 1230 and 1255 in Paris, which outlined a spherical universe in four short chapters., and later through a more comprehensive textbook, *Theorica planetarum*, whose anonymous author remedied the inadequate planetary theory of Sacrobosco's *De sphaera* with fundamental definitions and

spheres. Compare with Scot's Latin translation, *De motibus celorum* III, 25 (Carmody, p. 78).

elements of planetary astronomy.[14] The thirteenth-century *corpus astronomicum* additionally included Sacrobosco's *Algorismus vulgaris* and *Compotus*, Robert Grosseteste's *Calendar*, a minor work of Robert Anglicus (fl. 1272) on the so-called old quadrant, and most certainly the *Toledan Tables* with their canons, which enabled astronomers to calculate planetary positions.[15]

Rudimentary in nature, the aforementioned astronomical literature does not reflect the extent to which scholars in the second half of the thirteenth century were familiar with Arabic and Greek authorities in astronomy. It also conceals the fact that during this period there were disputes between supporters of Ptolemy and Aristotelian defenders of al-Bitrūjī. Several scholars – Albertus Magnus (ca. 1200–1280), Vincent of Beauvais (ca. 1190–1264),[16] Richard of Middleton (ca. 1249–1302), Bernard of Verdun (fl. latter part of the thirteenth century),[17] Giles of Rome (ca. 1243–1316), Pietro d'Abano (ca. 1257–1316), and John of Jandun (ca. 1285–1323) – spread al-Bitrūjī's ideas, regardless of the fact that they preferred the work of Ptolemy.[18] But some still adopted Aristotelian or Averroist views of

---

14 For Sacrobosco, see L. Thorndike, *The Sphere of Sacrobosco and Its Commentators*, and O. Pedersen, "In Quest of Sacrobosco." Sacrobosco's book was the principal textbook of elementary astronomy from the Middle Ages to the early Modern Period. On *De sphera* as a textbook, see O. Gingerich, "Sacrobosco as a Textbook." There is an English translation of *Theorica planetarum* by O. Pedersen in E. Grant, ed., *A Source Book in Medieval Science*, pp. 451–65. There are different hypotheses concerning whether or not Gerard of Cremona was really the author of this text. On *Theorica*, see O. Pedersen, "The Origins of the *Theorica planetarum*," and G. Federici-Vescovini, "Michael Scot et la *Theorica Planetarum Gerardi*."

15 For more details on this, see O. Pedersen, "Astronomy," p. 315, and "The Origins of the *Theorica planetarum*," pp. 113–5.

16 See *Speculum naturale* (ca. 1244), the first part of his *Speculum maius*. He spoke about "Avenalpetras" (most likely referring to al-Bitrūjī), who denied the existence of eccentrics and epicycles, since it would imply the divisibility of aether. A planet's distancing itself from the earth on an eccentric or an epicycle would suggest that the indivisible fifth element is divisible. In such a case, the vacuum left behind by the planet moving "up" or "down" would either be filled by aether or it would remain empty. As neither possibility was acceptable, Avenalpetras rejected the existence of epicycles and eccentrics. See P. Duhem, *Études sur Léonard de Vinci*, vol. 2, pp. 318–9, and *Le système du monde*, vol. 3, pp. 346–8. See also E. Grant, *Planets, Stars, & Orbs*, p. 287.

17 See his *Tractatus super totam astrologiam* (ca. 1300). On Bernard of Verdun, see P. Duhem, *Le système du monde*, vol. 3, pp. 442–60, and *Un fragment inédit*, pp. 65–9.

18 Another defender of the Ptolemaic astronomy was Guido de Marchia's or Guy de la Marche (fl. 1292–1310) in his treatise on *Planetarium*. For De Marchia, see M. Shank, "Rings in a Fluid Heaven," and "Mechanical Thinking in European Astronomy (13th–15th

astronomy. Such was the case of Henry of Langenstein, also known as Henry of Hesse (ca. 1325–1397), who appears to have drawn not so much on al-Bitrūjī as on Abū Ja'far al-Khāzin of the tenth century.[19]

The essential feature of Ptolemaic astronomy of the university was that it described and understood planetary motions in strictly mathematical terms. Both the anonymous *Theorica planetarum* and Sacrobosco's *De Sphaera* explained the planetary kinematics by geometric models derived from the *Almagest*, and neither led to the assumption that eccentrics and epicycles were real, material, three-dimensional orbs. Quite the contrary: according to Olaf Pedersen, the importance of the early *Theorica* is that it "showed how theoretical astronomy could be taught without any reference to cosmology and with no attempt to give actual dimensions or locations in space of the geometrical devices employed."[20] However, the status of astronomical models was understood differently already in the thirteenth century with the introduction of Bacon's *ymaginatio modernorum*,[21] which held and presented the Ptolemaic models as combinations of the real celestial spheres. The *Theorica planetarum* of Campanus of Novara (ca. 1220–1296), which appeared a few years after Bacon's *Opus tertium* (finished ca. 1267), considered the Ptolemaic astronomical system, to use Michel-Pierre Lerner's words, both "'realistic' and 'abstract' at the same time."[22] And finally, John of Sicily (fl. second half of the thirteenth century) stated in his *Scriptum Johannis de Sicilia super canones Azarchelis de tabulis Toletanis* of 1291, that geometric constructions contained in a special series of treatises known as the *Theorica planetarum* represented physical

Centuries)," pp. 15–9. De Marchia proposed elimination of the inner and outer orbs of the "three-orb system" of each planet.

19 On Henry of Hesse in general, see N. Steneck, *Science and Creation in the Middle Ages: Henry of Langenstein (d. 1397) on Genesis*. For his criticism of Ptolemy, see C. Kren, "Homocentric Astronomy in the Latin West: The *De reprobatione ecentricorum et epiciclorum* of Henry of Hesse," and "A Medieval Objection to 'Ptolemy'."

20 "Astronomy," p. 319.

21 It is not possible to determine with certainty how Bacon came across this *ymaginatio modernorum*. Some historians of medieval astronomy maintain that he adopted it from Ibn-al-Haytham's (Alhazen's) *De configuratione mundi*, others argue that such a system made its first appearance in the West about 1230 through the writings of William the Englishman, or in the *Compilatio de astrorum scientia* of Leopold of Austria. For him, see P. Duhem, *Le systéme du monde*, vol. 3, pp. 287–90, and 314.

22 See *Le monde des sphères*, vol. 1, pp. 118–20.

entities in the new *imaginatio*.[23] Obviously, he referred to the same *imaginatio* which Roger Bacon in his *Opus tertium* called *ymaginatio modernorum*.

Ymaginatio modernorum was a tentative response to the criticism of realistically understood eccentrics and epicycles – criticism which Bacon, as a supporter of the Alpetragian version of homocentric astronomy, found absolutely justified. On the other hand, those who tried to explain the phenomena by means of eccentrics and epicycles sought to strike a balance between the requirements of Aristotelian physics and Ptolemaic astronomy. The compromise between the Ptolemaic system of eccentrics and epicycles and the Aristotelian system of concentric spheres – also referred to as the "*Theorica* compromise," "Aristotelian–Ptolemaic compromise," or "three-orb compromise" in recent literature – was accomplished by embedding the mechanism of eccentrics and epicycles in a set of concentric spheres, so that the outermost convex and the innermost concave surfaces of individual orbs or spherical shells were concentric with respect to the centre of the earth. The fundamental principle of the Aristotelian system was thereby saved, while the region between the convex and concave bounding surfaces of every total orb could be subdivided according to the requirements of Ptolemaic astronomy into three or more orbs or spherical shells of various thicknesses. All this is clearly evident from Roger Bacon's description of the moon containing both an eccentric and epicycle (see *Figure 17*).

Point T is the centre of the earth and universe, as well as the centre of the total orb (*orbis*) of the moon.[24] The latter lies between convex surface ABCD and concave surface OQKP, which are both concentric with respect to T. Between these two surfaces are three orbs (a', b', and c') whose centre is at point V located towards the apogee of the moon. Around V as the centre are two surfaces, AGFE

---

[23] See *Scriptum Johannis de Sicilia super canones Azarchelis de tabulis Toletanis* (Pedersen, p. 135): "Huius vero novae positionis imaginationis, quae etsi forsan ab aliis visa est, a nullo tamen dearticulate particulariter explicata, defensionem ab hiis, quae possent obici, praetermitto, quia leviter omnia solvi possunt et alibi, videlicet in theorica planetarum, praecipue locum habent." Compare also with the following passage from which it is obvious that this imagination was invented to counter natural philosophical problems (ibid., p. 134): "Hoc autem sic imaginor: apello sphaeram planetae non unum solum corpus sphaericum, in quo planeta fixus existat, sed plures orbes ad motum planetae specialiter ordinatos [...]. Oportet autem ad hoc ut apparentia salventur et contra naturalem philosophiam inconvenientia non sequantur, in unoquoque planetarum ad minus 3 orbes sphericos assignare, quorum unus in concavitate alterius situatur [...]." See also H. Hugonnard-Roche, "Problèmes méthodologiques dans l'astronomie au début du XIVe siècle," pp. 62–3, and P. Duhem, *Le système du monde*, vol. 4, pp. 6–10.

[24] I am following closely E. Grant's description in *Planets, Stars, & Orbs*, pp. 279–80.

Figure 17. Roger Bacon's model for the moon. From P. Duhem, Un fragment inédit de l'Opus tertium de Roger Bacon, *p. 129.*

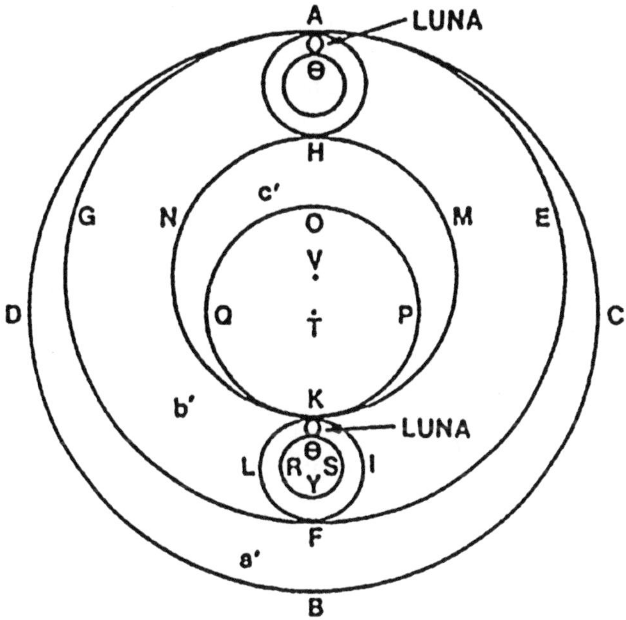

and HNKM, which enclose the lunar eccentric deferent (b'). Surrounding the eccentric deferent is the outermost orb (a') lying between surfaces ABDC and AGFE, whereas the innermost orb (c') lies between convex HNKM and concave OQKP. Between the surfaces of the middle or eccentric orb (b') is a concavity that contains a spherical epicycle. The latter may be understood in two ways: firstly, as a solid globe – Bacon calls it "convex sphere" (*spericum convexum*) – which resembles a ball (*pila*) because it lacks a concave surface; and secondly, as a ring with two surfaces – one convex (KLFI) and the other concave (RYSΘ) – whose central core belongs exclusively to the eccentric orb (b') and does not form part of the epicyclic sphere itself. The moon, which has only a convex surface and is therefore a solid body, is located in the concavity of the epicyclic sphere. The eccentric sphere moves around centre V, carrying the epicycle with it. The epicyclic sphere, however, also has its own simultaneous motion with which it carries the moon.

The basic characteristics of thirteenth-century astronomy can also be traced in the fourteenth and fifteenth centuries, with some smaller modifications and additions, such as the replacement of the Toledan Tables by the Latin version of the Alfonsine Tables, which constituted the canonical form of medieval mathematical astronomy until the Copernican reform.[25] After Ptolemaic mathematical astronomy prevailed in the thirteenth century, the position of the dominant system, accepted by all astronomers, was in the fourteenth century assigned to the *ymaginatio modernorum* described in Bacon's *Opus tertium*, which brought astronomy closer to natural philosophy with regard to planetary motion. The *Theorica planetarum* of about 1335 by John of Lignères (fl. first half of the fourteenth century) rendered a complete and highly systematic illustration of geometric and kinetic models of motion. These were the basis of not only the presentation of phenomena but also a model in which a physical interpretation of celestial motion agreed with Aristotelian natural philosophy, facilitating the construction of a model representing the celestial reality. According to John of Lignères, such a concept of the heavens complied with the Aristotelian assumptions about the immutability of celestial substance and impossibility of vacuum: "The aforementioned imagination saves all the apparent motions in the heavens with respect to the irregular motions of the sun, the moon and the five planets. It does not lead to condensation or rarefaction, nor does it cause any apparent motion of these bodies to leave behind a vacuum."[26] At the end of the thirteenth century Pierre d'Ailly provided a splendid description of this *imaginatio* in his *14 Questiones on the Sphere of Sacrobosco*, albeit using slightly different terms than Bacon.[27]

2. The Vienna school of astronomy, renewal and criticism of Ptolemy's *Almagest* and homocentric astronomy

Although the "three-orb compromise" between Ptolemaic astronomy and the requirements of natural philosophy eliminated several philosophical objections and was approved by several philosophers and astronomers in the fourteenth century, it was fully established only in the second half of the fifteenth century with Georg Peurbach's *Theoricae novae planetarum*. This text gradually

---

25 See H. Hugonnard-Roche, "Problèmes méthodologiques dans l'astronomie au début du XIVe siècle," p. 55. On the Alfonsine Tables, see E. Poulle, "The Alfonsine Tables and Alfonso X of Castille"; J. Chabás, "The Diffusion of the Alfonsine Tables: the Case of *Tabulae resolutae*"; and especially B. R. Goldstein and J. Chabás, *The Alfonsine Tables of Toledo*.

26 Translated from the Latin text quoted in H. Hugonnard-Roche, "Problèmes méthodologiques dans l'astronomie au début du XIVe siècle," p. 64, n. 21.

27 For more detailed exposition, see E. Grant, *Planets, Stars & Orbs*, pp. 284–6.

replaced the traditional *Theoricae planetarum (antiquae)* as a more or less standard university textbook.[28]

But Peurbach's interests were not wholly confined to astronomy. When he arrived at Vienna in 1446, its university had already been exposed to humanism, probably through the influence of Enea Silvio Piccolomini (1405–1464), then Chancellor to Emperor Frederick III (1415–1493).[29] In the 1450s several masters, including Peurbach, lectured on Vergil, Juvenal, Terence, Horace, Cicero and Lucan. Peurbach lectured from 1454 to his death in April 1561 on Latin poets (Vergil, Juvenal, Horace, pseudo-Cicero) and copied Latin poems. In 1458 he even published *Positio sive Determinatio de Arte Oratoria sive Poetica*. Peurbach only occasionally lectured on topics related to astronomy: in 1458 on sundials and four years earlier in 1454 at the Bürgerschule on planetary astronomy. These lectures became his most famous and influential work, *Theoricae novae planetarum*. Its main characteristic was that celestial orbs or spherical shells were imagined as real bodies rather than as mere illustrations used for educational purposes. This is clear not only from his models and expositions but also from the *incipit* of the manuscript preserved in Vienna, which does not appear in the printed version, but which reads as follows: "A new *theorica* presenting the *real* nature of the spheres and of motion with the vocabulary of the [astronomica] tables (*Theorica nova* realem *sperarum habitudinem atque motum cum terminis tabularum declarans*)."[30]

Peurbach, for instance, described the sun as follows (see *Figure 18*):

> The sun has three orbs, separated from one another on all sides and also contiguous to one another. The highest of them is concentric with the world on its convex surface, but is eccentric on its concave surface. The lowest, on the other hand, is concentric on its concave but eccentric on its convex surface. The third, however, situated in the middle of these, is eccentric to the world

---

[28] According to P. Barker and B. R. Goldstein, "Realism and Instrumentalism in Sixteenth-Century Astronomy: A Reappraisal," pp. 237–9, there were at least three reasons for the gradual replacement of circles by orbs: (1) educational and cognitive efficacy of such models, (2) concurrence of physical and astronomical curricula, (3) calculations of interplanetary distances.

[29] On Vienna curriculum in mathematics and humanism, see J. S. Byrne, "A Humanist History of Mathematics? Regiomontanus's Padua Oration in Context," pp. 42–4, and M. Shank, "The Classical Scientific Tradition in Vienna," pp. 119–26.

[30] Vienna, Codex 5203, fol. 2^r^. Latin text quoted from E. J. Aiton, "Peurbach's *Theoricae novae planetarum*: A Translation with Commentary," p. 8, n. 14. Emphasis added.

> on both its convex surface and its concave surface. Now an orb whose centre is the centre of the world is said to be concentric to the world, and an orb whose centre is other than the centre of the world to be eccentric. Therefore the first two are eccentric relatively (*secundum quid*), and they are called the deferent orbs of the apogee of the sun (*orbes augem solis deferentes*). The apogee of the sun varies according to their motion. The third is eccentric absolutely (*eccentricus simpliciter*) and is called the deferent orb of the sun (*orbis solem deferens*). The body of the sun is attached to it and moves indeed according to its motion. These three orbs take two centres. For the convex surface of the highest and the concave of the lowest have the same centre, centre of the world. From that fact, the whole sphere (*tota sphaera*) of the sun, just as the whole sphere of any other planet, is said to be concentric with the world. But the concave surface of the highest and the convex of the lowest, together with the surfaces of each side of the middle orb, share another centre, which is called the centre of the eccentric.[31]

Peurbach employed a standard terminology, with the exception of the term *orbis*, which he used exclusively in reference to partial orbs or spherical shells of a planetary model, and the term *sphaera*, which in his case denotes the entire model or a celestial region occupied by it. An *orbis* is therefore an orb or a spherical shell with two surfaces; the sun has three such orbs or spherical shells (*orbes*), where the outermost and the innermost orbs vary in thickness, while the middle (the eccentric orb or the sun's deferent enclosing the sun between its two surfaces) does not. The whole sphere (*tota sphera*) of the sun is concentric with the world, because the convex surface of the outermost orb and the concave surface of the innermost orb have the same centre, which is also the centre of the world. Therefore, the "total orb" or spherical shell, too, is of an even thickness and represents a planetary sphere as imagined by Aristotle.

As stated above, several early modern authors held that Peurbach's *Theoricae novae* was written in response to the Averroist criticism of the realistically understood Ptolemaic astronomy of eccentrics and epicycles.[32] According to Erasmus Reinhold, for instance, Peurbach enclosed the sun's eccentric orb within two spheres of uneven thicknesses for physical reasons already discussed by Bacon's *moderni*. In this way, he avoided the existence of a vacuum and the division of celestial substance: "The eccentric orb having been set up, they [astronomers]

---

31 *Theoricae novae* (Aiton, pp. 9–10).

32 See supra, p. 243, n. 3.

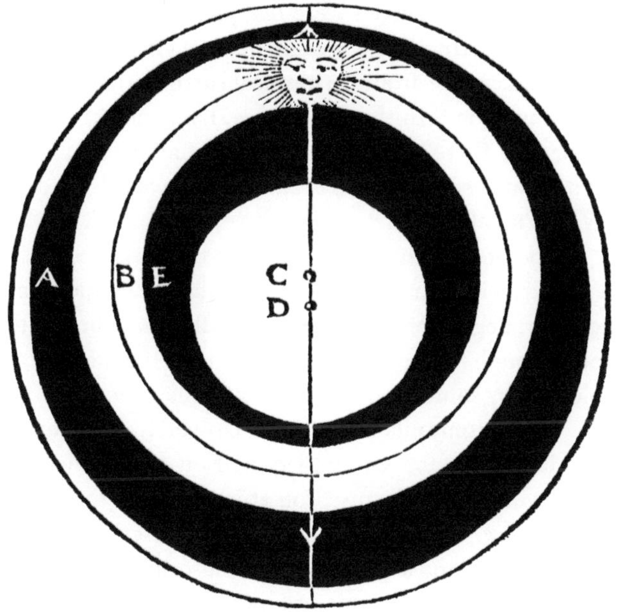

Figure 18. Peurbach's model for the sun. Theoricae novae planetarum *(Paris, 1534). Image courtesy of the History of Science Collections, University of Oklahoma Libraries.*

then gather together physical reasons whereby they attach to it two other orbs of unequal thickness – one above, the other below – so that the total sphere is made concentric with the centre of the world, lest it be necessary to assume vacuum or that celestial bodies are mutually torn apart."[33]

It seems very likely that Peurbach learned about the Averroist critique of Ptolemaic astronomy directly through al-Bitrūjī's *De motibus celorum*, available in Vienna no sooner than the 1450s.[34] The other Averroist critique of Ptolemaic astronomy accessible to Peurbach was Henry of Langenstein's *De reprobatione ecentricorum et epiciclorum*, which was published in Paris in 1364.[35] The treatise refuted epicycles and eccentrics on the basis of two general assumptions: if

---

33 *Theoricae G. Purbachii Germani ab Erasmo Reinholdo [...] illustratae scholiis*, Praefatio, fol. 3^{v}. Translation of R. S. Westman, *The Copernican Question*, p. 33.

34 Regiomontanus owned one copy and annotated it. See E. Zinner, *Leben und Wirken des Joh. Müller von Königsberg, genannt Regiomontanus*, pp. 61–2; M. Shank, "Regiomontanus and Homocentric Astronomy," p. 158.

35 On this, see the articles of C. Kren., "A Medieval Objection to 'Ptolemy'" and "Homocentric Astronomy in the Latin West: The *De reprobatione ecentricorum et epiciclorum* of Henry of Hesse."

epicycles and eccentrics were real physical spheres, they violated a number of Aristotelian principles, for example, that circular motion as defined by Aristotle should take place around a physical earth or that every simple body had a single simple motion.[36] In the early 1380s, Langenstein brought his work from Paris to Vienna. There he became an influential member of the university and his *De reprobatione ecentricorum et epiciclorum* was also read in the following century for its critical section.[37]

One of Langenstein's objections to eccentrics and epicycles was also often repeated by Peurbach's student and colleague, Regiomontanus. Unlike his mentor, however, Regiomontanus saw, as I shall explain below, the Averroist critique as a good enough reason to try to replace the Ptolemaic astronomy of eccentrics and epicycles by some kind of homocentric astronomy.[38]

Regiomontanus entered the University of Vienna in 1450, took his master's degree in 1457–58, and lectured on *perspectiva communis* and Euclid. He is believed to have lectured on at least one of the classical authors, Vergil and his *Bucolica*. Regiomontanus was a manuscript hunter, copyist, and reader of mathematical and astronomical manuscripts. Together with Peurbach, he made astronomical observations and intensely studied mathematical sciences (Euclid, Jordanus, Archimedes, Apollonius, al-Khwārzimī, Henry of Langenstein and

---

36 Unfortunately, the aforementioned articles of C. Kren do not cite all of Henry's objections to the Ptolemaic astronomy.

37 In Vienna he taught theology until his death in 1397 and became one of the leading figures of conciliarism. For more on the Viennese scientific tradition, see M. H. Shank, "The Classical Scientific Tradition in Vienna." M. Shank, "Mechanical Thinking in European Astronomy (13th–15th Centuries)," p. 19, suggests that Peurbach's three-orb compromise might be somehow inspired by Guido De Marchia's *Tractatus super planitorbium*, which surfaced in Vienna and its surroundings in the fifteenth century. De Marchia himself proposed the elimination of the inner and outer orbs of the "three-orb system" of each planet.

38 For more on Regiomontanus and his interest in homocentric astronomy, see M. H. Shank, "Regiomontanus on Ptolemy, Physical Orbs, and Astronomical Fictionalism: Goldsteinian Themes in the *Defense of Theon against George of Trebizond*"; N. Swerdlow, "Regiomontanus's Concentric-Sphere Models for the Sun and Moon." See also, M. Folkerts, M., "Regiomontanus als Mathematiker"; "Regiomontanus's Role in the Transmission and Transformation of Greek Mathematics"; "Regiomontanus's Role in the Transmission of Mathematical Problems"; C. Kren, "Planetary Latitudes, the *Theorica Gerardi*, and Regiomontanus"; F. J. Ragep, "Alī Qushjī and Regiomontanus: Eccentric Transformations and Copernican Revolutions."

others).[39] The course of their work was substantially altered by the arrival of Cardinal Bessarion in Vienna in May 1460.[40] Bessarion, a promoter of the Greek cultural heritage of united mathematical and philosophical sciences, and of Plato and Aristotle, was keen to see Ptolemy's *Almagest* properly translated into Latin. He was deeply disappointed with the recent translation of the *Almagest* from Greek by George of Trebizond (or Georgius Trapezuntius; 1395–1484) and his accompanying commentary.[41]

George of Trebizond was a Greek scholar commissioned by Pope Nicholas V (1397–1455) to make the Greek and patristic heritage available in Latin. He translated the works of several Greek Fathers including, for example, an incomplete translation of Eusebius of Caesarea's *De preparatione evangelica* (published in Venice in 1470),[42] Aristotle's principal works on natural philosophy, Demosthenes, Plato's *Laws* and *Epinomis*, and Ptolemy's *Almagest*. This later piece, accompanied by his commentary, began to circulate as a complete manuscript from 1451 onwards, but

---

39 On Peurbach and Regiomontanus, see P. L. Rose, *The Italian Renaissance of Mathematics*, pp. 90–117. For Regiomontanus, see also E. Zinner, *Leben und Wirken des Joh. Müller von Königsberg, genannt Regiomontanus*; G. Hamann, ed., *Regiomontanus-Studien*, and other articles, mostly by Shank and Swerdlow, cited infra.

40 On the University of Vienna, the intellectual and institutional home of Peurbach and Regiomontanus, and on Bessarion's role in the composition of the *Epytoma Almagesti*, see M. H. Shank, "The Classical Scientific Tradition in Fifteenth-Century Vienna." On Bessarion's search for manuscripts and his library, see S. Gentile, "Il ritorno delle culture classiche," pp. 83–5; On Bessarion, see J. Hankins, *Plato in the Italian Renaissance*, pp. 217–63; P. L. Rose, *The Italian Renaissance of Mathematics*, pp. 44–6. Cardinal Bessarion owned an immense library that held over 1000 volumes of Greek manuscripts and was as such instrumental in the transmission of Greek texts to the West. After his death it was bequeathed to the Venetian Republic.

41 On George of Trebizond, see J. Monfasani, *George of Trebizond: A Biography and a Study of His Rhetoric and Logic* and *Collectanea Trapezuntiana: Texts, Documents and Bibliographies of George of Trebizond*. See especially his "A Tale of Two Books: Bessarion's *In Calumniatorem Platonis* and George of Trebizond's *Comparatio Philosophorum Platonis et Aristotelis*." See also J. Hankins, *Plato in the Italian Renaissance*, pp. 165–92; C. Lohr, "Metaphysics," p. 561; C. S. Clenza, "The Revival of Platonic Philosophy," pp. 77–81. On George of Trebizond as a translator and on translations in Renaissance in general, see, for example, B. P. Copenhaver, "Translation, Terminology, and Style in Philosophical Discourse."

42 On Copernicus' possible acquaintance with the manuscript of *Preparatio evangelica* (but not in George of Trebizond's translation which was not complete; it ends with Book XIV), see A. De Pace, *Niccolò Copernico e la fondazione del cosmo eliocentrico*, pp. 25–7, and pp. 330–1, n. 93, 95, 97–9.

was unpublished until 1528. The controversy between George of Trebizond and Bessarion arose from their evaluation of the philosophies of Aristotle and Plato. In 1455 George of Trebizond attacked Plato and argued for Aristotle's superiority in *Comparatio phylosophorum Platonis et Aristotelis*. Bessarion intervened with a defense of Plato in his *In Calumnitatorem Platonis*, which appeared in 1569.

Upon his arrival in Vienna, Bessarion urged Peurbach to revise and comment on Ptolemy's *Almagest*. Since Peurbach did not know Greek, he began with the twelfth-century Arabic-Latin translation by Gerard of Cremona and the so-called *Almagesti minoris libri VI* (or *Almagestum parvum*), an unfinished book from the late thirteenth century that supplemented Ptolemy with new information derived from al-Battānī, Thābit ibn Qurra, Jābir ibn Aflah, al-Zarqāli, and the *Toledan Tables*. But Peurbach finished only the first six books before his death in April 1461. Regiomontanus took over the task and in late September 1461 he set out with Bessarion for Italy, where he visited astronomer Giovanni Bianchini (ca. 1410–1469) and lectured at the universities of Ferrara and Padua. By the end of 1462 the text of the *Epytoma Almagesti* was finished, but it did not make its way into print until 1496, when it appeared in Venice.

The first six books – originally written by Peurbach and later partly reworked by Regiomontanus with the help of Bessarion –[43] closely followed the *Almagestum minor* in its manner of exposition: the division of Ptolemy's chapters into propositions, showing the proof of a geometrical theorem, the derivation of a parameter, or the implementation of the process. The *Epytoma* gave astronomers an understanding of Ptolemy that they had not previously been able to achieve and was the true rediscovery of the ancient mathematical astronomy in the Renaissance.

Despite his extensive work on Ptolemy's *Almagest*, Regiomontanus revealed in a 1460 letter to János Vitéz that he was interested in the renewal of homocentric astronomy in addition to his restoration of Ptolemaic astronomy. Writing to Vitéz he briefly described his homocentric models for the sun and the moon, and promised that he would one day refute Ptolemy's "old theory of eccentrics and epicycles" and generate new tables.[44] While in Bessarion's service in Italy, Regiomontanus learned Greek, made astronomical observations, and also

---

[43] See A. Rigo, "Bessarione, Giovanni Regiomontano e i loro studi su Tolomeo a Venezia e Roma (1462–1464)."

[44] For his homocentric models for the sun and the moon, see N. Swerdlow, "Regiomontanus's Concentric-Sphere Models for the Sun and Moon." For his homocentric program described in a letter to Vitéz, see M. H. Shank, "Regiomontanus and Homocentric Astronomy," p. 158 and p. 164, n. 4.

composed works on trigonometry and the armillary sphere. He remained committed to his goal of restoring homocentric astronomy even after he completed the *Epytoma Almagesti* in 1462. In early 1464 he lectured on the Islamic scholar al-Farghānī at the University of Padua and in his inaugural address of which only *Oratio* remains, he renewed his vow.[45] In a letter to Bianchini of February 11, 1464, Regiomontanus repeated Henry of Hesse's "moon argument" – "If the moon had an eccentric and epicycle in the manner that is often alleged, it would be necessary for the moon in one position to be four times larger than it is at another […]" – and analyzed all the ways in which current astronomical theory disagreed with the observed phenomena. He reiterated his hope for a collaborative effort to restore the discipline.[46] Regiomontanus and Bianchini very probably met in Ferrara. Regiomonanus left at least two Ferrarese students in mathematics, Pellegrino Prisciani (1435–1510) and Domenico Maria Novara (1454–1504). The same critical attitude towards Ptolemaic astronomy is present also in his *Disputationes*, written after August 1464, in which he criticized the earlier *Theoricae planetarum antiquae*, also because it claimed that the moon's eccentric moved eastwards, while its centre moved westwards.[47] He continuously stressed that these hypotheses were philosophically unacceptable and contrary to nature.

During the same period, in the 1460s, Regiomontanus undertook his second major astronomical project: writing of *Defensio Theonis contra Georgium Trapezuntium*, a controversial attack on George of Trebizond's *Commentary on the Almagest*, with homocentric aspirations. The work was, at one point, offered to King Matthias Corvinus of Hungary (1443–1490), for whom Regiomontanus

---

45 On this lecture, see E. Zinner, *Leben und Wirken des Joh. Müller von Königsberg*, pp. 110–8; N. Swerdlow, "Science and Humanism in the Renaissance: Regiomontanus's Oration on the Dignity and Utility of the Mathematical Sciences," and J. S. Byrne, "A Humanist History of Mathematics? Regiomontanus's Padua Oration in Context."

46 A similar change in magnitude should have also been observed in Mars, Venus and Mercury. Regiomontanus copied Henry's *De reprobatione ecentricorum et epicyclorum*: "Iam venio ad rationes contra illam antiquam fantasiam eccentricorum, que si esset vera secundum approbatam sententiam ponentium eos, sequitur lunam in pleni lumine non plus quam sesquialtero esse remotiorem quam in quadraturis, id est quando est dimidiata in lumine, quod est falsum; quia sic in quadraturis deberet apparere notabiliter maioris diametri quam in pleni lumine." Quoted from M. H. Shank, "Regiomontanus and Homocentric Astronomy," p. 165, n. 14. The same argument is also mentioned in the *Epytoma Almagesti*. See also N. Swerdlow, "Regiomontanus on the Critical Problems of Astronomy."

47 Regiomontanus, again, based this criticism on Langenstein's *De reprobatione ecentricorum et epicyclorum*.

drafted a dedication and to whose court he also moved ca. 1467. The last books were finished in Nuremberg, where he moved in 1471, with the intention of implementing the observational reform of astronomy and establishing his own press to publish accurate editions of new and ancient texts. The program of 1474 lists forty-seven items, but only nine works were published, among them Peurbach's *Theoricae novae*.[48] In 1476, before much of his planned work could be completed, Regiomontanus died during his sojourn in Rome.

One of the books Regiomontanus never came to publish was his *Defensio Theonis*, written in the 1460s and 1470s, which, however, circulated in incomplete manuscript form among his and Bessarion's friends.[49] The *Defense of Theon against George of Trebizond* offers an account of Regiomontanus' thoughts during the time before his death. The essential feature of this work is his concern with philosophical (physical and cosmological) questions that ought to be addressed by every astronomer. He deemed it the duty of every astronomer to discuss not so much the plane circles of planets (*non tam de circulis planetarum superficialibus*) as their "corporeal orbs (*de orbibus corporeis*), namely, what their shapes, their order, and their magnitude are, about which poles they rotate, and other 'accidents' of this sort."[50] In his opinion, this duty arose from Ptolemy's failure to do just that – the very task he himself had set before astronomy. The other important information emanating from *Defensio Theonis* is that Regiomontanus was apparently familiar neither with Ptolemy's *Planetary Hypotheses* nor with any other medieval astronomical tradition of three-dimensional orbs, and considered the development of such theories as a fairly recent invention – he must have been ascribing this achievement to Peurbach's *Theoricae novae*.

In Regiomontanus' view – and Copernicus would have readily concurred – it was the duty of every astronomer to preserve the following two things in the celestial motions: the "primordial and intrinsic equality [or uniformity] (*equalitas videlicet primordialis et intrinseca*)" on the one hand and the "apparent inequality [or non-uniformity] (*inequalitas apparentia*)" on the other:

---

48 On this, see A. Wingen-Trennhaus, "Regiomontanus als Frühdrucker in Nürnberg."

49 *Defensio Theonis* has been preserved only in one 573-page autograph in St. Petersburg, Archive of the Russian Academy of Sciences. Richard L. Kremer and Michael H. Shank prepared a preliminary digital edition of the text available at http://regio.dartmouth.edu/index.html.

50 Quoted from M. H. Shank, "Regiomontanus on Ptolemy, Physical Orbs, and Astronomical Fictionalism: Goldsteinian Themes in the *Defense of Theon against George of Trebizond*," p. 192, n. 17.

> [t]he first is brought about by orbs, not by thin circles; the second by circles describable on planes, by the power of demonstration. The first pertains to the nature of celestial bodies, which can tolerate no motion other than the uniform; the second pertains to human observers, to whom these motions seem unequal [or nonuniform] and disorderly.[51]

According to Regiomontanus, Ptolemy completely neglected the first task, and did his utmost with the second:

> The first, if I may say so, Ptolemy completely neglected; the second, he pursued to the utmost, expressing the quantity of motions with numbers by means of foregoing demonstrations, their quality having been ignored, even though in book 3 [of the *Almagest*], concerning the sun, he adopts this almost as a principle, that the celestial motions are equal [or uniform] and regular [or ordinary], etc.[52]

Regiomontanus thus set before every astronomer the following two tasks: to save the "primordial and intrinsic equality or [uniformity]" of the celestial motions on the one hand and the "apparent inequality [or non-uniformity]" on the other. He treated the equality, uniformity or regularity of the celestial motions not as a phenomenon but as a physical principle which is – contrary to what can be seen in the heavens – connected to the nature of the celestial bodies. Hence, every astronomer ought to accomplish the first task in compliance with philosophical principles, that is, by applying some combination of uniformly moving orbs. As for the second task, preserving the apparent inequality of the celestial motions, Regiomontanus found circles to be an adequate solution. Ptolemy had done an excellent job in accomplishing this task in the *Almagest*; however, he had failed to fulfill the first one – to preserve the "primordial and intrinsic equality," even though he stated it as a principle. It was only the combination of these two tasks that could lead to a complete solution. As can be inferred from Regiomontanus' aspirations expressed in his works preceding the *Defensio*, the ideal solution for the first one was the homocentric system, by ensuring motions appropriate to the celestial bodies. The most appropriate system to save the anomalies was Ptolemy's *Almagest* and its circles.

Regiomontanus set the two approaches in a temporal framework and treated them as two stages in the history of astronomy that had been merged together by

---

51 Ibid., p. 194, n. 19.

52 Ibid.

recent developments. The older astronomical approach was two-dimensional and involved demonstrations in plane geometry that led to arithmetic calculations and facilitated predictions. The representatives of this approach used eccentric circles. According to Regiomontanus, this group was composed of astronomers from Hipparchus and Ptolemy to the Arabs. They all spoke of circles and used them for the sole purpose of saving the anomalies or deviations of the celestial motions by suitable mathematical techniques. Or, more precisely, they maintained that these anomalies might be numerically predicted. Regiomontanus found an indication of this in that "the princes of this art, and above all Ptolemy," immediately moved from linear demonstrations to computations, while "saying absolutely nothing about the poles or axes of the movements."[53] The fundamental shortcoming of this approach was its disregard for the physical aspect of astronomy.

The later astronomical approach was three-dimensional: it described corporeal orbs and concerned itself with their substance as well as such problems as *vacua*, collisions and penetrations of bodies. In contrast to the former approach, this method took into consideration the corporeal dimensions of the celestial bodies:

> The later [astronomers], penetrating most cleverly and intimately into the nature of celestial bodies, thought that carriers of the stars were not merely mathematical circles lacking a third dimension, but globular bodies that transport the stars by some multifarious and wonderful conveyance. Thus whereas the authority of Ptolemy had allotted the sun a single eccentric, they understood it to be not of a skinny circle, but an orb of equal thickness in every direction surrounding the entire elementary region, together with the part of celestial space below it [the sun]. They thought that the sun was fixed in this orb, enclosed by two spherical concentric surfaces, so that the sun itself was not wandering, as it were, through the aethereal region, but revolved about its centre [i. e. the centre of the orb] with some uniform speed under the constrained motion of the aforementioned orb that contained it. Now since this orb recedes from the middle of the universe, its parts necessarily will move away unequally from the centre of the universe, and therefore this orb in its motion will run into some celestial body placed near it, and will do violence to it by cleavage or compression (*per scissionem aut condensationem*); it will also leave a vacuum behind itself. Therefore they thought they should fill in, on either side, the space that was deficient in concentricity by means of two other orbs placed near the deferent of the

[53] Ibid., p. 195, n. 21.

> sun, orbs of uneven thickness, so that the entire congeries of the three aforementioned orbs would have a common centre with the world. So much for a cursory account of the sun.[54]

The contrast drawn by Regiomontanus between the two astronomical approaches, cast as historical periods, could not have been clearer. While ancient astronomers derived their predictions from two-dimensional demonstrations without referring to "poles or axes," later astronomers dealt with the "nature of celestial bodies," i.e. three-dimensional orbs possessing physical properties, including the ability to carry planets. Particularly interesting in this segment of the text, as Shank points out, was Regiomontanus' emphasis on the constraints that regulated the sun in the tradition of the later astronomy, which was in contrast to its hypothetical "wandering through the aethereal region" according to the tradition of the ancient astronomy: "His [i.e. Regiomontanus'] implicit criticism here is that the ancients', two-dimensional approach left the motion of the sun in causal limbo. The language of deferent 'carriers' (*delatores*) implicitly builds into the circles a causal expectation."[55] On the other hand, the three-dimensional approach, which relied on orbs, responded to the question what constrained the sun to move as it did. Regiomontanus further emphasized the physical component of the more "recent astronomers" by referring to physical entities (the "elementary region") that the sun's orb enclosed, as well as to the problems of vacua and condensation. He obviously examined the natural philosophical consequences of Ptolemy's two-dimensional model by converting it into a three-dimensional one. In his opinion, the sun's single eccentric orb, if understood as a real physical body, implied unacceptable physical consequences, such as expansion, division and condensation of the surrounding bodies (if the surroundings gave way) or collision and vacuum (if they did not). Therefore, to eliminate these philosophical problems, it was necessary to enclose the sun's eccentric orb within another two orbs of uneven thickness.

Regiomontanus did not identify this development as a compromise with Aristotle but rather as the progress of the Ptolemaic tradition itself. Even though he nowhere referred to the recent authors by their names, his accounts were

---

[54] Ibid., p. 197, n. 23. This is another confirmation that Peurbach devised – or revived – the three-orb system to answer the traditional Aristotelian objections to the realistically understood Ptolemaic astronomy.

[55] M. H. Shank, "Regiomontanus on Ptolemy, Physical Orbs, and Astronomical Fictionalism: Goldsteinian Themes in the *Defense of Theon against George of Trebizond*," p. 197.

very much similar to those found in the works of Bacon, d'Ailly, and above all in Peurbach's *Theoricae novae*.

Nonetheless, the "three-orb compromise," even in Peurbach's version, fell short of accommodating *all* demands of Aristotelian natural philosophy. The fact that it accepted orbs as real or that it was designed as a physical system rather than a strictly geometric instrument to "save the phenomena" continually raised new questions concerning its philosophical consistency.[56] The *ymaginatio modernorum*, indeed, assigned the earth a central position in the universe and eliminated the vacuum, which would occur if the eccentric deferent moved in celestial substance without being embedded between the outermost and innermost orbs. However, it flatly ignored certain objections to the Ptolemaic astronomy. Therefore, the eccentric celestial spheres in the *ymaginatio modernorum* moved in circular motion around centres not coinciding with the earth. The motion of the planets on epicycles should have led to a change in their observed magnitude. Orbs, carrying the epicycles, were not in direct contact; therefore the space between them was filled with celestial substance, which is neither divisible nor corruptible.[57]

Regiomontanus' homocentric attempts and aspirations may be considered in the following context: the "Aristotelian-Ptolemaic" three-orb compromise did not eliminate the disagreements between astronomy and natural philosophy. In his view, the problem of the Ptolemaic astronomy was a philosophical rather than astronomical one. Therefore, for instance, he maintained that Langenstein's "moon argument" was a problem to be discussed by those who wanted to philosophize (*philosophari*).[58] Regiomontanus' interest in the Ptolemaic mathematical astronomy presented in his most famous book, the *Epytoma Almagesti,* therefore also has its other side. Regiomontanus firmly believed that astronomy and philosophy belonged to one and the same realm. It was every astronomer's duty to incorporate physical considerations in constructing astronomy;[59] he considered astronomy as an enterprise that should integrate physical considerations and the task of producing numerical predictions. Nevertheless, the improvements in the

---

56 On this, see E. Grant, "Cosmology," p. 283.

57 On other cosmological problems raised by real eccentrics and epicycles, see also E. Grant, *Planets, Stars & Orbs*, pp. 286–308.

58 See M. H. Shank, "Regiomontanus and Homocentric Astronomy," p. 165, n. 17: "Talibus rebus sepenumero vexor et deflere cogor segnitiem et frigiditatem nostre etatis. Profecto materia copiosa est volentibus hodie *philosophari*. Habemus ante oculos vestigia maiorum nostrorum, quo fit ut cautius incedere possimus, modo ingenium huic rei accomodemus." Emphasis added.

59 See M. H. Shank, "Regiomontanus and Homocentric Astronomy," p. 162.

Ptolemaic tradition of physical astronomy made by his teacher Peurbach left him largely dissatisfied. In his opinion, the new astronomy, too, had fallen short of the ideal, because even in its philosophical form it continued violating a number of Aristotelian physical principles. Therefore, homocentric astronomy remained an ideal for Regiomontanus which he never fully attained. "It seems likely," states Shank, "that he went to his grave hoping for a homocentric solution to the inconsistencies he saw between Ptolemy's successful predictions of planetary positions and his neglect – as Regiomontanus perceived it – of physical considerations."[60]

60 M. H. Shank, "Regiomontanus on Ptolemy, Physical Orbs, and Astronomical Fictionalism: Goldsteinian Themes in the 'Defense of Theon against George of Trebizond'," p. 188.

CHAPTER TWELVE

How did Copernicus become a Copernican?

One of the most important questions which is still not completely satisfactorily illuminated is how Copernicus arrived at heliocentrism. What question was he trying to answer? Which elements, which factors shaped his decision to set the earth in motion and put it on a trip around the stationary sun in the middle of the universe? How did he become a Copernican? This is especially intriguing because in the period of his education and after his return to Warmia most of the fundamental characteristics of the astronomical landscape remained exactly the same as they were in the period of Peurbach and Regiomontanus. Copernicus learned astronomy from their works, from commentaries on their works, and from the familiar Averroist, homocentric criticism of Ptolemaic astronomy. What happened, then, what changed to incite and inspire him to write the *Commentariolus*?

In the *Commentariolus* Copernicus dismisses the concentric astronomy of Callippus and Eudoxus on the basis of its incapability to reproduce the varying distances of planets from the observer. Then he presents heliocentric astronomy as a consequence of his dissatisfaction with the Ptolemaic use of the equant, which contradicts the principle of regularity in explaining the apparent motion of the planets. A heavenly body, a perfect orb, could only move uniformly.

> I understand that our predecessors assumed a large number of celestial orbs principally in order to account for the apparent motion of the planets

> through regularity, for it seemed highly unreasonable that a heavenly body, which is absolutely round, should not always move uniformly. They had discovered that by the arrangement and combination of uniform motions in different ways it could be brought about that any body would appear to move to any position. // Callippus and Eudoxus, attempting to carry this out by means of concentric circles, could not by the use of these give an account of everything in the planetary motion, that is, not only those motions that appear in connection with the revolutions of the planets, but also that the planets appear to us at times to ascend and at times to descend in altitude, which concentric circles in no way permit. And for this reason a preferable theory, in which the majority of experts finally concurred, seemed to be that it is done by means of eccentrics and epicycles. // Nevertheless, the theories concerning these matters that have been put forth far and wide by Ptolemy and most others, although they correspond numerically [with the apparent motions], also seemed quite doubtful, for these theories were inadequate unless they also envisioned certain equant circles, on account of which it appeared that the planet never moves with uniform velocity either in its deferent orb or with respect to its proper centre. Therefore a theory of this kind seemed neither perfect enough nor sufficiently in accordance with reason *(Quapropter non satis absoluta videbatur huiusmodi speculatio, neque rationi satis concinna)*.[61]

It seems, at first sight, that in the following paragraphs Copernicus connects the problem of the equant directly to heliocentrism:

> Therefore, when I noticed these [difficulties], I often pondered whether perhaps a more reasonable model composed of circles could be found from which every apparent irregularity would follow while everything in itself moved uniformly, just as the principle of perfect motion requires. After I had attacked this exceedingly difficult and nearly insoluble problem, it at last occurred to me how it could be done with fewer and far more suitable devices than had formerly been put forth if some postulates, called axioms (*aliquae petitiones, quas axiomata vocant*), are granted to us, which follow in this order:
> First Postulate: There is no one centre of all the celestial orbs or spheres.
> Second Postulate: The centre of the earth is not the centre of the universe,

[61] *Commentariolus* (Swerdlow, pp. 432–4; Zekl, pp. 2–4).

but only the centre of gravity [i. e. centre towards which heavy things move] and the centre of the lunar orb.
Third Postulate: All orbs surround the sun as though it were in the middle of all of them, and therefore the centre of the universe is near the sun.
Fourth Postulate: The ratio of the distance between the sun and earth to the height of the firmament [i.e. the sphere of the fixed stars] is so much smaller than the ratio of the semidiameter of the earth to the distance of the sun that the distance between the sun and earth is imperceptible compared to the great height of the firmament [i.e. the sphere of the fixed stars].
Fifth Postulate: Whatever motion appears in the firmament [i. e. sphere of the fixed stars] belongs not to it but to the earth. Thus the entire earth along with the nearby elements rotates with a daily motion on its fixed poles while the firmament [i.e. the sphere of the fixed stars] remains immovable and the outermost heaven.
Sixth Postulate: Whatever motions appear to us to belong to the sun are not due to [motion] of the sun but [to the motion] of the earth and our orb with which we revolve around the sun just as any other planet. And thus the earth is carried by more than one motion.
Seventh Postulate: The retrograde and direct motion that appears in the planets belongs not to them but to the [motion] of the earth. Thus, the motion of the earth by itself accounts for a considerable number of apparently irregular motions in the heavens.
Now that these postulates have been set down, I shall attempt briefly to show how carefully the uniformity of the motions may be preserved. I have decided, however, for the sake of brevity to leave the mathematical demonstrations out of this treatise as they are intended for a larger book. [62]

---

[62] Copernicus' *petitiones quas axiomata vocant* have given some trouble to scholars. See E. Rosen, "Copernicus' Axioms," J. Dobrzycki, "Notes on Copernicus's Early Heliocentrism," and N. Swerdlow, "The Derivation and First Draft of Copernicus' Planetary Theory: A Translation of the *Commentariolus* with Commentary," pp. 437–8. What is Copernicus doing here? What is the nature of *axiomata*? They are not, as justly observed by Swerdlow, p. 437, "axioms since they are hardly self-evident." He argues that they take the place of the general description of the universe in the opening chapters of Ptolemy's *Almagest*, Regiomontanus' *Epytoma Almagesti*, and later of Copernicus' *De revolutionibus*, but he also maintains that "[w]hether Copernicus considers them absolutely true or merely working hypotheses is not made clear." While A. Goddu, *Copernicus and the Aristotelian Tradition*, pp. 251–6, makes a very intriguing suggestion linking their derivation with the method used by Plato in the *Parmenides*, I

> Nevertheless, the lengths of the semidiameters of the orbs will be set down here in the explanation of their circles, from which anyone not ignorant of mathematics will easily understand how very precisely such an arrangement of circles agrees with computations and observations. // In the same way, in case anyone believes that we have asserted the movement of the earth for no good reason along with the Pythagoreans, he will also receive considerable evidence [for this] in the explanation of the circles. And in fact, [the evidence] by which natural philosophers (*physiologi*) attempt so very hard to confirm the immobility of the earth (*stabilitatem eius*) depends for the most part upon appearances. All [their evidence] falls apart here in the first place, since we explain [or consider] it [i. e. the immobility of the earth] as being also due to an appearance.[63]

Copernicus' apparent connection of the problem of the equant with the movement of the earth around the sun in the *Commentariolus* led some scholars to propose interpretations of his transition from the equant to heliocentrism in more or less purely technical, astronomical terms.[64] Copernicus' statements

---

believe Copernicus' manner of exposition and the nature of *petitiones quas axiomata vocant* find their explanation in Proclus' *Commentary on Plato's Timaeus* II, 3 (Runia and Share, pp. 77–80). Proclus explains that Plato is not an empiricist: Plato will not start with experiences and then draw conclusion from them. Plato's *methodos* is hypothetical, or, rather, Plato uses the method of the hypothesis. He sets out fundamental *axiômata* and *hypotheseis* and draws conclusions from them. Proclus presents first a list of five *axiomata*, and then follows another list of seven *axiômata*. Describing Plato's "hypothetical method" Proclus does not refer to Plato's own description of hypothetical method but explicitly refers to the *methodos* used by geometers. They first postulate, define and name their key principles before proceeding to their demonstrations based on them. And he cites an example from *Euclid*. On the basis of the fundamental principles or hypotheses Plato's *Timaeus* then proceeds, in Proclus' reading of the text, to a number of 'demonstrations' (*apodeixeis*) based on them and required in order to solve the problems. See also Runia and Share's very concise summary of that issue on pp. 18–20. Copernicus' method in the *Commentariolus* is highly reminiscent of Proclus: he first establishes seven *petitiones quas axiomata vocant* and then promises to provide mathematical *demonstrationes* in a larger book.

63 *Commentariolus* (Swerdlow, pp. 435–9; Zekl, pp. 4–8). The last sentences are extremely difficult to translate. Consequently, each scholar translates it differently. For the explanation of my changes to Swerdlow's translation, see infra, pp. 384–5.

64 See N. Swerdlow, "The Derivation and First Draft of Copernicus' Planetary Theory: A Translation of the *Commentariolus* with Commentary," pp. 423–511, and M. Clutton-Brock, "Copernicus' Path to His Cosmology: An Attempted Reconstruction." In my opinion B. R. Goldstein, "Copernicus and the Origin of his Heliocentric System," who

in the *Commentariolus* are, indeed, slightly misleading insofar as they seem to imply that the problem of the equant *alone* could bring him to heliocentrism. On the other hand, a closer reading of the above quotation and other relevant passages reveals enough to conclude that this was not the case; it was not *only and exclusively* the problem of the equant that led Copernicus to heliocentrism and geokinetism. Upon a closer reading of the *Commentariolus*, one can see that Copernicus talked about "every apparent irregularity" that should be accounted for by "a more reasonable model composed of circles," and that he explained that "the motion of the earth by itself accounts for a considerable number of apparently irregular motions in the heavens," which means that he was concerned not only with the non-uniform motion implied in the concept of equant but also with other "apparent irregularities."

This was confirmed as well by another of Copernicus' elliptic but still illuminating explanations in Chapter 2 of Book V of *De revolutionibus*, where he first criticized the equant for permitting circular motion to be uniform with respect to an extraneous centre not its own – which is "something Cicero's Scipio would hardly dream of" – and then said that "such and similar things" gave him "the occasion to consider the motion of the earth and other ways of preserving uniform motion and the principles of the science, as well as of making the computation of the apparent nonuniform motion more enduring."[65] Not only the equant but also *similia*, or "similar things," provided him with an opportunity to consider the ways to "preserve uniform motion *and* the principles of the science." Therefore there must have been other characteristics of Ptolemaic astronomy, other than just the equant, that annoyed Copernicus and ultimately led him to heliocentrism.[66]

What, then, were these *simila* which bothered him? Or, if we look at the equation from the other side, which were the *principia artis* he was trying to preserve?

Let us look at the *similia* first. The seventh *petitio* of the *Commentariolus* reads: "The retrograde and direct motion that appears in the planets belongs not to them but to the [motion] of the earth. Thus, the motion of the earth by itself accounts for a considerable number of apparently irregular motions in the

stresses the importance of the distance–period relationship, and A. Goddu, "Reflections on the Origin of Copernicus's Cosmology," who adds some new elements and considerations to the question, are basically on the right track, but they do not tell the whole story. What follows is a more complete story.

65 *De revolutionibus* V, 2 (Rosen, p. 240; Gansiniec, p. 246).

66 A. Goddu, *Copernicus and the Aristotelian Tradition*, p. 41, is, in my opinion, absolutely right, claiming that the "rejection of the equant did not lead directly (and certainly not by itself) to a heliostatic cosmology."

heavens." This could provide us with the first provisory explanation for the connection of the equant with heliocentrism. Copernicus was troubled not only by the first anomaly (the equant) but also by the second anomaly, that is, the apparent progressive and retrograde motion of the planets. But on the other hand this could be just Copernicus' afterthought, coming to him when he already had the heliocentric order of the planets. There is at least one other apparent irregularity of the Ptolemaic astronomy which he mentions in passing in the *Commentariolus* and in *De revolutionibus*: the size of the lunar epicycle which predicted a doubling of the size of the moon in quadrature.[67] And one could also add to the list the fact that Venus and Mars have the largest epicycles of the planetary models.[68]

But, as it is obvious from the *Commentariolus* and the *De revolutionibus*, the biggest and the most important issue was the (dis)order of the spheres. In the *Commentariolus* he linked this question to the distance–period relationship and solved it as follows:

> The heavenly spheres surround each other in the following order: The highest is the immovable sphere of the fixed stars which encloses and fixes the positions of all [the others]. Under it is the sphere of Saturn which the sphere of Jupiter follows, then the sphere of Mars, under which is the sphere in which we are carried around, then the sphere of Venus, last the sphere of Mercury. The sphere of the moon revolves around the centre of the earth, and is carried with it like an epicycle. In this order one [planet] exceeds another in rapidity of revolution, according as they traverse the larger or smaller circles

---

67 The size of the lunar epicycle is mentioned already by Codro and Regiomontanus in the *Epytoma Almagesti* V, 22. See *Commentariolus* (Swerdlow, p. 461; Zekl, p. 18): "Those who hold the opinion that this is brought about by means of an eccentric circle, in addition to the improper non-uniform motion in the circle itself, have fallen into two obvious errors. For it follows by a mathematical proportion that when the moon in quadratures rests at the lowest part of the epicycle, it should appear about four times larger (if only the whole moon were illuminated) than the new and full moon, unless in addition an increase and decrease in the size of its body is rashly maintained." See also *De revolutionibus* IV, 2 (Rosen, p. 176; Gansiniec, p. 176): "Secondly, experience and our senses themselves show us that the lunar parallaxes are different from those indicated by the ratio of the circles. [...] Thus, the moon would generally look four times larger, on the supposition that it shone with its full disc, in quadrature, when nearest to the earth, than when in opposition to the [sun]. See also P. L. Rose, *The Italian Renaissance of Mathematics*, p. 126.

68 See A. Goddu, "Reflections on the Origin of Copernicus's Cosmology," p. 41. He refers to *De revolutionibus* I, 10.

(*eodem quoque ordine alius alium revolutionis velocitate superat; secundum quod maiora minorave circulorum spatia emetiuntur*). Thus, Saturn returns to the same position in the thirtieth year, Jupiter in the twelfth, Mars in the second, the earth in the annual revolution; Venus completes a revolution in the ninth month, Mercury in the third.[69]

If we look at these matters from the perspective of the *principia artis* – and take into account also what he says in his *Preface* to *De revolutionibus*, written in 1542, and *Book I of De revolutionibus* where these matters were more fully and articulately expressed – it is clear he was trying to rescue (at least) the following *prinicipia artis [astronomicae]*:

* the sphericity of the earth (Chapter 1 of Book I of *De revolutionibus*),
* the sphericity of the universe (Chapter 2 of Book I of *De revolutionibus*);
* the uniform, eternal and circular motion of the heavenly bodies (*Commentariolus* and Chapter 4 of Book I of *De revolutionibus*);
* the immensity of the heavens compared to the size of the earth (*Commentariolus* and Chapter 6 of Book I of *De revolutionibus*);
* the revolution in a smaller circle is faster than that in a bigger one (*quae breviori clauduntur ambitu, revolvantur citius, iis quae maiorem circulum ambient*) (Chapter 6 of Book I of *De revolutionibus*), which equals the principle that the planets should be arranged "in accordance with the duration of the revolution," which "assumes that of the objects moving equally fast, those further away seem to travel more slowly" (Chapter 10 of Book I of *De revolutionibus*).[70] The two principles above are essentially a restatement of the claim from the *Commentariolus* that "one [planet] exceeds another in rapidity of revolution, according as they traverse the larger or smaller circles."

In sum: Copernicus was obviously annoyed not just by the equant but by several other apparent irregularities. He was also keen to save some of the generally accepted astronomical principles. This eventually led him to the discovery of his own heliocentric and geokinetic system. But this provisional and very general answer raises – or transforms itself into – another four important questions.

---

69 *Commentariolus* (Swerdlow, p. 440; Zekl, p. 8).

70 Compare with J. Evans, *The History and Practice of Ancient Astronomy*, p. 76, where five fundamental propositions of Greek astronomy from Eudoxus on are enumerated: "[1] The earth is a sphere, [2] which lies in the centre of the heaven, [3] and which is of negligible size in relation to the heaven. [4] The heaven, too, is spherical [5] and rotates daily about an axis that passes through the earth." Numbers in square brackets added.

Firstly: Where *all* of these issues really operative in the development of the helio-centrism? Secondly: How, when and where did Copernicus become aware of all of these problems? Thirdly and most importantly: Why was he annoyed by all those issues, which – as we shall see shortly – were more all less known to other astronomers and philosophers of his time? And fourthly: How, if at all, did these considerations lead him to geokinetism and heliocentrism?

Copernicus' own narrative about the motivation for the astronomical reform: a second look. In search for an answer to these questions, we may derive some basic orientation from Copernicus' own narrative about his discovery in combination with his sociocultural context – his education; the books he owned, annotated, or disposed of; manuscripts he may have copied; as well as discussions or controversies he may have heard of or even taken part in. First let us take yet another look at Copernicus' own story about his discovery of heliocentric astronomy, this time from the chronological perspective he provides in the *Preface*, where he narrates it most exhaustively, in the *Introduction*, and in the *Commentariolus*. Seen from the perspective of when the *Preface* was written, the entire story can be divided into the first part that led to the *Commentariolus*, and then the part that led to *De revolutionibus*. The part that led to the *Commentariolus*, which concerns us most here, can again be divided into three logical steps, although it is not likely that the process was completed in a chronological or linear fashion at the time.

The first step was Copernicus' confrontation with the many problems of astronomy. This "divine rather than human science," was full of perplexities (*Preface*, *Introduction*). On the one hand, astronomers related to "different principles, assumptions and explanations" in establishing heavenly motions (*Preface*). These "principles and assumptions" were the source of disagreement. There were two incompatible traditions: homocentric and eccentric astronomy (*Introduction* and *Commentariolus*). On the other hand, each of these two traditions had its own problems. Homocentric astronomy could not provide models in accordance with the phenomena (*Preface* and *Commentariolus*). Eccentric, that is, Ptolemaic astronomy, was in good enough agreement with most of the phenomena (*Preface*), but it suffered from the equant problem, whereby it contradicted the first principle of uniform motion. Its major problem was its inability to establish *formam mundi*, that is, the certain *symmetria* of the world. Another critical problem of Ptolemaic astronomy was the length of the tropical year. Astronomers were not certain about the motion of the sun and the moon (*Preface*); there were different opinions about the sun's tropical year (*Introduction*).

In the second step, as the *Preface* reveals, Copernicus' dissatisfaction with eccentric astronomy was narrowed to one particular problem, that of the order of planetary spheres. He reflected for a long time on the confusion in the astronomical traditions concerning the derivation of the motions of the universe's spheres. Philosophers, who occupied themselves with trifles, did not know how the universe was constituted. This confusion was annoying, because the best Artisan created the world machine for our sake, and philosophers were still uncertain about its constitution (*Preface*).

In the next, third, step, Copernicus started reading all available philosophers to see whether anyone had ideas about the motions of the universe's spheres different from those espoused by astronomers in the schools. He found that several authors argued that the earth moved not only around its axis but also around the sun. This enabled him to attain certainty in ordering the planetary spheres, since in a heliocentric universe its parts cannot be moved from one place to another, without disrupting the remaining parts and the universe as a whole (*Preface*).[71]

These three texts reduce the number of problematic issues in existing astronomy to basically three essential problems: the calendar problem, the equant problem and the *forma mundi* problem. They show, I believe, that most of the points in the above list of several possible apparent irregularities which could have troubled Copernicus, were either Copernicus' afterthoughts or played only a minor role in the development of the heliocentrism. What really concerned him were the equant and the structure of the universe.

1. A general outline of Copernicus' education

Copernicus' own outline of his path to heliocentrism is, despite some important clues, obviously still too general and leaves out some crucial details that would enable us to answer the questions raised above. We can discern additional information and draw a clearer picture, however, if we put his narrative into the socio-cultural context of his education in Cracow, Bologna and Padua.

Cracow. In 1491 Copernicus, then aged eighteen, enrolled as a student of Liberal Arts at the University of Cracow and left after four years without a degree in

[71] The next step, or rather the absence thereof, after the *Commentariolus* was Copernicus' "Pythagorean" silence about the movement(s) of the earth (see *Lysis' letter to Hipparchus*), and the final step began when he was persuaded by his friends to change his mind and publish *De revolutionibus*.

1495.[72] In the second half of the fifteenth century Cracow University was highly praised for its achievements in astronomy, which were partly a result of its ties with the University of Vienna.[73] Several Cracow figures belonged to the same circle as Peurbach and Regiomontanus. Martin Krol of Žuravica was a friend of Peurbach's and Martin (Marcin) Bylica of Olkusz, also known as Stanislaw of Ilkush (ca. 1460–1514) met Regiomontanus in Padua, became his friend and worked with him on *Tabulae directionum*.[74] He also revised Regiomontanus' *Tabulae ac problemata primi mobilis* and introduced Peurbach's *Theoricae novae* at Cracow University. At the end of the fifteenth century, around seventy-five professors taught at the university, twenty of them in the mathematical *quadrivium*.

Copernicus' exact curriculum is not known, but he will have encountered many important teachers and been exposed to many intellectual currents. Special attention ought to be paid to preeminent astronomy and philosophy teachers Albert of Brudzewo (or Wojciech Brudzewski; ca. 1445–1495) and John of Glogów (or Jan of Glogów; ca. 1445–1507). Another important figure was the humanist and teacher of the liberal arts Laurentius Raabe Corvinus (ca. 1462/5–1527), who wrote the eulogy in the preface to Copernicus' first publication, a Latin translation of Theophylactus Simocatta's Greek *Letters* on ethics, rustic topics, and love. We should also consider the often-overlooked or underestimated humanist and Platonist influences.

---

72 Copernicus' motives for entering the University of Cracow, the school's masters, texts, logic, and natural philosophy, Cracow humanism, and Cracow's general and astronomical curriculum in the years 1491–1495 are very thoroughly discussed in A. Goddu's book, *Copernicus and the Aristotelian Tradition*, pp. 13–167. I do not share his views, however, on a number of important issues. See also a very condensed overview of the Cracow University in the fifteenth and sixteenth centuries by M. Kokowski, in *Copernicus's Originality*, pp. 169–83. See also G. Rosińska, *Scientific Writings and Astronomical Tables in Cracow: A Census of Manuscript Sources (XIVth-XVIth Centuries)*; P. W. Knoll, "The Arts Faculty at the University of Cracow at the End of the Fifteenth Century"; R. Lemay, "The Late Medieval Astrological School at Cracow and the Copernican System." Although out of date in many respects, I. Polkowski, *Żywot Mikolaya Kopernika*, pp. 114–23; L. A. Birkenmajer, *Stromata Copernicana*, pp. 54–134; and L. Prowe, *Nicolaus Copernicus*, vol. 1, pp. 117–58, are still worth consulting.

73 See, for example, M. Markowski, "The Earliest Unknown Excerpts from Nicholas Copernicus' *De revolutionibus*," p. 6. On the influence of Vienna on Cracow, see J. Babicz, "Die exakten Wissenschaften an der Universität zu Krakau und der Einfluss Regiomontans auf ihre Entwicklung."

74 On Martin (or Marcin), see J. Zathey, "Marcin Bylica of Ołkusz."

Albert of Brudzewo's most important work was his commentary on Peurbach's *Theoricae novae*, *Commentariolum super Theoricas novas planetarum Georgii Purbachii*. Brudzewo started lecturing on the *Theoricae novae* in 1482, and wrote at least two versions of his manuscript before the text was published in Venice in 1494 and in Milan the following year.[75] Copernicus may have attended Brudzewo's lectures on Aristotle's natural philosophy, but the timing is wrong for him to have studied astronomy with him. However, since Brudzewo was the most famous astronomer among Cracow teachers and his *Commentariolum* would have been used in teaching and circulated among students, even when he was not teaching it himself, it is certain that Copernicus was familiar with the *Commentariolum super Theoricas novas* from the very beginning of his studies.

Another very important figure at Cracow University was John of Glogów, the *maximus sagax vir in omni scientia decus singular Universitatis Cracovensis* who, in his forty years at the university (1468–1507) trained most of the masters who taught logic, grammar, metaphysics, astronomy and geography in the 1490s.[76] He wrote on Peter of Spain's *Summulae logicales*, old and new logic, commentaries on *Analytica priora* and *Analytica posteriora*, *Quaestiones* on Aristotle's *Physica*, *De anima* and *Metaphysica*, as well as commentaries on Aristotle's *Metaphysica* and on Thomas Aquinas' *De ente et essentia*. He also authored *Introductorium compendiosum in tractatum sphere materialis magistri Ioannis de Sacrobusto*, and many practical works in astronomy.[77] Although John of Glogów appears to have mostly drawn on the Aristotelian traditions, he was well-versed in other

---

75 See L. A. Birkenmajer's edition *Commentariolum super Theoricas novas planetarum Georgii Purbachii*. For a brief summary of Brudzewo's *Commentary*, see A. Goddu, *Copernicus and the Aristotelian Tradition*, pp. 162–7. I do not share, however, his view expressed on p. 164, that "Brudzewo at times expresses himself agnostically on the actual existence of eccentrics in the spheres of the planets, yet elsewhere he speaks as if they really exist, although he seems to be quoting Peurbach when he says so." For a brief explanation of this point, see infra, p. 295, n. 129. On Brudzewo, see also B. Pawlikowska Brożek, "Wojciech of Brudzewo."

76 On John of Glogów, see M. Zwiercan, "Jan of Glogów"; B. Biliński, *Il Pitagorismo di Niccolò Copernico*, pp. 26–7, and A. Goddu, *Copernicus and the Aristotelian Tradition*, pp. 35–6. Goddu examines Glogow's theory of consequences on pp. 76–82, and in "Consequences and Conditional Propositions in John of Glogowia's and Michael of Biestrzykowa Commentaries on Peter of Spain and their Possible Influence on Nicholas Copernicus."

77 Some of his views are discussed by P. Barker, "Albert of Brudzewo's *Little Commentary on George Peurbach's 'Theoricae novae planetarum'*." See also N. Jardine, "The Significance of the Copernican Orbs," p. 193, n. 95.

philosophical schools of thought as well. He knew and referred to Pierre d'Ailly's *Imago mundi*, for example, and very often quoted Pliny the Elder. He was also familiar with Macrobius' *In somnium Scipionis*, as well as with Pythagoras, Hermes Trismegistus (whose name he wrote as Trimegistus), and Plato.[78]

Apart from various predominantly "Aristotelian" currents, including especially Buridanism,[79] Cracow also hosted scholars who were strongly influenced by humanist and Platonist views. Already in 1485 one of the university reading rooms was called Platonist, and perhaps it should be underlined that Albert of Brudzewo is mentioned in that connection.[80]

---

78 On this, see his *Disputationes in Metaphysicam* (or *Quaestiones super duodecim libros Metaphysicae Aristotelis magistri Joannis Glogoviensis*) preserved as a manuscript: "Introductio in primam philosophiam Aristotelis per magistrum Johannem Gloger [...] et ille Hermes antiquus Trimegistus scribens de Deo ad Asclepium collegam suum, inquit, quod homo est nexus Dei et mundi"; "[...] opinio Platonis et Pythagorae [...]"; "[...] quod pulcre dicit Plato ille divinus in *Timaeo*, cum dixit [...]; secundum Platonicos et Pytagoricos [...]"; "[...] ergo anima humana non ingreditur de uno corpore in aliud corpus, ut dicebat Pitagoras [...]." There is an especially interesting passage regarding the importance of sun: "Sol est dignissimus planeta, ergo etiam eius influentia est nobilior [...]. Ille planeta est dignior, qui omnes planetarum motus regit, dirigit et mensurat, quarum Sol est huiusmodi, quia regens est dignior recto. Declarat Ptholomeus in sapiencijs Almagesti (est liber magne utilitatis et est dictio Arabica), qui ostendit, quod omnes motus planetarum mensurantur et inventi sunt per motum Solis [...]. Unde argumenta concludunt quod uno modo Sol est dignior ceteris stellis et planetis et alio modo aliae stellae digniores sunt Sole, ut dictum est sufficiente in notabilibus et argumento quarto [...]." All quotations are from L. A. Birkenmajer, *Stromata Copernicana*, pp. 122–3, and B. Biliński, *Il Pitagorismo di Niccolò Copernico*, pp. 27–8.

79 See M. Markowski, *Burydanizm w Polsce w okresie przedkopernikańskim*.

80 See W. Wislocki, ed., *Acta rectoralia Almae Universitatis studii Cracoviensis inde ab anno 1469*, vol. 1., p. 213, n. 970: "Die vero Saturni nona mensis Aprilis, anno Dni 1485 [...] Noveritis, quomodo lectorium, dictum Platonis, in Collegio Maiori alias habitacio diucius est occupata et clausa in preiudicium et gravamen Universitatis et presertim domus Collegii Artistarum, et hactenus manet clausum et occupatum [...] quatenus accedentes personaliter Vener. mgros. Albertum de Brudzew, prepositum domus Colegii Artistarum, drem de Zawada et Andream de Labischin, ipsos salubriter et efficaciter inducentes moveatis et requiratis [...] ut infra diem naturalem monitionem sequendo mgr. Albertus de Brudzew, prepositus domus Colegii Artistarum, ex officio suo iubeat lectorium apreiri [...]." See also B. Biliński, *Il Pitagorismo di Niccolò Copernico*, p. 145. I. Polkowski, *Żywot Mikolaya Kopernika*, p. 114, gives the names of other reading rooms: *Aristotelis, Ptolomei, Maronis, Socrattis, Theologorum, Prope Valvam* (*Prope Vawam*). Compare with M. Kokowski, *Copernicus's Originality*, p. 176.

Other teachers at the university, too, were interested in humanism and Platonism, which is evident from the content of their libraries. Matthew of Miechów, for example, professor of medicine, who had studied in Bologna, listed in his 1514 library catalogue "Sexternus Theorice asserentis Terram moveri, Solem vero quiescere," Marsilio Ficino's *Opera*, Filippo Beroaldo the Elder's *Orationes et poemata*, Pomponio Leto's *Historiae imperatorum romanorum*, and *De mysteriis Aegyptiorum, Chaldaerum, Assyriorum* translated by Ficino. This volume included the following entries: Iamblichus, *De misteriis*; Proclus, *De sacrificis*; Porphyrius, *De divinis et daemonibus*; Synesius, *De somniis*, Alcinous, *De doctrine Platonis*; Speusippus, *De diffinitionibus Platonis*; *Pitagorae aurea verba*; *Simbola pitagoricae*; Xenocrates, *De Morte* and Marsilio Ficino, *De voluptate*.

The Platonist current at Cracow University was probably the fruit of the activities of Italian humanist immigrant Filippo Buonaccorsi (called Callimachus Experiens; 1437–1496), a figure with great influence on Cracow's intellectual life.[81] Callimachus, once a member of the Accademia Romana established by Pomponio Leto (ca. 1428–1498), came to Poland in 1470 and was appointed secretary to the king in 1472. He constantly traveled back and forth from Cracow to Italy and Constantinople either as a member or the head of Polish diplomatic missions: in 1477 he was in Venice and Rome; in 1483 in Venice, Padua and Florence; in 1486 in Padua and Venice, where he met Ermolao Barbaro (1453/1454–1493) and a group of scholars that included Giorgio Valla (1447–1500) and Nicholas Leonicus Tomaeus (1456–1531); in 1490 he was in Rome meeting with Italian humanists such as Angelo Poliziano (1454–1494), Giovanni Pico della Mirandola (1463–1494) and above all Marsilio Ficino (1433–1499), who called Callimachus "my fellow Platonist" and "a follower of our Plato."[82] Ficino even supplied Callimachus and his friends with his works. There is a very high possibility that Ficino sent a copy of the 1484 edition of *Opera Platonis* to him[83] and on April 1, 1494, Marsilio wrote a letter to

---

[81] On Callimachus, see, for example, H. B. Segel, *Renaissance Culture in Poland*, pp. 36–82; G. Avezzù, "Le fonti Greche di Copernico," p. 137; A. Goddu, *Copernicus and the Aristotelian Tradition*, p. 141; P. L. Rose, *The Italian Renaissance of Mathematics*, p. 119, and B. Biliński, *Il Pitagorismo di Niccolò Copernico*, pp. 125–9.

[82] See Ficino's letter to Callimachus, in *Epistolarum liber* VIII (Members of the Language Department of the School of Economic Science, vol. 7, letter 5, pp. 7–8; *Opera omnia*, pp. 865–6). For his correspondence with Ficino, see J. Domański, "La fortuna di Marsilio Ficino in Polonia nei secoli XV e XVI."

[83] See Ficino's letter dated April 18, 1485. *Epistolarum liber* VIII (Members of the Language Department of the School of Economic Science, vol. 7, letter 16, p. 19; *Opera omnia*, p. 870): "I would very willingly discuss this matter with you further were it not that

Callimachus to which he attached a catalogue of his works as well as three copies of his *De sole* and an accompanying piece *De lumine.* [84]

Callimachus' friends, to whom Ficino referred in the letter, might have been Albert of Brudzewo and Laurentius Corvinus, two of the most active members of the humanist society *Sodalitas Litteraria Vistulana.* The Vistulan Literary Sodality was established by Callimachus, Albert of Brudzewo and Conrad Celtis (1459–1508), after the latter moved from Italy to Cracow in 1489.[85] Having completed his studies in Germany, Celtis set out for Italy, where he visited Venice,

---

our Plotinus may call me back at any moment to translate him. For this reason, while Plotinus, the follower of Plato, is detaining me, let Plato, who is on his way to you, speak to you in Latin through the work done here. Let him also give you gracious thanks for your gifts, since my own gratitude is undying." It is not certain whether Callimachus actually received *Opera Platonis.* In fact, Callimachus wrote to Ficino's after the latter's house and library burned down, a letter dated May 15, 1488 (Members of the Language Department of the School of Economic Science, vol. 7, p. 98), inquiring whether he had foreseen the fire and had for this reason *not* sent him his Plato: "But I am coming back to the fire, which I believe you predicted would happen, and this is why you did not send your Plato." D. Knox, "Ficino and Copernicus," p. 406, n. 27, speculates that Ficino had *not* sent his copy and dismisses the possibility that the reference could be to the second copy of Ficino's Plato sent to Callimachus as "less likely." I would not exclude this possibility. There was a three year gap between the two letters and since Callimachus sent him generous gifts – according to H. B. Segel, *Renaissance Culture in Poland*, p. 81, Ficino received a coat of marten fur, a smaller garment woven of feathers, and a sword whose hilt was made of horn resembling jasper – he may have expected something in return. Callimachus' and Ficino's correspondence began in the early 1480s. The most substantive letter of Callimachus to Ficino related to the issue of demons.

84 See Ficino, *Epistolarum liber* XII (not yet translated in *Letters of Marsilio Ficino*; *Opera omnia*, p. 957: "Quid autem hoc iamdiu circuitu fecerim, rursusque refecerim atque retexam, intelliges ex catalogo librorum meorum, quem una cum libro *De Sole* legendum ad te mitto, ut clarissimo Phoebi muneris testimonio amicitia nostra praesentibus posterisque refulgeat. Mitto autem triplicem *De sole et lumine* codicem, ut non tibi solum ego gratificer, sed tu quoque duobus tuis amicis gratificari possis. Quae igitur componatur a nobis, ex his ipse perspicies. Ego vero te iam res tuas composuisse ex Octaviano tuo et Tedaldis accepi, quod quidem mihi contigit quam gratissimum." See D. Knox, "Ficino and Copernicus," p. 406, n. 27; B. Biliński, "Intorno ai problemi del Copernico padovano," p. 98, and *Il Pitagorismo di Niccolò Copernico*, pp. 102–3, and 127–8.

85 On Celtis, see L. Spitz, *Conrad Celtis: The German Arch-Humanist*; H. B. Segel, *Renaissance Culture in Poland*, pp. 83–106; B. Biliński, *Il Pitagorismo di Niccolò Copernico*, pp. 129–33, and p. 145; L. A. Birkenmajer, *Stromata Copernicana*, pp. 80–1; H. Rupprich, *Der Briefwechsel des Conrad Celtis.*

Padua, and Ferrara, where he met Baptista Guarinus (ca. 1434–1503). From Ferrara he moved to Bologna, where he made contacts with Filippo Beroaldo the Elder (1453–1505) and from there to Florence, where he met Ficino in person. His journey continued to Rome, where he made acquaintance with members of the Accademia Romana. From Rome, perhaps on the advice of the head of the Roman Academy, Celtis came to Cracow.

The *Sodalitas Litteraria Vistulana*, supported not only by the university and town elite but also by members of the royal court, provided undeniable impetus to a Renaissance humanist literary and philosophical tradition in Cracow. Its members gathered in private residences and discussed literature and humanist learning, the sources for which were provided from Celtis' vast library, which included, among other items, Ficino's translation of Plato's *Opera omnia*. Celtis became a close friend of his teacher in *astrorum studio*, Albert of Brudzewo, and maintained a correspondence with him after his leaving Cracow in 1491.[86] In his *Inaugural oration* on August 31, 1492 at Ingolstadt University, where he moved from Cracow, Celtis expressed his views on the reform of university education, which he believed should be based on the development of a new cultural course of studies. This would be a synthesis of mathematical and astronomical studies, revised grammatical instructions, geography, history (classical and German), as well as rhetoric, poetry and music.[87] Celtis' poetic philosophy was a broad system of thought that embraced both human and divine knowledge, just as Plato and Pythagoras combined "the light of nature with that of grace." This method of knowledge included, according to Celtis, revealing the work of nature and the wisdom of its director by mathematical truth.

Other members of the Vistulan Literary Sodality were: astronomer, mathematician and physician Stanislaw Selig (called Statilius Simonides; d. 1512), doctor of medicine and law and lecturer on Salust; Jan Ursyn (Ursinus; ca. 1460–ca. 1503); Sigismundus Fusilius (ca.1465–1504), lecturer on Vergil's *Georgica*; Johannes

---

86 Celtis dedicated several odes to the members of the *Sodalitas Litteraria Vistulana*. See, for example, *Odarum liber* I, 17, devoted to Albert of Brudzewo, who was praised not only for his knowledge in astronomical matters but also for his competence in astrology: "You know, father, the secret signs of the oblique heaven that change the shapes of things and by their gifts preside over births in the elevated constellation: Aries and Taurus and Gemini, Cancer and Leo, Virgo, Libra and Scorpio, Sagittarius and Capricorn, and the Fishes [Pisces] born in the flowing river. Everything in the heavenly order is known to you." Translation of H. B. Segel, *Renaissance Culture in Poland*, p. 94.

87 See, L. Spitz, *Conrad Celtis: The German Arch-Humanist*, p. 31.

Sommerfeld (or Johannes Aesticampianus the Elder; 1457–1520), who owned Ficino's translation of Plotinus (Florence, 1492) and his *Commentaria quinque in Platonem perpetua* (Florence, 1492), and lectured on Aristotle; Nicholas Czepiel (or Czepel; 1452–1518), who owned Ficino's translation of Plato's *Opera omnia*; Jan Heydecke Mirica (ca. 1443–1512); Piotr Tomicki (1464–1525), who received a doctorate at Bologna in 1498 (during Copernicus's studies there), owned Pythagoras' *Carmina aurea*; *Centiloquium Hermetis in sexternis* and *Opera* of Giovanni Pico della Mirandola (Bologna, 1496);[88] and especially Laurentius Corvinus, who was the most active member of the Literary Sodality after Celtis left the town in late summer of 1491, that is, exactly in the year of Copernicus' enrollment.

Laurentius Raabe Corvinus enrolled at Cracow University in 1484; he received his bachelor's degree in 1486 and his master's degree in 1489, when he started lecturing at the university as *extraneus non de facultate*.[89] Corvinus was a dedicated advocate of Italian humanism, but he also had mathematical interests. He taught literature (e.g. Vergil's *Bucolica*), philosophy (in 1492 he lectured on *De ente et essentia* of Thomas Aquinas, and in 1493 Aristotle's *Analyticae libri posteriorum*, Boethius' *De consolatione philosophiae*), astronomy and geography until 1494, when he left Cracow and moved to Świdnica; he later continued to Wrocław. In 1496 Corvinus published *Cosmographia dans manuductionem in tabulas Ptolemei*, which appeared in Basel.[90] Between 1506 and 1508 he lived in Torun/Thorn and in 1508 – the year when the *Commentariolus* may have already been written – helped to publish Copernicus' Latin translation of Theophylactus Symocatta's Greek *Epistolae morales, rurales et amatoriae*.

Probably while still in Cracow, Corvinus wrote the ode *Carmen de novem musis*, which takes place on the banks of the Vistula River and is thematically very close to his later ode *Carmen de Apolline et novem musis* published in Wrocław in 1503. The poem exhibits a strong influence of Platonic and Neo-Platonic philosophy. He presented the celestial harmony of the Platonic and Pythagorean tones with citations from Marsilio Ficino and Plato. One can trace the thoughts derived from Ficino's *De lumine* and *De sole*, and from Ficino's *argumentum* to his translation

---

88 See B. Biliński, *Il Pitagorismo di Niccolò Copernico*, p. 138.

89 On Corvinus, see A. Goddu, *Copernicus and the Aristotelian Tradition*, pp. 43–4; J. Krókowski, "Laurentius Corvinus und seine Beziehungen zu Polen"; J. Domański, "La fortuna di Marsilio Ficino in Polonia nei secoli XV e XVI," pp. 579–81.

90 In *Cosmographia* Celtis referred to Macrobius' *Saturnalia* thus demonstrating his familiarity with the author of *In somnium Scipionis*. See B. Biliński, *Il Pitagorismo di Niccolò Copernico*, p. 27.

of *De republica*. Corvinus' Platonism is traceable in all of his publications, but above all in his *Dialogus carmine et solute oratio conflates de mentis saluberrima persuassione ad honesta ingenarum artium studia* (1516). *De mentis saluberrima persuassione* was written in the style of Boethius' *De consolatione philosophiae* as a mixture of prose and verse. Corvinus explained how *Mens* accompanied by Muses appeared to him and told him about the meaning and goal of life. Plato was selected as the best leader on the way to virtue, because his philosophy was closest to Christianity. Corvinus revealed a good knowledge of some Neo-Platonists (for instance, Plotinus) and Ficino. Besides, he was also influenced by Hermes Trismegistus and other hermetic and Neo-Platonic sources. He knew almost all Plato's dialogues in the translation of Marsilio Ficino. In *Mens*' eulogy to Plato, the *Epinomis*, the *Timaeus*, the *Parmenides* and the *Laws* are quoted. In the *Epinomis*, says *Mens*, Plato speaks about *logos*, The Word of the Father, and about the world created by *logos*, that is, by the Divine Mind.[91]

While in Cracow, Copernicus bought several books that are now bound together in two volumes. One volume contains Euclid's *Elementa geometriae* with the commentary by Johannes Campanus, published in Venice in 1482, and Albohazen Haly, *In iudiciis astrorum*, also published in Venice in 1485. The second volume contains *Alphonsi Tabulae astronomicae*, published in Venice in 1492, and Regiomontanus' *Tabulae directionum et profectionum*, published in 1490. This volume also contains sixteen subsequently handwritten pages by Copernicus known as "Uppsala Notebooks" or "Uppsala Notes."[92]

Bologna. In 1496 Copernicus continued his studies in Italy, probably on the advice of his uncle Lucas Watzenrode or, the latter being on friendly terms with Callimachus, on the advice of both his uncle and Callimachus. At the age of twenty-three he enrolled at the University of Bologna and studied canon and civil law as a member of German nation until 1500.[93]

---

91 See J. Domański, "La fortuna di Marsilio Ficino in Polonia nei secoli XV e XVI," p. 580, n. 44.

92 Both volumes have Cracow bindings and Euclid is annotated by Copernicus. On books that Copernicus bought, consulted and annotated, see P. Czartoryski, "The Library of Copernicus," and A. Goddu, "Copernicus's Annotations: Revisions of Czartoryski's 'Copernicana'."

93 For Copernicus' studies in Italy in general, see C. Vasoli, "Copernic et ses études italiennes." Natural philosophy in these Italian universities is treated by A. Poppi "La filosofia naturale del primo Cinquecento nelle Università di Padova, Bologna e Ferrara."

At that time Bologna was not only famous for its studies of law and medicine, it also hosted an important humanist school, with teachers of grammar, rhetoric, poetry and Greek, such as Filippo Beroaldo the Elder and Antonio Urceo Codro (1446–1500),[94] as well as Angelo Poliziano, a friend of Venetian publisher Aldo Manuzio (1449–1515) and Giovanni Pico della Mirandola. It is possible that Copernicus started learning Greek with Codro: Aldo Manuzio dedicated his edition of a collection of Greek letters, including *Epistolae morales, rurales et amatoriae* of Theophylactus Symocatta, to Codro with the suggestion that he should show them to his students of Greek.[95] Copernicus went on to translate this work (published in 1509), for which Corvinus wrote the *encomium*.

Bologna also hosted at least one famous Averroist: Alessandro Achillini (1463–1512).[96] He taught there from 1484, when he received his degree (he taught logic until 1487, then natural philosophy from 1487 to 1494; he became a morning professor of medical theory from 1494 to 1497, and afternoon ordinary professor of natural philosophy from 1497 to 1500) to 1506, when he left for Padua.[97] It was in Bologna in 1498 that Achillini published his *De orbibus*.

The most important figure Copernicus encountered in Bologna was, as is well known, Domenico Maria Novara, a doctor in liberal arts and medicine, mathematician, astronomer, astrologer and physician, who continued the tradition of technical, mathematical astronomy and considered himself a student of

---

See also L. Olivieri, ed., *Aristotelismo veneto e scienza moderna*; A. Poppi, ed., *Scienza e filosofia all'Università di Padova nell Quattrocento.*

94 On Codro, see L. Prowe, *Nicolaus Coppernicus*, vol. 1, p. 249 et seq.; C. Vasoli, "Copernic et ses études italiennes," pp. 163–5; on Filippo Beroaldo the Elder, pp. 162–3. On humanism in Bologna, see P. F. Grendler, *The Universities of the Italian Renaissance*, pp. 216–9, on Beroaldo and Codro, pp. 218–9. On Beroaldo, see also E. Garin, "Filippo Beroaldo il Vecchio: un universitario inquieto."

95 The aforementioned suggestion reads as follows: "Aldus Manutius Antonio Codro Urceo S. P. D. // Collegimus nuper, Codre doctissime, quotquot habere potuimus Graecas epistolas, easque typis nostris excusas duobus libris publicamus, praeter multas illas Basilii, Gregorii, et Libanii, quas, cum primum fuerit facultas, imprimendas domi servamus! // Auctores vero, quorum epistolas damus, sunt numero circiter quinque et triginta, ut in ipsis libris licet videre. Has ad te, qui et Latinas et Graecas literas in celeberrimo Bononiensi Gymnasio publice profiteris, muneri mittimus, tum ut a te discipulis ostendantur tuis, quo ad cultiores literas capessendas incendantur magis, tum ut apud te sint Aldi tui *mnemosynon* – et pignus amoris. Vale."

96 On Achillini, see B. Nardi, "Appunti sull'averroista bolognese Alessandro Achillini."

97 See P. F. Grendler, *The Universities of the Italian Renaissance*, p. 13.

Regiomontanus.[98] Copernicus resided in Novara's house.[99] According to Rheticus, he "made observations with the utmost care at Bologna, where he was not so much a student as he was an assistant and witness to the astronomical observations of the learned Dominicus Maria (*non tam discipulus quam adiutor et testis observationum doctissimi viri Dominici Mariae*)."[100]

Padua. After Copernicus' brief visit to Rome in 1500, where he held some lectures on mathematical subjects,[101] he went to Warmia, and in 1501 asked the Warmian Chapter for a permission to continue his education for two more years. He obtained it and promised to study medicine.[102] In the same year he returned to Italy, this time to the University of Padua.[103]

---

98 See E. Zinner, *Leben und Wirken des Joh. Müller von Königsberg*, pp. 240–1; R. S. Westman, *The Copernican Question*, p. 97.

99 See R. S. Westman, *The Copernican Question*, p. 528, n. 67. On Copernicus and Novara, see ibid., pp. 87–99.

100 *Narratio prima* (Rosen, p. 111; Hugonnard-Roche and Verdet, p. 43).

101 Rheticus, *Narratio prima* (Rosen, p. 111; Hugonnard-Roche and Verdet, p. 43) describes him as "a public lecturer on mathematical subjects before a large audience of students and also to a cycle of great men and to craftsmen skilled in this kind of learning." B. Biliński, "Intorno ai problemi del Copernico padovano," pp. 98–9, believes that Alessandro Farnese, later Pope Paul III, attended these lectures and became acquainted with Copernicus, which could explain Copernicus' later boldness in dedicating *De revolutionibus* to him.

102 This means that although Copernicus has permission to study medicine, he was not obliged to do so. See the text of the Warmian Chapter: "[…] in Die Pantaleonis martyris [July 27, 1501] comparuerunt coram Capitulo domini canonici Nicholaus et Andreas Coppernick fratres: desideravit ille ulteriorem studendi terminum, videlicet ad biennium, qui iam tres annos, ex licentia Capituli, peregit in studio. […] Post maturam deliberationem Capitulum votis utriusque condenscendit: maxime, ut Nicholaus medicinis studere promisit consulturus olim Antistiti nostro Reverendissimo, ac etiam dominis de Capitulo medicus salutaris et Andreas pro literis capescendis abilis videbatur." Quoted from B. Biliński, "Intorno ai problemi del Copernico padovano," p. 85. See also M. Biskup, *Regesta Copernicana*, p. 43, n. 38.

103 On the University of Padua, see P. F. Grendler, *The Universities of the Italian Renaissance*, pp. 21–40. For a brief overview of the history of astronomy at University of Padua, see M. di Bono, *Le sfere omocentriche di Giovan Battista Amico nell' astronomia del Cinquecento*, pp. 45–7. Many important themes and subjects related to the University of Padua in the fifteenth and sixteenth centuries are thoroughly discussed in A. Poppi, ed., *Scienza e filosofia all'Università di Padova nell Quattrocento*. On Copernicus in Padua, see B. Nardi, "Copernico studente a Padova"; B. Biliński, "Intorno ai problemi del Copernico padovano," and "Il periodo padovano di Niccolò Copernico (1501–1503)." Compare with A. Goddu, *Copernicus and the Aristotelian Tradition*, pp. 198–203.

There were several elements that may have motivated him to choose the study of medicine at the University of Padua, apart from the wish of the Warmian Chapter and his uncle Lucas Watzenrode. Since medicine was closely related to the study of astronomy,[104] he may have chosen medicine on the advice of Novara, who pursued a similar path studying *artes liberales* and medicine in Ferrara. This motive may have been further strengthened by the important astronomical tradition in Padua. As we have seen, Georg Peurbach visited and possibly lectured there, as did his collaborator Johannes Regiomontanus in 1464. Other figures connected to Padua included Paul of Middelburg, teacher of astronomy in 1480 and later Bishop of Fossombrone, who invited Copernicus to collaborate on calendar reform. Similarly Francesco (Giambattista) Capuano da Manfredonia, a fierce anti-Averroist, was master of astronomy in Padua ca. 1494–1495. He was a doctor of arts and medicine, as well as author of the *Commentary on Johannes de Sacrobosco's Sphaera* and the second commentary on *Commentary on Peurbach's Theoricae novae planetarum* after Brudzewo's; both were for the first time printed in Venice in 1499.[105] After 1498 the chair of mathematics and astronomy was held by Benedetto del Tiriaca (fl. 1495/98), who worked on *Questio de astronomiae veritate* and denied the movement of the eighth sphere.[106] In the last fifteen years of the fifteenth century the University of Padua attracted other important figures such as Pietro Pomponazzi (1462–1525), who obtained a doctorate in arts there in 1487 and became extraordinary professor of natural philosophy in 1488. He was promoted to the first ordinary professor of natural philosophy in 1495 and had Nicoletto Vernia (1420–1499) as his competition in the second position. He left in 1496 for Ferrara to pursue private studies, but returned in 1499 to the same position. This time his rival was Agostino Nifo (ca. 1470–1538), who taught at Padua from 1492 to 1499. In 1501 Pomponazzi discussed – *secundum imaginationem* – a possible rotation of the earth while commenting on Aristotle's *De caelo* and mentioned Nicole Oresme in that context.[107] Pomponazzi left Padua in 1509.[108]

Copernicus enrolled in medicine which was in the period of our interest deeply connected with astronomy and astrology. It is questionable whether he ever

---

[104] On the relationship between medicine and astrology or astronomy, see, for instance, R. S. Westman, "The Astronomer's Role in the Sixteenth Century," pp. 118–21.

[105] On Manfredonia, see M. Shank, "Setting up Copernicus."

[106] See B. Biliński, "Il periodo padovano di Niccolò Copernico (1501–1503)," pp. 246–7.

[107] See A. De Pace, *Niccolò Copernico e la fondazione del cosmo eliocentrico*, pp. 178, 523 and 524.

[108] See P. F. Grendler, *The Universities of the Italian Renaissance*, p. 29.

studied natural philosophy (but if he did, he must have encountered Pomponazzi).[109] Paduan physicians and philosophers particularly occupied themselves with reading Galen and with the problems of the scientific *methodos*.

According to Goddu, Copernicus studied only theoretical medicine.[110] The most known teachers of theoretical medicine from 1501 to 1503 were Bartholomaeus Montagnana and Marco Antonio (Marcantonio) Della Torre (1481–1511). His younger brother, Giovanni Battista (Giambattista) Della Torre (d. 1534), was later praised by Girolamo Fracastoro in his *Homocentrica* published in 1538 in Venice as his inspiration for the revival of homocentric astronomy. It is a curious and perhaps significant coincidence that Fracastoro was in Padua at about the same time as Copernicus: first as a student and then as a teacher of logic. Fracastoro, whose name appeared for the first time in *Acta graduum academicorum Gymnasii Patavini* in 1496, studied literature, mathematics, astronomy, philosophy (the latter under the guidance of Pietro Pomponazzi and Nicholas Leonicus Tomaeus), knew Tiberio Bacilieri (1461–1511), doctor *in artibus* and *medicine lector* (a disciple and friend of Alessandro Achillini), and lived in the house of Girolamo Della Torre, father of Marcantonio Della Torre and Giambattista Della Torre.[111]

On June 9, 1502, Fracastoro was present as a witness (together with Antonio Fracanzani (d. 1506), teacher of natural philosophy *in secundo lougo* – and hence a competitor to Pomponazzi – and Giovanni Mondela, student of arts) at the conferment ceremony at which Giambattista Della Torre received his doctoral insignia from the hands of his father Girolamo Della Torre, teacher of medicine. Giambattista Della Torre's other promoters were his brother Marcantonio Della Torre, Giovanni dell'Aquila, Gabriele Zerbi, Pietro Trapolin, Simone Marini and

---

109 B. Nardi, "Copernico studente a Padova," argues for such scenario, B. Biliński, "Intorno ai problemi del Copernico padovano," disagrees.

110 See *Copernicus and the Aristotelian Tradition*, pp. 198–203.

111 According to G. Ongaro, "Girolamo Fracastoro e lo Studio di Padova," p. 35, Girolamo della Torre organized in his residence some kind of boarding house for students. See an anonymous biography of Fracastoro in *Opera Omnia*, fols. 2^a–5^b, written very probably by the editor of the Fracastoro's *Opera* Paolo Ramusio (1532–1600), son of geographer Giovanbattista Ramusio (1485–1557), Fracastoro's friend from the student days. See fol. 2^a: "[Fracastoro] Studiarum socius et sodales habuit M. Antonium, Ioannem Baptistam, et Rhamundum, Turrianos nobilissimos fratres, quibuscum quotidiano fere in convictu Patavij versabatur." See also the beginning of Fracastoro's *Naugerius sive de poetica dialogus. Ad Ioannem Baptistam Rhamusium* I, 1–10 (Peruzzi, p. 44), where Fracastoro reminds G. B. Ramusio about their friendship from the days of their youth and friends, all of them (including Ramusio) students at Padua: Andrea Navagero, Giambattista Della Torre, Raimondo Della Torre, and Marcantonio Della Torre.

Girolamo Polcastro.[112] Fracastoro received his doctorate *in artibus* on November 2, 1502. His conferment ceremony was witnessed by Tiberio Bacilieri, Francesco Burana and Marco Faella, and there is a note regarding Fracastoro's future obligation: *ad lecturam logice publice deputatus*. Fracastoro's promoters were Giovanni dell'Aquila, Girolamo della Torre, Pietro Trapolin, Marcantonio Della Torre, *qui dedit insignia*, and Gabriele Zerbi (1435–1505), a professor of theoretical medicine and humanist, who discovered several medieval scientific manuscripts and had contacts with Turkey and who later died on the way back home from Turkey. Immediately after receiving his doctoral degree *in artibus* in 1502, Fracastoro became an instructor in logic, but began studying medicine already the following year. He received his doctorate in medicine between June and September 1505.

It is very likely that Copernicus continued – or started – learning Greek in Padua. Padua had had a professorship of Greek since 1463, when Athenian Demetrius Chalcondyles (1423–1511) was appointed to teach Greek, but a decisive shift was brought about by Nicholas Leonicus Tomaeus in 1497. Tomaeus received his degree *in artibus* in 1485 – his promoters were Pietro Roccabonella, Paolo dal Fiume, Girolamo Polcastro, Nicoletto Vernia and Girolamo Della Torre. In 1486 – as already mentioned – he visited the house of Filippo Buonaccorsi (Callimachus) in Venice. Upon his return to Padua in 1497 he started reading Aristotle's and – according to some – Plato's books in Greek and did so until 1506.[113]

---

112 See *Acta graduum academicorum Gymnasii Patavini, vol. 1: Ab anno 1501 ad annum 1525*, n. 136, p. 50. See also G. Ongaro, "Girolamo Fracastoro e lo Studio di Padova," p. 38, n. 44, and G. M. Varanini, "Giralomo Fracastoro nel contesto. La famiglia e gli studi," p. 22. Compare with Francesco Patrizi, later also a student at Padua, who in an overview of the history of astronomy in his *Nova de universis philosophia* IV, 12, lists Giambattista Della Torre and Fracastoro as representatives of the homocentric astronomy from Verona; Patrizi makes no mention of Amico. See also J. M. Lattis, *Between Copernicus and Galileo*, p. 90, and his "Homocentrics, Eccentrics and Clavius's Refutation of Fracastoro," pp. 704–5.

113 On Nicholas Leonicus Tomaeus, see a concise summary in A. De Pace, *Niccolò Copernico e la fondazione del cosmo eliocentrico*, pp. 130–4, and relevant notes. See also P. F. Grendler, *The Universities of the Italian Renaissance*, p. 237. On Leonicus and Plato, see J. Facciolati, *Fasti Gymnasii Patavini I*, Pars II, p. 110: "MCDXCVII. Nicolaus Leonicus de Tomaeis Epirota, adoptione Venetus, pridie kal. april. munus suscepit, et quidem primus in hoc Gymnasio, Philosophiae ex Graecis Aristotelis et Platonis scriptis explicandae, ita postulante Universitate." On the other hand, Facciolati in the same work, *Fasti Gymnasii Patavini*, Pars I, p. LV, suggests that only Aristotle was read in Greek: "Hoc tempore tantus fuit animorum ardor, ac paene impetus erga litteras Graecas, ut se Artium Rector valde comitatus Venetias contulerit, egeritque Universitatis nomine de

Tomaeus also read texts of the Greek and Byzantine commentators on Aristotle (Johannes Philoponus, Alexander of Aphrodisias, Michael Ephesius and others) in the original and was a passionate manuscript hunter. In the spirit of a lot of Paduans, such Agostino Nifo,[114] Tomaeus also tried to reconcile Aristotle and Plato.

Tomaeus was a translator of pseudo-Aristotle's *Quaestiones mechanicae*, published in Venice in 1525, and *Parva naturalia*, and of Ptolemy's *Inerrantium stellarum significationes*, published in the Aldine edition of Ptolemy in 1515. He read and annotated the famous codex of Plato's *Timaeus* (Par. 1810). His *lectiones* are included in the Aldine edition of Plato's works in Greek in 1513. Tomaeus also translated a portion of Plato's *Timaeus* 35a–36e (on the World-Soul and its division into harmonic intervals; on the construction of the circles of the Same and the Different; and on the world's body that fits its soul), with Proclus' commentary on the same passage.[115] It is no wonder that he was called *doctus disciplinae Platonicae* by one of his contemporary professors at Padua and *vir integer et in platonica philosophia feliciter exercitatus* by Erasmus of Rotterdam (1466–1536). He tried to establish the original text of Plutarch's *Moralia* and translated Galen,

---

schola instituenda, in qua Graeci Aristotelis libri, et ea Omnia, qua a Graecis fontibus petita, ad Philosophiam et Medicinam pertinerent, Graeca eruditione, explicarentur. In ejus verba Senatus consultum factum est pridie non. April. MCCCCXCVII quo munus tam grave Nicolao Leonico de Thomaeis Epirotae delatum est. Primus igitur omnium Aristotelicos libros, ut habet inscriptio, quam illi Petrus Bembus in aede S. Francisci posuit, Graeco sermone Patavii publice docuit usque ad quintum saeculi proximi annum, magnamque eruditione novitiate admorationem movit." See also J. Hankins, *Plato in the Italian Renaissance*, vol. 1., p. 98, n. 174; D. De Bellis, "Niccolò Leonico Tomeo interprete di Aristototele naturalista", p. 74, and D. J. Geanakoplos, "The Career of the Little-known Renaissance Greek Scholar Nicholas Leonicus Tomaeus and the Ascendancy of Greco-Byzantine Aristotelianism at Padua University (1497)", p. 363. Compare with A. De Pace, *Niccolò Copernico e la fondazione del cosmo eliocentrico*, pp. 131–2, n. 399. Compare also with Pietro Bembo's sepulchral inscription in St. Francis' church in Padua: "Leonico Thomaeo Veneto mitioribus in literis pangendisque carminibus ingenio amabili, philosophiae vero in studiis, et academica peripateticaque disciplina longe praestanti; nam et *Aristotelicos libros Graeco sermone Patavii primus omnium publice docuit*, scholamque illam a multis Latinis interpretibus inculcatam perpolivit, et Platonis maiestatem nostris hominibus iam prope abolitam restituit, multa praetera scripsit, multa interpetatus est, multos claros erudiit viros, praeter virtutem bonasque artes tota in vita nullius rei appetens." Emphasis in original. Quoted from, D. De Bellis, "La vita e l'ambiente di Niccolò Leonico Tomeo", p. 48.

114 On that, see, for example, E. De Bellis, "Monographia introduttiva," in A. Nifo, *La filosofia nella corte*, p. 24.

115 See J. Hankins, *Plato and the Italian Renaissance*, vol. 2, p. 799.

who was, according to him, full of valuable comments on Aristotle's logic and on Plato. Tomaeus, furthermore, translated certain writings of Hippocrates and Proclus' *Commentary on Plato's Parmenides*. His *Dialogi decem*, published in 1524 and composed *more Academicorum*, suggest that he was interested in Cicero's *Queastiones academicae* and that he may have lectured on it while Copernicus was in Padua.[116]

Besides Nicholas Leonicus Tomaeus, there were other figures at Padua interested in humanism and Plato. One of them – perhaps an indication of a broader movement – was Federicus Chrysogonus (1472–1538). He began his legal studies in Padua in 1498/1499, but did not complete them. Instead, he turned to "divine mathematical sciences," left Padua, traveled across Italy and Europe, fighting wars for different masters, "following the doctrine of the divine Plato and Pythagoreans."[117] He returned to Padua in 1501 and was promoted to doctor of philosophy and medicine in 1507. In the same year he published his *Speculum astronomicum* in Venice. In the *Epistola*, found at the beginning of the *Speculum*, written by Marco Antonio Contarini – then *philosophiae et sacrae theologiae candidatus* and later a correspondent of Fracastoro – Chrysogonus is praised for his keen mind and expertise in mathematical sciences.[118] The *Speculum* reveals that, while in Padua, Chrysogonus was inspired by somebody or something to follow *iter pytagorica et platonica*, to work and lecture on Ptolemy's *Almagest* and *Quadripartitum*, and to discuss the problem of *metabasis*. In *Oratio* he characterized mathematical

116 See B. Biliński, *Il Pitagorismo di Niccolò Copernico*, p. 56 et seq.

117 See *Speculum astronomicum*, "Oratio clarissimi artium et medicinae doctoris Federici Chrysogoni Iadertini in Academia Patavina publice habita" (Girardi-Karšulin and Perić, p. 8): "Et me divinis mathematicis ex toto dicavi, amore quarum accensus, non solum per totam Italiam, sed ferme per totam Europam peregre profectus sum sub diversis militando ducibus dogmata divi Platonis et Pythagorae secutus."

118 *Speculum astronomicum*, "Epistola" (Girardi-Karšulin and Perić, pp. 4–6): "Sed quid etiam dicam de tuis divinis mathematicis in quibus divina integritas maxime suam bonitatem et excellentiam humano generi communicavit, quum nos (quantum potentiae ordinatae et fragilitati humanae conveniebat) sibi similes fecerit, ut subtilissime in hac tua doctissima oratione in principium studii Patavini doctissime ostendisti? Quarum vere princeps vel magis parens esse videris, in manu cuius destructio et constructio ipsarum sita est, quia te dubitante omnes convenit dubitare. Quae tu non solvis, non erunt soluta. Aliena video tibi nullam facere difficultatem, sed tua quidem omnibus. Hoc tuae publicae disputationes, lectiones, commentaria et additiones super *Quadripartitum* et *Almagestum* Ptolemaei clare significant."

sciences as the most certain (Aristotle, Boethius, Averroes are listed as authorities on this point) and those which make a man most godlike.[119]

Another indication that Plato's works were seriously studied in Padua may have been Fracastoro's attitude towards Plato. He believed in his philosophical dialogues that it was possible to reconcile Plato and Aristotle, discussed typically Platonist topics such as the philosophical investigation of beauty and revealed his extensive knowledge of Plato.[120] What is more, he also reported that the homocentric revival initiated by Giambattista Della Torre was somehow related to Plato's *Timaeus*. In his dedication to the Pope Paul III in *Homocentrica* he explains that Della Torre on his deathbed told him to recall the circles from the *Timaeus* in the shape of letter X. Fracastoro refers here to *Timaeus* 36b–c, which is included in the part translated by Leonicus Tomaeus.[121]

Besides, there is ample evidence that important Paduan philosophers, usually labeled as rigid Averroists, such as Nicoletto Vernia (ca. 1420–1499), Agostino Nifo and Marcantonio Zimara (ca. 1460–1532), read Plato in Ficino's translation, used Plato in their discussions and, at least Vernia and Nifo, even accepted the reconciliation of Plato and Aristotle through interpretations of Greek commentators, such as Themistius, Simplicius and Philoponus. Nifo, in particular,

---

119 See *Speculum astronomicum*, "Oratio clarissimi artium et medicinae doctoris Federici Chrysogoni Iadertini in Academia Patavina publice habita" (Girardi-Karšulin and Perić, p. 10): "[…] sed unica ratione volo esse contentus, excellentes scilicet mathematicas sua nobilitate omnes scientias humanitus acquisitas excedere, quia per eas solas Deo optimo maximo similes esse possumus, cum ipsi (Deo benedicto) et praeterita et futura sint praesentia."

120 See, for instance, E. Peruzzi, "Introduzione," in G. Fracastoro, *Navagero. Della Poetica*, pp. 7–33, and, "La poetica del *Naugerius* tra platonismo e aristotelismo."

121 See *Homocentrica*, "Sanctissimo Pavlo Pontifici Maximo": "Moriturus autem quum iam esset (qualem Trimegistum Mercurium ferunt morientem astantes amicos rogasse, ut memores forent illius navis quam in medio aethere perpetuo ascendere et descendere docuisset), ita ille at nos conversus, qui astabamus amici, quum multa alia dixisset, tum me rescpiciens, – Quidam – inquit – habeo, Fracastori, quod te supremum rogem –. Hortante vero me ut diceret, – memorem – inquit – te fore velim eorum circulorum, *quos Timaeus ad figuram X literrae primum secat*, mox ita retorquet, ut linearum capita inter se coeant (significabat autem a se inventos motus). Tum, quod mihi nunc invidet mors, si te praestaturum sperem ac coeptum per me negotium perfecturum, magna morienti mihi consolatio futura est –. Cui ego memorem fore me, si opus fuisset, et desiderio illius satisfacturum promisi." Emphasis added. Fracastoro's *Homocentrica* is full of references to Plato's *Timaeus* and Proclus' *Commentary on Timaeus*. See E. Peruzzi, *Il nave di Ermete*, pp. 1–38, 45–54, and 97–108. Fracastoro refers to the same *Timaeus* passage also in his *Turrius*.

showed profound knowledge of the *Timaeus*, the *Parmenides*, the *Republic* and the *Phaedo*.[122]

Greek, which, as already said, Copernicus very probably learned with Leonicus Tomaeus, opened new horizons for him.[123] He was able to read the Greek authors in original and from Padua he had easy access to the publishing activity in Venice, where Aldo Manuzio, for whom Tomaeus worked and prepared editions, was the leading printer and publisher. Copernicus must have followed the standard learning process, which in its initial stage comprised copying of Greek texts – Bessarion, for example, in 1461 persuaded the twenty-four-year old Regiomontanus to copy the New Testament – and then progressed to autonomous translation (Isocrates, Xenofon, Lucan).[124] For that purpose Copernicus bought Johannes Crastonus' *Dictionarium graecum cum interpretatione latina*, which was published by Dionysius Bertochus in Modena in 1499 or, more probably, in 1500. Copernicus left a great many notes in Greek and Latin; some of them, not necessarily from the Padua period, indicate his interest in the reconstruction of the calendar, while others show his work on the translation of Symocatta. He also bought three books that are nowadays bound together in a single volume; all annotations contained therein are Copernicus'; some annotations are in Greek. The first book in the volume is Giovanni Pontano's *Opera*, published by Bernardinus Vercellensis in Venice in 1501. The other two books were published in Venice by Aldo Manuzio: Bessarion, *In calumniatorem Platonis* (1503), and Aratus,

---

122 See E. P. Mahoney, "Marsilio Ficino's Influence on Nicoletto Vernia, Agostino Nifo and Marcantonio Zimara." One of Mahoney's conclusions, p. 530, is that with Ficino's translation, "there is [...] a dramatic expansion of the range of Platonic dialogues available for study by philosophers in the Aristotelian tradition." And that "[...] even a rather cautious and traditional thinker like Zimara was not adverse to reading and citing works like the *Laws*."

123 On this, see B. Biliński, "Il periodo padovano di Niccolò Copernico (1501–1503)," pp. 252–3, and 256. Copernicus debt to Nicholas Leonicus Tomaeus is very likely expressed in *De revolutionibus* IV, 7 (Rosen, p. 191; Gansiniec, p. 194), where Copernicus defined the meridian of Cracow with the meridian of Frombork and Dyrrhachium in Epirus, the original town of Nicholas Leonicus Tomaeus' family: "All these values [are reduced] to the meridian of Cracow. For Gynopolis, which is commonly called Frombork, where I generally made my observations, is located at the mouth of the Vistula River and lies on the meridian of Cracow, as I learn from lunar and solar eclipses observed simultaneously in both places. Macedonia's Dyrrhachium, which was called Epidamnus in antiquity, is also located on this meridian."

124 See G. Avezzù, "Le fonti greche di Copernico," p. 126.

Phenomena with Theon's commentary (1499). Both the text and commentary are in Greek, attesting to Copernicus' interest in astronomy *and* Greek.

Warmia. After receiving his doctorate *in iure canonico* at Ferrara on May 31, 1503, Copernicus returned to Warmia, where he first went to live with his uncle in the episcopal palace in Lidzbark and then, in 1510, took up residence in his Cathedral Chapter in Frombork. He remained there to the end of his life, with the exception of occasional trips. During the Lidzbark period he was very likely in contact with Tiedemann Giese, later Bishop of Fossombrone, who wrote marginal comments on the works of Aristotle in editions with commentaries by Johannes Versoris (d. ca 1485).[125]

2. Copernicus' encounter with the problems of astronomy

During his education in Cracow and Italy, Copernicus had ample opportunity to learn about each and every astronomical problem he listed in his *Commentariolus, Introduction* and especially *Preface* as motives for his astronomical renovation. These questions concerned the basic principles of astronomy (homocentric astronomy as opposed to eccentric astronomy) and the problems of homocentric and eccentric astronomy respectively. That is, the inability of homocentric astronomy to provide accurate mathematical models that could predict the celestial phenomena on the one hand, and the inability of the Ptolemaic astronomy to determine the length of the tropical year, its problematic use of the equant, and finally and most importantly, its inability to resolve the problem of the *forma mundi* on the other.

Averroist criticism of Ptolemaic astronomy and possible discussions on the revival of homocentric astronomy. According to his own words, Copernicus started his renovation project because of various disagreements among astronomers. The first disagreement regarded the principles of astronomy: while some defended homocentric astronomy, others built astronomy on eccentric models. In the *Commentariolus* Copernicus mentioned two ancient representatives of homocentric astronomy, Callippus and Eudoxus, which suggests that at the time he was writing this short text, he was not aware of any *contemporary*

[125] See A. Goddu, "Reflections on the Origin of Copernicus' Cosmology," p. 50, n. 36. On Johannes Versoris, see also A. Goddu, *Copernicus and the Aristotelian Tradition*, pp. 47, 72–4, 95–6, 122–8, 134, 322, 333, and 461–70.

revival of homocentric astronomy; if he had been, he would have probably mentioned it. This might well be the case, but the mere mention of homocentric astronomy indicates that he must have encountered some discussions or disputes regarding homocentric astronomy before he set out to write about it. He actually did have a number of opportunities to hear and learn – if not about the revival of homocentric astronomy, which occurred later in the 1530s with the work of Giovanni Battista Amico (*De motibus corporum coelestium iuxta principia peripatetica sine eccentricis et epicyclis*, Venice, 1536 and 1537, Paris, 1540) and Girolamo Fracastoro (*Homocentrica*, Venice, 1438) when he was practically done with *De revolutionibus* – then at least about the attacks from Averroist philosophers against the Ptolemaic astronomy and discussions about the homocentric astronomy of Eudoxus and Callippus.[126]

In Cracow Copernicus may have learned about Eudoxus and Callippus reading Aristotle's *Metaphysica*, but he also must have encountered classical Averroist objections to the eccentrics and epicycles in Albert of Brudzewo's *Commentariolum super Theoricas novas planetarum Georgii Purbachii*.[127]

Brudzewo listed classical Averroist arguments against eccentrics regarding the "tearing" or "splitting" of celestial substance and the introduction of vacuum (*scissio sphaerarum et commixtio vacui*), which the Averroists believed would be required to accommodate eccentric orbs. He recognized the worth of the Averroists' position, but denied them the right to dismiss mathematical astronomy. Averroes as a philosopher rightfully refuted eccentrics; however, it was not a philosopher's task to discuss individual "partial orbs" but to consider "the total orb" of a planet, that is, a spherical shell concentric with the earth. On the other hand, it was the duty of astronomers to examine "partial orbs" of a "total orb," which were eccentrics and epicycles according to the Ptolemaic tradition, and concentric, like the "total orb," according to the Averroists. Quoting Richard of Wallingford's *Albion*, Brudzewo maintained that without eccentrics and epicycles it would be impossible to make exact predictions of celestial positions.[128] The best

---

[126] On this, see also P. Barker, "Copernicus and the Critics of Ptolemy," and "The Reality of Peurbach's Orbs," pp. 16–7; M. A. Granada and D. Tessicini, "Copernicus and Fracastoro: The Dedicatory Letters to Pope Paul III, the History of Astronomy, and the Quest for Patronage," pp. 433–7.

[127] For a more detailed analysis, see P. Barker, "Albert of Brudzewo's *Little Commentary on George Peurbach's 'Theoricae novae planetarum'*."

[128] See J. North, *God's Clockmaker: Richard Wallingford and the Invention of Time*, pp. 351–79.

response to the Averroist critique was offered by the embedding of an eccentric partial orb into a concentric total orb.[129]

In Bologna Copernicus worked with Novara, his direct link to Bianchini and Regiomontanus, and through them to Peurbach and Bessarion. This leaves ample space for speculation about the transmission of the astronomical material and preoccupations from Bessarion's circle and Vienna's astronomical school to Copernicus: Did Novara read or know anything about Regiomontanus' *Defensio* and his homocentric aspirations?[130]

Besides, Copernicus must have heard of or read Averroist arguments against Ptolemaic astronomy in Alessandro Achillini's *De orbibus*.[131] Achillini composed

---

[129] See *Commentariolum super Theoricas novas planetarum Georgii Purbachii*, "Theorica Solis" (Birkenmajer, pp. 22–7). In this passage Brudzewo also clearly affirms his belief in realistic understanding of partial and total orbs. He says that without the help of the spirits "no mortal" can know which eccentrics and epicycles truly exist, and whether or not those proposed by astronomers are merely figments of their imagination. But there are real eccentrics and epicycles in the heavens; we, the mortals, just do not know exactly which ones of the many possible are real. Compare with Melanchthon's position in *Initia Doctrinae Physicae*, analyzed by P. Barker and B. Goldstein, "Realism and Instrumentalism in Sixteenth Century Astronomy: A Reappraisal," and P. Barker, "The Lutheran Contribution to the Astronomical Revolution: Science and Religion in the Sixteenth Century." There is a clear affirmation of the reality of particular orbs in Brudzewo, for example, in *theorica solis*. According to Brudzewo, the total sphere of the sun is divided into three real partial orbs, and then there is also an explanation of those movements by imaginary circles. If anything, this proves that circles are imaginary, not particular orbs; particular orbs are real (Birkenmajer, p. 22): "Theorica Solis principali divisione dividitur in tres partes. In prima parte ponit Magister divisionem totius sphaere solaris in orbes reales partiales ostendendo quot sunt et dispositionem quoad eorum superficies uniuscuisque centrum assignando. In secunda determinat de motibus illorum orbium, qua scilicet velocitate aut tarditate quilibet eorum moveatur, in quibus polis et axibus et cum regularitatem et irregularitatem eorum [...]. In tertia, aptando illos orbes ad circulos imaginarios, definit circulum ecentricum et ipsius consequentia declarat, videlicet: quid sit longitudo longior? Et sic de aliis [...]." P. Barker has very recently arrived to the same conclusion in his "Albert of Brudzewo's *Little Commentary on George Peurbach's 'Theoricae novae planetarum'*."

[130] See E. Zinner, *Leben und Wirken des Joh. Müller von Königsberg, genannt Regiomontanus*, pp. 239–41; P. L. Rose, *The Italian Renaissance of Mathematics*, p. 120; P. Barker, "Copernicus and the Critics of Ptolemy," p. 348.

[131] On Achillini, see also P. Barker, "Copernicus and the Critics of Ptolemy," pp. 349–50, and "The Reality of Peurbach's Orbs," pp. 16–7. For an overview of *De orbibus*, see also A. Goddu, *Copernicus and the Aristotelian Tradition*, pp. 238–43. On some of Achillini's cosmological reflections in *De orbibus*, see A. Poppi "La filosofia naturale del primo

three sets of arguments against Ptolemaic astronomy in *dubium tertium: vtrum eccentrici sunt ponendi.* He presented the three-orb compromise as a solution offering simpler hypotheses for saving the celestial appearances and as something that avoided void, penetration of the bodies, rarefaction, and condensation. He drew on the authority of Averroes to claim that there was not a single eccentric body in the heavens, and concluded that eccentrics and epicycles were *extra naturam.* Accordingly, he declared that "the motions that Ptolemy proposes are built on two foundations that do not agree with the natural sciences, the eccentric and epicycle." He presented three arguments to prove it.

The first argument was based on the idea of a unique centre of the world. If there were many centres of the world, a distinction between the natural centre of the world (= the centre of the element earth) and the mathematical centre of the world (= the centre point of the gravity of the element earth), that is, if the centre of the magnitude differed from the centre of the gravity, then something impossible would follow, for example, the existence of many earths. The second argument followed from the nature of the celestial bodies. If orbs were to suit epicycles and eccentrics they could not be absolutely and perfectly spherical. They would also necessitate the introduction of physically superfluous bodies to prevent the creation of *vacua.* The third group of arguments began with the consideration that circular motion was defined by a single centre. The first of these, for example, was based on the postulate that for one simple body there is only one movement that is naturally suited for him. If there were eccentrics and epicycles there were also many motions, of which some would be necessarily violent, that is, *praeter naturam.* And in the fourth of these, Achillini stated that circular motion could be performed only around the centre of the whole. In short, there were too many centres in Ptolemaic astronomy.[132]

Cinquecento nelle Università di Padova, Bologna e Ferrara," pp. 56–7. On Achillini as astronomer, see M. di Bono, *Le sfere omocentriche di Giovan Battista Amico nell' astronomia del Cinquecento*, pp. 62–4.

132 See *De orbibus* (*Opera Omnia*, fol. 29rb–29vb). It seems that Copernicus' first *petitio* of the *Commentariolus* – "There is no one centre of all the celestial circles or spheres [...]"– is directly related to Achillini's objections against many centres of the celestial movements, which makes Copernicus's second *petitio* completely logical: "The centre of the earth is not the centre of the universe, but only of gravity and of the lunar sphere." I agree with E. Rosen, "Copernicus' Axioms," pp. 46–7, that Copernicus *petitiones quas axiomata vocant* are logical and respond to historical situation, but in my view Copernicus' first *petitio* could be a reaction to Achillini and not to Eudoxus and Callippus. It is also possible that Achillini's elaboration on two centres, that of the

Padua, which hosted, as Swerdlow puts it, "some sort of Aristotelian school of astronomy,"[133] was an even more interesting place from the perspective of Copernicus' possible knowledge of Averroist attacks on Ptolemaic astronomy and perhaps even of some indications of a future revival of homocentric astronomy. Copernicus was in Padua at the same time as Fracastoro and Giambattista Della Torre, whom Fracastoro praised in his "Sanctissimo Paulo III pontifici maximo" in a later version of *Homocentrica* as the reviver of homocentric astronomy (even though he never published anything).[134] Fracastoro and Della Torre remained friends also after their student years, corresponding and even working together.[135] *Homocentrica* was the fruit of their later collaboration, but one can easily imagine that some discussions about homocentric astronomy must have already taken place when they were students. Note, for example, the similarity between Copernicus' description of the problems of homocentric astronomy in the *Commentariolus* and Fracastoro's in the *Preface*. They both maintained that the main problem facing homocentric astronomy was its inability to provide demonstrations for the celestial appearances. But while Fracastoro believed that his version of homocentric astronomy, based on Della Torre's ideas, could amend that, Copernicus looked for a solution in a heliocentric version of eccentric astronomy.

The Averroist criticism of Ptolemaic astronomy at Padua continued after Copernicus' departure in 1503. Amico studied there and in 1536, at the age of twenty four, published *De motibus*, his version of homocentric astronomy. The second, improved edition of *De motibus* was published in 1537.[136] Agostino Nifo,

---

magnitude and that of the gravity, stimulated Copernicus to reflect about this issue in more detail in *De revolutionibus* I, 3.

133 "Aristotelian Planetary Theory in the Renaissance: Giovanni Battista Amico's Homocentric Spheres," p. 36.

134 The first version of Fracastoro's *Homocentrica* was completed already in 1530. He sent one of the early versions of *Homocentric*a to his friend, Cardinal Gasparo Contarini (1483–1542), who responded with a short treatise, *De homocentricis ad Hieronymum Fragastorium*, containing doubts and questions. Fracastoro, in turn, responded to the cardinal. Both treatises were published in Gasparo Contarini, *Opera* (Paris, 1571).

135 Fracastoro, among other things, dedicated to Giambattista Della Torre one of his poems (*Opera*, fols. 199^r–200^v: *Carmen XIII*) about the death of his brother Marcantonio Della Torre (died in 1512) and named the second book of his trilogy of philosophical dialogues *Turrius* after Giovanni Battista Della Torre; the dialogue is actually Fracastoro's long monologue.

136 On Amico, see M. Di Bono, *Le sfere omocentriche di Giovan Battista Amico nell' astronomia del Cinquecento*; N. M. Swerdlow, "Aristotelian Planetary Theory in the Renaissance: Giovanni Battista Amico's Homocentric Spheres," and "Amico's Homocentric Spheres."

also a Paduan philosopher, attacked Ptolemaic astronomy *In Aristotelis libros De coelo et mundo commentaria*, which appeared in 1517.[137]

In sum: Both in the time before and during Copernicus' education – and even later, when he was compiling *De revolutionibus* – the most basic controversy was between two kinds of astronomy. One was homocentric astronomy based on the approved Aristotelian natural philosophy as promulgated by Averroes and his partisans, which tried – unsuccessfully – to provide homocentric models in agreement with observations. The other was Ptolemaic astronomy (*Almagest*; *Epytoma Almagesti*), whose models facilitated relatively adequate predictions of celestial positions that were consistent with observations, but failed to fully satisfy the requirements for compliance with Aristotelian natural philosophy even with the aid of the "Aristotelian-Ptolemaic" compromise (*Theoricae novae*). The Averroist critique of Ptolemaic astronomy was based on strict insistence to Aristotelian principles of natural philosophy as individual authors interpreted them. But this placed restrictions on astronomy. Philosophy or, more accurately, natural philosophy, was considered a superior science to astronomy and as such determined the principles from which astronomy was to follow. Since, to repeat one example, the Aristotelian natural philosophy in its Averroist interpretation only allowed circular motion which had the earth as its physical centre, eccentrics and epicycles were considered unacceptable. The problem facing the Averroists and their refutation of eccentrics, epicycles and associated real orbs was that homocentric spheres alone did not provide astronomy capable of producing correct and exact positional calculations. But that, of course, did not deter them from repeatedly chastising the Ptolemaic astronomy and making ever new attempts to introduce a homocentric – and philosophically acceptable – astronomy.

Problems within the Ptolemaic astronomy. At one level Copernicus' work may therefore be seen as a unique Ptolemaic response to the Averroist program. But this is, in my opinion, neither the whole story nor the most important part of it. He, like the majority of his astronomical contemporaries, deemed the Ptolemaic astronomical tradition the only satisfactory technical astronomy. But this does not mean that the Ptolemaic astronomy itself was, as Copernicus made very clear as well, without problems.

For a comparison between Fracastoro and Amico, see M. Di Bono, "Teme e fonti dell'*Homocentrica* di Fracastoro"; and "Copernicus, Amico, Fracastoro and Ṭūsī's Device: Observations on the Use and Transmission of a Model."

[137] See Book II, fols. 23–6.

(i) The equant. One of them was the equant. When Copernicus set out to write the *Commentariolus*, the equant was relatively well known to the Latin West. It was described several times by Peurbach in his *Theoricae novae planetarum*, in Albert of Brudzewo's *Commentariolum super Theoricas novas planetarum Georgii Purbachii* and in Peurbach and Regiomontanus' *Epytoma Almagesti*, which were all available to Copernicus during and after his studies. In *Theoricae novae planetarum* and the *Epytoma Almagesti* it was described without any qualifications,[138] so that Jerzy Dobrzycki and Richard Kremer rightfully claim that they "know of no extant text by Peurbach or Regiomontanus in which the Ptolemaic models are criticized explicitly on the grounds that they violate uniform, circular motion."[139] Brudzewo also described it as imaginary, but did not do anything more than that.[140] Upon closer reading of Brudzewo's text one can hardly claim that he was a vehement critic of the equant mechanism or that he, as Goddu claims, "clearly recognizes a physical problem with the equant model," that is, "it cannot refer to the motion of an orb."[141]

In sum, there was universal acceptance of the postulate of the uniform motion of the celestial bodies. Regiomontanus in his *Defensio* defined the task of astronomy as accounting for *equlitas primordialis et intrinseca*, which belonged to the nature of the celestial bodies by means of orbs;[142] Brudzewo called it in *Commentariolum super Theoricas novas* a philosophical principle according to which a celestial object must move in its orb with uniform and simple movement (he even revealed the source of his claim, *Almagestum parvum*, and its maxim that "all movements of the celestial bodies are simple and truly uniform").[143] Therefore

---

[138] See, for instance, *theorica lunae* (Aiton, p. 13): "From this it follows, first, that although the eccentric deferent of the epicycle moves about the axis and also its poles, it still does not move uniformly about them […]"; and especially *theorica trium superiorum* (Aiton, p. 18): "The motion of the deferent of the epicycle about its centre and its poles is nonuniform. Nevertheless this nonuniformity has this rule of regularity: that the centre of the epicycle moves uniformly about a certain point in the line of the apogee, as far from the centre is distant from the centre of the world. Consequently that point is called the centre of the equant, and the circle imagined about that point, the same size and in the same plane as the deferent, is called the eccentric equant."

[139] "Peurbach and Marāgha astronomy? The Ephemerides of Johannes Angelus and their Implications," p. 211.

[140] See *Commentariolum super Theoricas novas planetarum Georgii Purbachi, III:* "De tribus superioribus" (Birkenmajer, pp. 79–80).

[141] A. Goddu, *Copernicus and the Aristotelian Tradition*, p. 157.

[142] See, supra, pp. 260–1.

[143] See *Commentariolum super Theoricas novas planetarum Georgii Purbachii,* "Theorica

Copernicus' strong critical attitude towards this concept is surprising since none of these scholars explicitly criticized it and even less attempted to get rid of it and replace it with something else. It seems that Copernicus was the first in the Latin West to object to Ptolemy in specific regard to the equant.[144]

(ii) The order of the planetary spheres. Another problem facing the Ptolemaic astronomy of eccentric and epicycle models was that of the order of the planetary spheres. Copernicus was not the first to address this, however. Bits and pieces of discussions regarding the order of the planets were already presented by different authors who worked and published in various places.

Albert of Brudzewo discussed it in his *Commentariolum super Theoricas novas planetarum* and finally decided in favor of Ptolemy's ordering: Saturn, Jupiter, Mars, the sun, Venus, Mercury, the moon and the earth.[145] The situation in Bologna was more interesting. Two years before Copernicus arrived there, Codro expressed in his inaugural lecture for the academic year 1494–95 (now *Sermo I*) a deep mistrust in the reliability of liberal arts, including astronomy. "I shall pass over," he said, "the controversy concerning the size of the moon; they disagree also concerning the place and arrangement of the planets." Plato and those who followed the Egyptians located the sphere of the sun as the second after the moon and below Mercury. Archimedes and Chaldeans believed it occupied the fourth place among the seven, that is, in the middle of the planets. And he concluded: "Yet it is no wonder that the astronomers are inconstant and cannot agree among themselves when the very universe which they measure is the most inconstant of all things."[146]

Solis" (Birkenmajer, p. 23): "Et cum ipse Sol diverso motu et inaequali moveatur in zodiaco, senserunt ipsum – ex pricipiis philosophiae – debere moveri in suo orbe, aequali motu et simplici. Et hoc quidem dicit communis acceptio in Abbreviatio Almagesti (qui Alberto ascribitur) libro tertio, in haec verba: 'Omnem motum coelestis corporis simplicem et verum aequalem esse,' hoc est, super aequos angulos in centro motus consistentes et in arcus cadentes aequales, aequalibus fieri temporibus." *Abbreviatio Almagesti*, also called *Commentario Ptolomaei* or *Almagestum parvum* and ascribed to Albert the Great, begins with the following words: "Omnium recte philosophantium [...]."

144 See also F. J. Ragep, "Copernicus and his Islamic Predecessors," p. 67: "In particular Copernicus objected explicitly to Ptolemy's use of the equant, an objection that had been a staple of Islamic astronomy for some five centuries (but which seems not to have been made by earlier European astronomers)." See also pp. 70–1.

145 R. S. Westman, *The Copernican Question*, p. 54, underlines the astrological context of his account.

146 See Antonio Urceo Codro, *Orationes, Epistolae, Sylvae, Satyrae, fols.* VIIII^{v}–X^{r}: "De

Perhaps Codro's *Sermo I* reflected the thoughts of his friend Giovanni Pico Della Miarandola that first circulated from mouth to mouth and finally found the light of the day in its written form as his fierce attack on astrology in *Disputationes adversus astrologiam divinatricem* published in Bologna in July 1596.[147] One of the major arguments of Giovanni Pico Della Mirandola against astrology, discussed in Chapter 4 of Book X of *Disputationes*, was the lack of consensus among its practitioners regarding the order of all celestial spheres.[148]

According to Pico, the ancients and the moderns disagreed on the order of the planets. It was an ancient quarrel whether the sun was next to the moon (the earth, the moon, the sun, Mercury, Venus, Mars, Jupiter, and Saturn) or in the middle of the planets (the earth, the moon, Mercury, Venus, the sun, Mars,

---

magnitudine itidem lunae altera discordia quam praetereo et de situ sive ordinem planetarum etiam dissentiunt. Nam Plato a luna sursum secundam sphaeram solis locat, et aegyptios secutus uidetur qui inter lunam et mercurium solem ponunt. Archimedes cum chaldeis quartum de septem in medio locatum dicunt. [...] Sed non mirum est astrologos esse inconstantes et inter se non convenire cum caelom ipsum quod metiuntur, omnium sit inconstantissimum." See also, P. L. Rose, *The Italian Renaissance of Mathematics*, p. 121, and p. 135, n. 42.

[147] The unfinished manuscript of Giovanni Pico, who died on November 17, 1494, was prepared for publication by his nephew Gian Francesco Pico; the edition was pirated in 1597 by Venetian printer Bernardino Vitali. See more on *Disputationes* in M. Bertozzi, ed., *Nello specchio del cielo. Giovanni Pico della Mirandola e le 'Disputationes contra l'astrologia divinatoria'*, especially C. Vasoli, "La polemica antiastrologica di Giovanni Pico"; O. Pompeo Faracovi, "In difesa dell'astrologia: risposte a Pico in Bellanti e Pontano"; and S. Caroti, "Le fonti medievali delle *Disputationes adversus astrologiam divinatricem*." Compare with, M. Ficino, *Scritti sull' astrologia*. Ficino does not discuss the order of the spheres.

[148] See R. S. Westman, *The Copernican Question*, pp. 86–7, and N. Swerdlow, "Copernicus and Astrology," pp. 363–4. Swerdlow's translation of Chapter 4 of Book X of Pico della Mirandola's *Disputationes adversus astrologiam divinatricem*: "That the arguments of astrologers by which they strengthen their doctrines are for the most part reduced to five weak and ineffective kinds" is readily available at: http://www.mitpressjournals.org/doi/pdf/10.1162/ POSC_a_00071. The first to respond to Pico's attack on astronomy were Pontano and Bellanti. On this, see O. Pompeo Faracovi, "In difesa dell'astrologia: risposte a Pico in Bellanti e Pontano." Pontano was obviously an interesting author for Copernicus. Copernicus bought his *Opera*, and Rheticus, *Narratio prima* (Rosen, p. 143; Hugonnard-Roche and Verdet, pp. 58–9), also quoted from his *Urania* or *De stellis*: "Then, in harmony with these arrangements, God stationed in the centre of the stage His governor of nature, king of the entire universe, conspicuous by their divine splendor, the sun 'To whose rhythm the gods move, and the world // receives its laws and keeps the pacts ordained'."

Jupiter, Saturn). According to the Egyptians, the sun was next to the moon, whereas according to the Chaldeans, Ptolemy and a host of the moderns, it was in the middle of the planets. In Pico's reconstruction, others disagreed, including Jābir ibn Aflah, Theon of Aleksandria (ca. 335–405 AD), a Greek commentator on Ptolemy, as well as Plato and, surprisingly, according to Pico even Aristotle. They all placed the sun immediately above the moon: "However, not only Jābir, the acute mathematician, but also Theon himself, the Greek interpreter of Ptolemy, dissenting from the master, believe the sun is to be placed immediately above the moon, which Plato and Aristotle also affirm."[149]

The disagreement among authors comes as no surprise to Pico. There were many different criteria used in the dispute, none of them certain. The only certain criteria for ordering the spheres are parallaxes; without them there can be no decision on the arrangement of planets:

> And it is no wonder that the best authors disagree in this matter, concerning which many reasons have been handed down on both sides, none of which can be considered certain. For when you have not (add: *non*) offered parallaxes (parallaxes, text: *parallelares*), that is, differences of view (*obtutus discrimina*, parallaxes), mathematics in no way shows which of two stars is higher, and not only is there doubt concerning the sun but also concerning Mercury and Venus.[150]

Pico is, like Ptolemy before him, adopting parallaxes as criterion for determination of the order of the planets. There is no other certain and mathematical criterion that could do the job. This means that not only was the traditional order of Mercury and Venus under question but also that of Mars, Jupiter, and Saturn.[151] The same lack of consensus applied to the positions of Mercury and Venus in relation to the sun. Ptolemy placed the sun conveniently amidst the planets because of their elongation. While Saturn, Jupiter and Mars can be at any longitude away from the sun and reach opposition, Venus and Mercury are always close to it. But Pico is not satisfied with Ptolemy's criterion of placing the sun in the middle of the planets on the account of the difference between those planets that have every elongation (i.e. reach opposition as do Saturn, Jupiter

---

149 *Disputationes adversus astrologiam divinatricem* X, 4 (Swerdlow, pp. 4–5; Garin, pp. 372–3). According to Swerdlow, "Copernicus and Astrology," p. 363, the reference to Jābir ibn Aflah is to his *Correction of the Almagest*.

150 *Disputationes adversus astrologiam divinatricem* X, 4 (Swerdlow, p. 5; Garin, pp. 372–3).

151 See ibid.

and Mars) and those which have only limited elongation (Venus and Mercury). There is one planet, the moon, which reaches every elongation, just like Saturn, Jupiter and Mars, but is still believed to be closest to the earth. For Pico, this is a ridiculous and incoherent conjecture, because the moon, like the superior planets, also digresses from the sun, which means that the sun is not situated between the planets which do and do not digress:

> That well-known sort (of reason) of Ptolemy's, of the sun suitably placed in the middle between the planets, since some move away from it to a place at a great distance (i.e. reach opposition), which Saturn and Jupiter and Mars do, some always accompany it, as Venus and Mercury, is a worthless conjecture, not effective, since the moon departs from the sun as far as the superior planets, so the sun is not in the middle between the planets departing and not departing from it.[152]

Similarly weak, according to Pico, is the argument of al-Bītrūji, in which Venus was placed above the sun and Mercury below it.[153] Pico then considers the position of those who place Mercury and Venus above the sun, which means that the sun is placed next to the moon. Their reasons seem more reasonable, since the fact that no transit of Venus and Mercury has been observed seems to confirm this ordering. But Pico is still not satisfied and goes on to find different explanation for this fact. The absence of any observed transit of Mercury is explained on account of its smallness. The absence of Venus' transit is explained by the possibility that Venus, when very close to the sun, "is fallen upon and intercepted by the [sun's] rays" and is therefore unable to obstruct the path of the sun's rays toward the observer on the earth. The moon, on the other hand, which is much more distant from the sun, can obstruct the sun's rays. There is yet another explanation for why Mercury and Venus's transit is not observed. In contrast to the moon they could be translucent. The last possibility Pico discusses is that they have their own light, that is, they could be self-luminous and thus replace the sun's light they have taken away by their transit. Then follows account on Averroes' report from *Paraphrase of the Great Treatise of Ptolemy* of the observation of two spots on the sun and his calculation that suggested it was transit of Mercury.[154] Pico

---

[152] Ibid.

[153] See ibid. (Swerdlow, p. 5; Garin, pp. 373–5). See also Regiomontanus' explanation in *Epytoma Almagesti* IX, 1, in N. Swerdlow, "Copernicus and Astrology: Appendix," p. 9, quoted, infra, p. 305, n. 158, and Swerdlow's explanation, pp. 10–1.

[154] See *Disputationes adversus astrologiam divinatricem* X, 4 (Swerdlow, pp. 5–6; Garin, pp.

concluded this passage alluding to the authority of Maimonides that the order of all planets, not only of the sun, Mercury and Venus, remain uncertain.[155]

Pico's book was definitely one of Copernicus' major sources in writing Chapter 10 of Book I of *De revolutionibus*,[156] but there were two other books that appeared in Bologna during his stay there that discussed the arrangement of the celestial spheres. In the same year as Pico's *Adversus*, Regiomontanus' *Epytoma Almagesti* was published. In Proposition 1 of Book IX, "To show in what order the celestial spheres are to be arranged," Regiomontanus, following Ptolemy's *Almagest*, draws attention to different orderings of the planets and uses the term *controversio* in reference to the lack of consensus regarding the position of Mercury and Venus in relation to sun. He first reports, in contrast to Pico, that there is agreement in regard of the order of the sphere of the fixed stars, Saturn, Jupiter and Mars. There is also agreement on the lowest place of the moon on the basis of its parallax:

> [o]ur predecessors held different opinions concerning this, yet they all agreed in this one: that the sphere of the fixed stars is higher than all the other spheres of the planets, under which in mutual agreement they placed in order the sphere of Saturn, next the sphere of Jupiter, and under this the sphere of Mars. Likewise, they assigned the sphere of the moon the lowest place, and sensibly so whether because (otherwise) the condition for a solar eclipse is precluded or because among all the stars the moon has a parallax more apparent to perception.[157]

According to Regiomontanus the problem is with the ordering of the sun, Venus and Mercury. Some, "the most ancient," adopted the arrangement Mars, the sun, Venus, Mercury, and the moon. Later astronomers, lacking evidence of transit by Venus and Mercury, placed them above the sun. Al-Bitrūjī, for his

---

374–5). The work of Averroes, *Paraphrase of the Great Treatise of Ptolemy*, was available only in Hebrew, but Pico studied Hebrew with Elia del Medigo, who had Averroes' books. See also Swerdlow's explanations, "Copernicus and Astrology: Appendix," pp. 7–8, accompanying his translation.

[155] See *Disputationes adversus astrologiam divinatricem* X, 4 (Swerdlow, p. 6; Garin, pp. 374–5).

[156] See also Rheticus, *Narratio prima* (Rosen, pp. 126–7; Hugonnard-Roche and Verdet, pp. 49–50): "If such an account of the celestial phenomena had existed a little before our time, Pico would have had no opportunity, in his eighth and ninth books, of impugning not merely astrology but also astronomy."

[157] *Epytoma Almagesti* IX, 1 (Swerdlow, p. 9).

own reasons, placed Venus below the Mars, that is, above the sun, and Mercury below the sun:

> But concerning the other three there was disagreement. For the most ancient placed in order under Mars the sun, under which Venus, and above the moon Mercury. Those coming later, who directed more frequent sightings to conjunctions of the sun with Venus and Mercury, since they never observed an eclipse of the sun at the appropriate conditions of Venus and Mercury, they considered those to be placed above the sun.[158]

In Regiomontanus' opinion the moderns approve the ancient order according to which Venus and Mercury are below the sun. Then he considers the absence of the observed transits of the Mercury and Venus. In addition to Pico's reasons, he mentions two further explanations.

When Mercury and Venus are in conjunction they may not be in the line that connects the eye with the centre of the sun. Their latitudes may, as Swerdlow puts it, "exceed the apparent radius of the sun." Besides, Venus and Mercury are extremely small and therefore their transit should be undetected by the eye.[159] Regiomontanus then brings into discussion another reason why Mercury and Venus should be placed below the sun. There is an immense distance of about 1006 earth's radii between the moon and the sun and nature does not want this enormous space to be left vacant. It is perfect that it is occupied by the spheres of Venus and Mercury. However, it is impossible to determine which of these two planets is above and which below, since their parallaxes cannot be determined precisely enough.[160]

Another text available and very likely known to Copernicus, although there seems to be no textual evidence that he used it when writing Chapter 10 of Book I of *De revolutionibus*, was Alessandro Achillini's already mentioned *De orbibus* of 1498, also published in Bologna. It also critically discussed the order of the planets, especially Mercury and Venus. The question was whether Mercury and Venus were above the sun since they never eclipsed it. Achillini accused Ptolemy of contradicting himself in his treatment of Mercury and Venus in respect to the

---

[158] Ibid. He continues: "However al-Bitrūjī, who believed that the inequalities of motions and apparent velocities of the planets occur through a kind of falling behind (*quadam incurtatione*), placed under Mars Venus, under which the sun, then Mercury, for Venus falls behind (*incurtat*) from the first motion less than the sun, as he said, in fact on account of the epicycle, but Mercury more than the sun."

[159] See ibid.

[160] See ibid.

sun in Chapter 1 of Book IX of the *Almagest*.[161] Ptolemy first stated that he could not determine whether Venus and Mercury ever eclipsed the sun, and concluded that they must be above it. But subsequently he admitted that they were in the same line that connected the sun and our eyes. Achillini's conclusion was that they were below the sun.[162]

In writing Chapter 10 of Book I of *De revolutionibus*, Copernicus did also use, as mentioned, Proclus' *Hypotyposis* in Giorgio Valla's *De expetendis et fugiendis rebus*, published in Venice by Aldo Manuzio in 1501.

(iii) The Calendar. The last and the least problem of Ptolemaic astronomy mentioned by Copernicus in the *Introduction* and the *Preface* was that of the calendar and the tropical year. It seems that Copernicus became aware of it through Pico della Mirandola's *Disputationes*.[163] There Pico not only accused astronomers of being unable to locate the planets with certainty, but also pointed to their disagreement regarding the length of the tropical year. In Chapter 9 of Book IX, there is an account on Hipparchus, Ptolemy, al-Battānī and Thābit ibn-Qurra, who disagreed on that matter. Some believed that all years were of the same length, while others held them to be unequal and attributed this inequality to the non-uniform motion of the eighth sphere.[164]

3. Platonism as a common denominator

These were, according to Copernicus, major critical moments of the Ptolemaic astronomy: the tropical year and the calendar, the equant and especially its inability to determine with certainty the order of the planetary spheres. As stated above, Copernicus might have been aware of some other problems or perceived some of the characteristics of Ptolemaic astronomy as problematic as well – the retrograde motion of the planets; the size of the lunar epicycle, which predicted a doubling of the size of the moon in the quadrature; the fact that Venus and Mars had the largest epicycles of the planetary models – but he explicitly mentioned only the tropical year, the equant and the order of the planetary spheres.

---

161 See also R. S. Westman, *The Copernican Question*, p. 99.

162 See *De orbibus* (*Opera Omnia*, fol. 29rb–fol. 29vb). R. S. Westman, *The Copernican Question*, p. 531, n. 139, explains that since there is no such discussion of Mercury and Venus in Chapter 1 of Book III of the *Almagest*, Achillini might have been using Geber's *Correction of the Almagest* (*Islah al-Majisti*).

163 See also R. S. Westman, *The Copernican Question*, pp. 85–6.

164 See Pico Della Mirandola, *Disputationes adversus astrologiam divinatricem* IX, 9 (not in Swerdlow; Garin, pp. 327–33).

Now it is time to answer the third question we posed: Why were all these problems so critically important to him?

All the aforementioned issues were familiar to both the astronomers before Copernicus and to his contemporaries. The problem of the equant was known to Muslim astronomers and in the Latin West to Peurbach and Regiomontanus. Brudzewo knew and discussed it, but none of the Latin authors seemed that concerned about it and would make even less effort to abolish it. Copernicus was apparently the first in the Latin West to address the problem of the equant with such vigor and to present it as the stimulus to overthrow the traditional geocentric cosmology. Why was he, of all capable astronomers, so upset over it? And why was the disagreement among astronomers and philosophers regarding the ordering of the universe's spheres, or rather, the real form of the world, an issue also known to Copernicus' contemporaries and predecessors, all of a sudden so immensely important to him?

The arrangement of the planetary orbs was, as we have seen, indeed, not unproblematic even before Copernicus: Pico della Mirandola attacked the lack of consensus on the matter, and Regiomontanus even characterized it as a controversy. But no one was so alarmed by it as to call for a radically new arrangement of the celestial spheres. No one before Copernicus went to such great lengths to solve it. Why? Or, as Westman asks:

> what historical circumstances could have made the ordering of Venus and Mercury into a matter of such urgency that it motivated a new look at the application of Euclid's or Aristotle's version of the period–distance principle? And why was Copernicus motivated to adopt the sun rather than the earth as the centre of all, rather than some, celestial motions?[165]

Or, in my own words: Why did *all* of these problems become so important to Copernicus? Were all of these issues somehow related, and if so, how?

We can narrow these questions even further. How is the equant problem related to the *forma mundi* problem? Several centuries of Arabic astronomy, which dealt critically with that concept, clearly prove that the solution to the equant problem did not imply any move in the direction of geokinetism or heliocentrism. Copernicus used heliocentrism to solve the problem of the order of the spheres. Was there a hidden common denominator, a missing link that connected all these issues into an inherently coherent logic?

---

[165] *The Copernican Question*, p. 61.

Copernicus himself, again, provided a tiny clue to the answer to these questions. According to his own words in the *Preface*, he reflected for a long time on the confusion in the astronomical traditions concerning the derivation of the motions of the universe's spheres and then went on to explain what upset him so much: "I began to be annoyed that the movements of the world machine, created for our sake by the best and most systematic Artisan of all, were not understood with greater certainty by the philosophers, who otherwise examined so precisely the most insignificant trifles of this world."[166] He expressed a similar view, with a slightly different undertone, in the *Introduction*. There he spoke about "things established in the finest order and directed by divine management" that led a man to "the admiration for the Maker of everything, in whom are all happiness and every good."[167] It is in this conception of the universe, the *machina mundi* created by the best Artisan in the finest order, and managed by him *for our sake*, to provide us with a visible guide through which we would be led to supreme Good – and thus to the supreme happiness – that we must in my opinion seek the ultimate reason for Copernicus' deep dissatisfaction with the state of astronomy of his time.

But whose conception was it? Copernicus' own? Or was it a general Christian view? This seems to be suggested by the Psalm Copernicus quotes in the next line of the *Introduction*: "For would not the godly Psalmist in vain declare that he was made glad through the work of the Lord and rejoiced in the works of His hands, were we not drawn to the contemplation of the highest good by this means, as though by a chariot?"[168] The view expressed in these few quoted sentences – and some others as we shall see shortly – is, of course, easily adaptable to the Christian understanding of the universe and its purpose for mankind. But it also has a very specific pagan origin, a source which Copernicus himself reveals in the sentence that follows his paraphrase of the Psalms. Here Copernicus points out how advancements in astronomy serve the greater community: "The great benefit and adornment which this art confers on the commonwealth (*vtilitatem et ornamentum Reipublicae conferat*) – not to mention the countless advantages to individuals – are most excellently observed by Plato in the *Laws, Book VII.*"[169] It is my thesis that Copernicus' critical attitude towards the state of astronomy, which ultimately resulted in his geokinetism and heliocentric arrangement of the planetary orbs, was founded upon Plato's views on the order and arrangement of the universe created by the supreme Artisan, and on the status and

[166] *Preface* (Rosen, p. 4; Zekl, p. 72).

[167] *Introduction* (Rosen, p. 7; Zekl, pp. 81–2).

[168] Ibid. (Rosen, p. 7; Zekl, p. 82). The reference is to *Psalms* 92, 4.

[169] *Introduction* (Rosen, p. 7; Zekl, p. 82).

role of astronomy in discovering that order developed in his dialogues (mainly the *Laws*, the *Epinomis*, the *Republic* and the *Timaeus*) and then subsequently repeated and reformulated by different Platonists and commentators on Plato. Those Platonist conceptions not only played a negative role in the criticism of Ptolemaic astronomy but were at the same time to certain extent instrumental in Copernicus' discovery of heliocentric cosmology.

There are several different classes of evidence, of different importance and weight, which support this claim.

First, during his education in Cracow and Italy, Copernicus moved in circles that evaluated Plato positively. Second, his humanist leanings and positive attitude towards the Platonist (and Pythagorean) conception of world harmony presided over by the sun are expressed in his private seal. But in addition to this circumstantial and indirect evidence, there are several more important classes of direct evidence. Third, Copernicus bought and annotated a book of Cardinal Bessarion, *In calumniatorem Platonis*, in which he read the praise of Plato as a mathematician. Fourth, and even more importantly, he annotated at least one of Plato's dialogues, the *Parmenides*, if not more, in Marsilio Ficino's translation of Plato's *Opera omnia*, first published in 1484. He certainly read several others. Fifth, Platonic influence is corroborated by his references in the holograph of *De revolutionibus* and in its printed edition of 1543, which proves that Copernicus had Plato constantly at hand and referred to him when he needed to. Sixth, Plato's influence on Copernicus as well as the fact that Copernicus had access to his work is also evident from Rheticus' references to Plato in *Narratio prima*. And finally, the seventh and most important piece of evidence: there is philosophical correspondence between the views of Plato and Copernicus in regard to the universe's arrangement and the role of astronomy in discovering that order.

Plato and Platonism during Copernicus' education. Copernicus was surrounded by humanists and Platonists and had easy access to the works of Plato and others throughout his studies. From Cracow to Padua, and even later, upon his return to Warmia in 1503, first to Lidzbark and afterwards to Frombork in 1510, he would have been surrounded by works of Plato, and of the Pythagorean, Neo-Platonist and Hermetic traditions.

At the University of Cracow, the student Copernicus had close contact with scholars who evaluated Plato positively and were up-to-date with the latest Neo-Platonist developments. The presence of a Plato Reading Room at the university indicates a strong interest in Plato, which was also shared by two most important teachers, Albert of Brudzewo and John of Glogów. They had contacts with humanist circles and read Plato. Plato's works were also widely available: the content of the libraries of some teachers show that there were several copies of his *Timaeus* with Chalcidius' *Commentary* already in the 1470s and 1480s,[170] and there were at least four complete copies of Plato's works (*Opera Platonis*) available at the turn of the sixteenth century.[171] Among them was a copy that served as the university copy (*Opera Platonis Universitatis*).[172] Conrad Celtis, for example, had one copy in his own library and another one was owned by Jan Heidecke (Mirica). Callimachus corresponded with the leading Florentine humanists, including Ficino, who sent him his books. Ficino's works, as well as those of the Hermetic and Pythagorean traditions, were also owned by some other Cracow intellectuals.

One humanist and Platonist at Cracow was especially close to Copernicus. Although there is no decisive evidence to support that Copernicus was ever a student of Laurentius Corvinus, it seems that they established a close relationship, maybe even a friendship, after they met at the university and remained in contact up to at least 1509.[173] Corvinus, whose work shows Platonic and sometimes even Hermetic influences, derived from Ficino's translations, helped to publish Copernicus' Latin translation of Theophylactus Symocatta's Greek *Letters* and wrote a eulogy for it, describing the translator as a man "who discusses the swift course of the moon and the alternating movements of its brother as well as the stars together with the wandering planets – the Almighty's marvelous creation – and he knows how to seek out the hidden causes of phenomena by the aid of wonderful principles."[174]

---

170 These copies belonged to Andrzej Grzymala (d. 1466), Sędziwój z Czechla (d. 1476) and Jan Dąbrówka (d. 1472). See M. Kokowski, *Copernicus's Originality*, p. 176.

171 On the date of the first publication of Ficino's translation, see P. O. Kristeller, "The First Printed Edition of Plato's Works and the Date of its Publication (1484)."

172 See also L. A. Birkenmajer, *Stromata Copernicana*, pp. 80–1, and n. 6; B. Biliński, *Il Pitagorismo di Niccolò Copernico*, p. 145; J. Domański, "La fortuna di Marsilio Ficino in Polonia nei secoli XV e XVI," pp. 569–70; D. Knox, "Ficino and Copernicus," p. 405.

173 See E. Rosen's explanations, in Copernicus, *Minor works*, pp. 19–22. See also J. Dobrzycki, "Notes on Copernicus's Early Heliocentrism," p. 223, and A. Goddu, *Copernicus and the Aristotelian Tradition*, pp. 43–4.

174 Theophylactus Simocatta, *The Moralia. His Ethical, Rustic and Love Letters in a Latin Translation*, in Copernicus, *Minor works* (Rosen, p. 27): "Qui celerem lunae cursum alternosque meatus Fratris cum profugis tractat et astra globis, Mirandum Omnipotentis opus, rerumque latentes Causas scit miris quaerere principiis." Latin text is taken from M. Markowski, "The Earliest Unknown Excerpts from Nicholas Copernicus' *De revolutionibus*," p. 6, and B. Biliński, *Il Pitagorismo di Niccolò Copernico*, p. 61. For

Copernicus' humanist and Platonist interests were also well served during the time he spent in Italy, first in Bologna, where he lived with Novara, a student of Regiomontanus. This relationship linked him directly to the Plato–Aristotle concordist Bessarion. Later, probably in Padua, he bought Bessarion's *In calumniatorem Platonis*. Despite its being an Averroist stronghold, not even Padua could escape humanist and Platonist influences. Apart from Nicholas Leonicus Tomaeus, who very probably read Plato in Greek and translated portions of the *Timaeus*, there were apparently other figures associated with the *Studio Patavino* who followed the "Platonic way." These included Federicus Chrysogonus, and more importantly a group which included Fracastoro and other philosophers usually labeled rigid Averroists: Nicoletto Vernia, Agostino Nifo and Marcantonio Zimara.

Apollo insignia. Copernicus' humanist, Pythagorean and Platonist attitude is also revealed thorough his insignia used in his private letters: Apollo playing a lyre (see *Figure 19*).

According to a very persuasive analysis of Stanislaw Mossakowski,[175] his choice of Apollo playing a lyre may have been linked to his thorough humanistic education, admiration for Antiquity and membership in the humanistic *respublica literaria*. But it may also have a deeper meaning. Apollo occupied a special place in the astronomy and astrology of Antiquity, as well as that of the Middle Ages and Renaissance. Apollo-Phoebus, a sun-god, appeared frequently in classical Latin literature related to the Pythagorean theory of the harmony of the universe, a theme taken up by Plato. According to the Pythagorean theory of the harmony of the universe, the celestial spheres in their constant motions produced certain clearly definable sounds which constituted a full and perfect harmony, inaudible to human beings. This was interpreted as a song or a play of nine Muses, who at the same time became the movers of the spheres. Apollo-Phoebus, the planetary god Helios, was a synonym for a sun-god, the ruler of the cosmos. The sun as

different interpretations of this passage, see M. Markowsky, "The Earliest Unknown Excerpts from Nicholas Copernicus' *De revolutionibus*," pp. 6–7; Rosen's comment on p. 24, and Dobrzycki, "Notes on Copernicus's Early Heliocentrism," p. 223; M. Clutton-Brock, "Copernicus's Path to his Cosmology," p. 203; A. Goddu, "Reflections on the Origin of Copernicus's Cosmology," pp. 37–8.

175 S. Mossakowski, "The Symbolic Meaning of Copernicus' Seal." Conrad Celtis had an image of Apollo and Hermes, and Fracastoro had an image of an altar dedicated to Aesculapius, Apollo, and Minerva. See also two studies of K. Hutchinson, which add some additional dimensions to Mossakowski: "Copernicus, Apollo, and Heracles," and "Harmony and Authority: The Political Symbolism of Copernicus' Political Seal."

Figure 19. Copernicus' seal: Apollo playing a lyre.

the leader of the Muses (*Dux Musarum*) was "thought to be the ruler of the planetary spheres and the seven tones of his musical instrument, represented by the seven-corded ancient Greek lyre, were held to be the parallel of the sounds of the seven spheres from which the world harmony arose."[176] This means that the whole Cosmos was ruled by Apollo with a lyre in his hand (see *Figure 20*).

The ancient theme of Phoebus and his playing as a source of the harmony of the world could not have been unknown to Copernicus. He probably knew of it from Macrobius' *In somnium Scipionis*, which was obviously, as indicates John of Glogów's familiarity with it, available to him already in Cracow:

> In a discussion in the *Republic* about the whirling motion of the heavenly spheres, Plato says that a Siren sits upon each of the spheres, thus indicating that by the motion of the spheres divinities were provided with song; for a singing Siren is equivalent to a god in the Greek acceptance of the world.

---

176 S. Mossakowski, "The Symbolic Meaning of Copernicus' Seal," p. 456. See also his explanation, pp. 454–6, of Franchino Gafurio's frontispiece in his *Practica musicae* (Milan, 1496). Gafurio was born in 1451 and died in 1522.

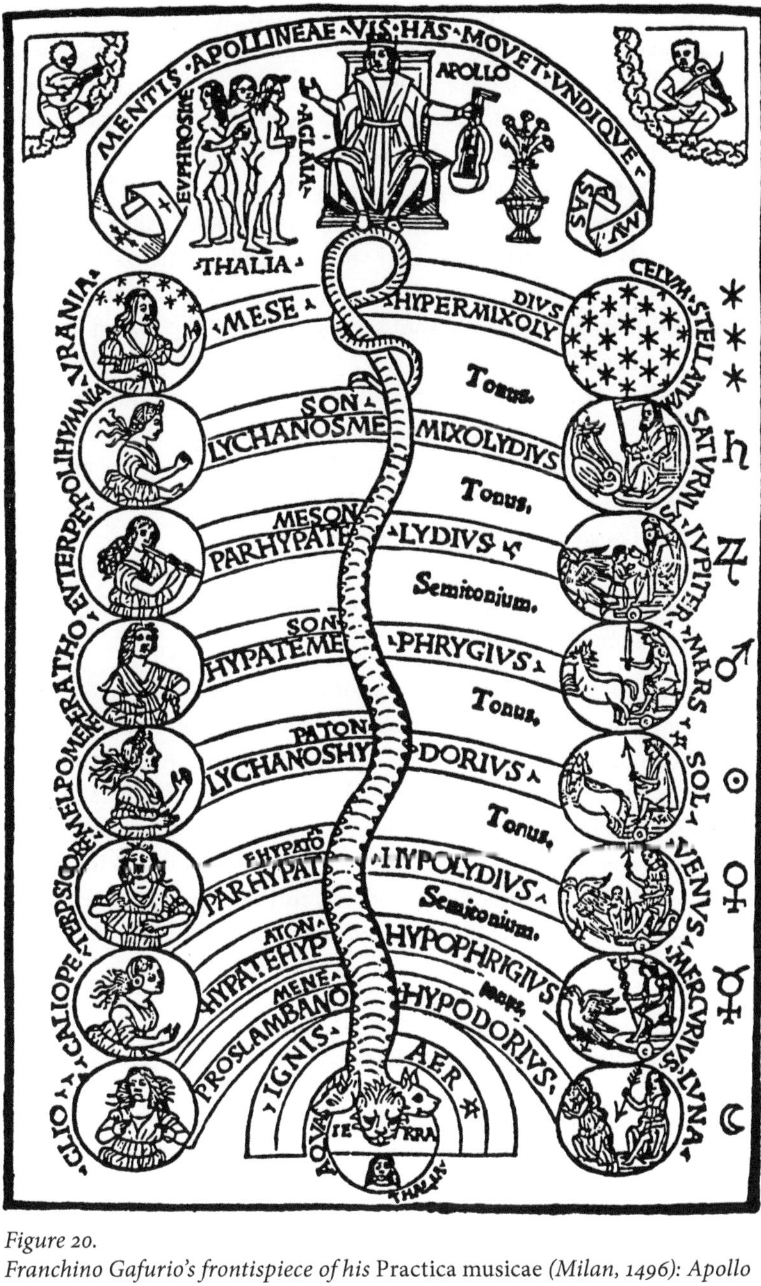

Figure 20.
Franchino Gafurio's frontispiece of his Practica musicae *(Milan, 1496): Apollo with a lyre rules the universe.*

> Moreover, cosmogonists have chosen to consider the nine Muses as the tuneful song of the eight spheres and the predominant harmony that comes from all of them. In the *Theogony*, Hesiod calls the eight Muse Urania because the eight sphere, the star-bearer, situated above the seven errant spheres, is correctly referred to as the sky, and to show that the ninth was the greatest, resulting from the harmony of all sounds together, he added 'Calliope, too, who is preeminent Among all.' The very name shows that the ninth Muse was noted for the sweetness of her voice, for Calliope means 'best voice'. In order to indicate more plainly that her song was the one coming from all the others, he applied to her a word suggesting totality in calling her 'preeminent among all'. Then, too, they call Apollo, god of the sun, the 'leader of the Muses', as if to say that he is the leader and chief of the other spheres, just as Cicero, in referring to the sun, called it leader, chief of the other planets, mind and moderator of the universe. The Etruscans also recognize that the Muses are the song of the universe, for their name for them is *Camenae*, a form of *Canenae*, derived from the verb *canere*.
> That the priests acknowledge that the heavens sing is indicated by their use of music at sacrificial ceremonies, some nations preferring the lyre or cithara, and some pipes or other musical instruments. [177]

These topics of Apollo, the lyre (not necessarily always his) and the harmonious arrangement of the world was very much also on Marsilio Ficino's mind, but as far as I know there is not a single passage that links all the three themes together.[178] Apollo's theme was treated by at least two members of the *Sodalitas Litteraria Vistulana*. Copernicus's friend Laurentius Corvinus wrote the poem *Carmen elegiacum de Apolline et novem Musis* and Conrad Celtis, who wrote an *Ode to Apollo*,[179] also ordered a woodcut representing Apollo-Phoebus playing the lyre surrounded by the Muses, produced probably in 1502 but certainly before

---

177 *In somnium Scipionis* II, 3, 1–5 (Stahl, pp. 193–4; Neri, pp. 452–4). Compare with Cicero, *De re publica* VI, 17 (= *Somnium Scipionis*, 4–5; Stahl, pp. 72–3), and Pliny the Elder, *Historia naturalium* II, 3–4, 6–13 (Rackham, vol. 1, pp. 172–9) and 20, 84 (Rackham, pp. 226–9).

178 As far as I am aware *Theologia platonica* II, 9, 7, is closest to it. See infra, p. 351, for the quotation. For a summary of Ficino's thoughts on these subjects, see C. L. Joost-Gaugier, *Pythagoras and Renaissance Europe*, pp. 80–2, and 86–7. For more on Ficino's Pythagoreanism, see also Ch. Celenza, "Pythagoras in the Renaissance: The Case of Marsilio Ficino."

179 See English translation, in L. W. Spitz, *Conrad Celtis. The German Arch-Humanist*, p. 10. The ode is from Celtis' *Ars versificandi*, p. 47.

1508. The woodcut, titled *Melopoiae Sive Harmoniae Tetracenticae*, was probably made by Hans Suess von Kulmbach (ca. 1480–ca.1522) and illustrated by Petrus Tritonius (Treybenreif; 1465–ca. 1525). It was published in 1507 under the auspices of Celtis and the literary sodality in Augsburg.[180]

Bessarion: *In calumniatorem Platonis.* As may be inferred from the evidence above, Copernicus was constantly exposed to Platonist influences and moved in humanistic circles who favored Plato. But in addition to this contextual evidence attesting to his Platonist philosophical orientation, there are several even more important types of direct evidence, the first being represented by Bessarion's book *In calumniatorem Platonis*, published for the second time in 1503 by Aldo Manuzio in Venice.[181] Copernicus bought it a few years after he became acquainted with the most vigorous attack on the certainty of astronomy by Pico della Mirandola in his *Adversus astrologiam divinatricem* and its purchase suggests that he was, very likely through Novara, acquainted with Bessarion's role in the renovation of astronomy, his influence on Regiomontanus, and his broader humanist project that included the rehabilitation of Plato and Platonism.

In calumniatorem Platonis represents philosophical aspect of Bessarion's polemics with George of Trebizond. While Bessarion's dissatisfaction with George's translation and commentary on the *Almagest* brought about Peurbach and Regiomontanus' *Epytoma Almagesti* and Regiomontanus' *Defensio Theonis*, his *In calumniatorem Platonis* was Bessarion's response to George's evaluation of Plato's philosophy in his *Comparatio philosophorum Platonis et Aristotelis* of 1458.[182] This particular work was in turn provoked by Georgios Gemistos Pletho's book comparing Plato and Aristotle, in which the author expressed his preference

---

[180] See, L. W. Spitz, *Conrad Celtis: The German Arch-Humanist*, p. 81, and p. 130, n. 7.

[181] The full title of the book reads: *In calumniatorem Platonis: Correctio librorum Platonis De legibus Georgio Trapezuntio interprete. De natura et arte adversus eundem Trapezuntium tractatus.*

[182] The controversy over the relative superiority of Plato and Aristotle was initiated by Georgius Gemistus, known as Pletho (ca. 1355–1452/1454), in his famous treatise *Peri hon Aristoteles pros Platona diapheresthai (De differentiis Platonis et Aristotelis).* See an overview of the debate in J. Hankins, *Plato in the Italian Renaissance*, pp. 193–217; C. H. Lohr, "Metaphysics," pp. 558–68. For a historical overview of the "comparison genre," see F. Purnell, Jr.," The Theme of Philosophic Concord and the Sources of Ficino's Platonism," pp. 401–15. What should be underlined for our purposes is that, in addition to Bessarion, Copernicus read at least two more "concordists," Alcinous and Cicero.

for Plato. In contrast, George tried to show that Aristotle was a better philosopher than Plato. What is more, while Plato only seemingly agreed with Christianity and was actually a polytheist who believed in creation from uninformed matter and in pre-existence of souls, Aristotle, though not a monotheist, believed in creation *ex nihilo* and immortality of the human soul. George also denounced Plato's immorality and bad influence on political life. Bessarion's *In calumniatorem Platonis*, written and rewritten several times in Greek before being published for the first time in Latin in 1469, was, according to John Monfasani "the most important Platonic text in the Renaissance before Ficino's *Theologia Platonica*."[183] Copernicus bought a new Aldine edition published in 1503, in which he read several interesting things regarding Plato's views on astronomy.[184]

In the first book Bessarion responded to the George of Trebizond's claim that Plato's learning was deficient. Actually, according to Bessarion, Plato and Aristotle agreed on natural philosophy. Plato also made a great contribution to the liberal arts, which Aristotle neglected. Plato's role was especially important in mathematical sciences and in theology. In Chapter 1 of Book I, Bessarion explained his reasons for defending Plato against the attack of George of Trebizond: the Latin West only disposed of a few of Plato's works which were only rarely read. In Chapter 2, he claimed that Plato spoke most sublimely of divine things and gave reasons why he either wrote nothing about them or wrote enigmatically. In this regard, he held Plato to be in agreement with the Pythagorean manner of keeping things secret, and added *Lysis' letter to Hipparchus*, which Copernicus later translated himself and intended to include in his *De revolutionibus* (he ultimately decided to omit it; his translation drew on Bessarion's). In Chapter 2 of Book I, Bessarion claimed that even in natural philosophy some people would occasionally believe Plato over Aristotle. In Chapter 3 of Book I, he defended Plato's rhetorical style. In Chapter 4, he tried to show that Plato used all kinds of syllogisms and that the demonstration in the *Timaeus* was based on syllogisms. In Chapter 6, he maintained that Plato and Aristotle did not differ on the principles of natural things, but that Plato had a more perfect doctrine on the heavens than Aristotle. In Chapter 8, Bessarion claimed that Plato was a most profound student of mathematical disciplines and that his work could not be understood without mathematics. He explained the relevance of geometry for the education

---

183 "George of Trebizond's *Comparatio Philosophorum Platonis et Aristotelis*," p. 2. See also, "A Tale of Two Books: Bessarion's *In Calumniatorem Platonis*."

184 For an overview of the book, see also A. Goddu, *Copernicus and the Aristotelian Tradition*, pp. 220–5.

of legislators, referring to the *Laws* and the *Epinomis*, in which Plato ascribed four mathematical arts to genuine wisdom, while relegating all others to a lesser status.

In Books II and III, Bessarion addressed the question of each authors compatibility with Christianity and defended the view that Plato was in a better agreement with it. In Chapter 1 of Book II, he referred to the Church Fathers who quoted Plato more than any other gentile author. In Chapter 3 of Book II, he discussed Plato's philosophy regarding the doctrine of super-substantial divinity exposed in the *Parmenides*. According to Plato, mind governed heaven and earth. Chapter 6 of Book II contains a summary of Plato's views on the creation of the world from the *Timaeus* 28a–37e. Bessarion pointed to the goodness of God (constantly referring to Him as *opifex*, *auctor* and *creator*, who looked at the eternal exemplar while creating the world (several times referred to as *machina mundi*).[185] In Chapter 7 of Book II, Bessarion explicated Plato's views on the soul and utility of mathematical disciplines in bringing the humans to recollect the divine. In Chapter 10 of Book II, he exposed Plato's doctrine on the principles of nature, and the resolution of the elements into plane and linear figures. Book III was more or less devoted to theological and metaphysical issues, only at the end of it, in Chapter 26, Bessarion discussed Plato's opinions on mathematical disciplines, the abstractions of which prepared a rational soul to understand the divine.

In Book IV, Bessarion responded to the charge of Plato's immorality. In Chapter 12, for example, he defended Plato against the accusation that mathematics was to be taught to those who wanted to become divine. He declared that, according to Plato, mathematics was truly most worthy of a free man and continued, paraphrasing the *Epinomis*, that the easiest way to ascend to the divine was through mathematics. He concluded the chapter by referring the reader to Books VII and X of the *Laws*, to the *Epinomis* as well as to Books V, VI and VII of the *Republic*.

Copernicus' reading of Plato. It seems that Copernicus was impressed by Bessarion's report on Plato's praise for mathematical disciplines. He also took Bessarion's advice from *In calumniatorem* seriously and read the dialogues in Ficino's Latin translation or, better, reread them, since he could have read Ficino's Latin translation of Plato at any stage of his career, before, during or after his sojourn in Italy.[186] There were two editions of Ficino's translation of Plato before 1515, the *datum ante quem* of the *Commentariolus* (Florence, 1484;

---

[185] See also A. Goddu, *Copernicus and the Aristotelian Tradition*, p. 223, n. 50.

[186] See D. Knox, "Ficino and Copernicus," p. 405.

Venice, 1491), and four additional ones after that date (Paris, 1518; Basel, 1522; Paris, 1533; Basel, 1539).[187] In Cracow, Copernicus had access to several copies of Ficino's translation of Plato already at the turn of the century, one of which belonged to the university. There was a copy of Ficino's translation in Lidzbark, and Plato's work was also available to Copernicus in Frombork.[188]

Besides library inventories, Plato's availability to Copernicus and his influence on him is also corroborated by Rheticus' *Narratio prima* which was written in the library of Frombork (*in Bibliotheca Warminense*) in the summer of 1439.[189] Rheticus referred to or quoted from the *Republic*, the *Epinomis*, the *Georgias*, the *Phaedrus*, the *Phaedo* and the *Timaeus*.[190] According to Edward Rosen, Rheticus was using Simon Grynaeus' revision of Ficino's translation, which was published in Basel in 1532,[191] but since he also quoted Plato in Greek, he – and Copernicus – apparently also had access to a Greek version of Plato.[192]

Copernicus annotated at least one of Plato's works, the *Parmenides*, in the 1484 edition of Ficino's translation.[193] The *Parmenides* was included in Part I of

---

187 See J. Hankins, *Plato in the Renaissance*, "Census of Printed Editions," pp. 739–59. Ficino's *Plato*, pp. 740–2, n. 5 and 8; pp. 749–51, n. 30, 32, and 37; pp. 753–6, n. 43, 46, and 52; D. Knox, *Ficino and Copernicus*, p. 403.

188 See A. Goddu, *Copernicus and the Aristotelian Tradition*, pp. 224–6.

189 See *Narratio Prima* (Hugonnard-Roche and Verdet, p. 87): "Ex Musaeo nostro Varmiae, IX Calend. Octobris, Anno Domini M.D. XXXIX." Rosen, p. 196, apparently mistranslated the date as: "From my library at Frauenburg, September 23, 1539."

190 Rheticus quotes from *Republic* 533b–c in Greek (Rosen, p. 142; Hugonnard-Roche and Verdet, p. 110); evokes *Timaeus* 40 b–d without mentioning it (Rosen pp. 147 and 150; Hugonnard-Roche and Verdet, pp. 60 and 62. He quotes again in Greek from *Epinomis* 990b (Rosen, 162; Hugonnard-Roche and Verdet, p. 68) and paraphrases *Epinomis* 989d–990a in Latin (Rosen p. 163; Hugonnard-Roche and Verdet, p. 68). He refers explicitly to *Georgias* 458a (Rosen, p. 165; Hugonnard-Roche and Verdet, p. 69) and quotes again in Greek from *Phaedrus* 266b (Rosen, p. 168; Hugonnard-Roche and Verdet, p. 71). The last refrence is the Latin paraphrase of explicitely mentioned *Phaedo* 86b–c and 92a–95a (Rosen p. 196; Hugonnard-Roche and Verdet, p. 86).

191 See *Three Copernican Treatises*, p. 165, n. 198.

192 The other possibility – not very likely, in my opinión – is that Rheticus brought a Greek Plato with him to Frombork.

193 On this, see P. Czartoryski, "The Library of Copernicus," p. 382, and especially A. Goddu, "Copernicus's Annotations: Revisions of Czartoryski's 'Copernicana'," pp. 208–20. For the analysis of Copernicus' handwriting, annotation to the *Parmenides* and related matters, see pp. 209–15. Goddu, p. 215, concludes that "Copernicus did not own the book but he very likely annotated it." See also his, *Copernicus and the Aristotelian Tradition*, pp. 209–10, and D. Knox, "Ficino and Copernicus," p. 406.

Ficino's translation of Plato's *Opera omnia*. The copy in question belonged to the library of Frombork (*Liber Bibliothecae Varmiensis*), but other copies were also available, as mentioned above, in Cracow and in Lidzbark. According to Goddu, Copernicus wrote a small note – "what needs to be known about hypotheses (*quid aduerti oporteat circa hypotheses*)" – on the following passage of the *Parmenides*: "If you like, said Parmenides, take as an example this hypothesis that Zeno entertained: if many are, what must the consequences be both for the many themselves in relation to themselves and in relation to the one, and for the one in relation to itself and in relation to the many."[194] In Part II of Plato's *Opera omnia*, containing very few annotations, there is also a small note to the *Laws* that may have been, according to Goddu, made in Copernicus' hand, while the annotation to the *Timaeus* is more difficult to evaluate.[195] Extant annotations by Copernicus– and there may have been others now lost – show indisputably that Copernicus read and studied Plato with great care.

Copernicus also referred to Plato or his works on several occasions, and he explicitly mentioned two of Plato's dialogues.[196] I have already pointed to his paraphrase of Plato's *Laws* in the *Introduction*, and there are two further explicit references to the *Timaeus* in his *De revolutionibus*. The first one is in Chapter 10 of Book I and relates to Plato's ordering of the planets Venus and Mercury above the sun.[197] The other one is in the introductory passage to the Book V, referring to the *Timaeus* as the authority on the Greek names of the planets.[198]

---

194 *Parmenides* 136a–136b (Gill and Ryan, p. 370). For Ficino's translation of *Parmenides* 134b–137d, see A. Goddu, *Copernicus and the Aristotelian Tradition*, pp. 471–2.

195 Goddu discusses the destiny of Part II of Plato's *Opera omnia* on pp. 215–8 of his "Copernicus's Annotations: Revisions of Czartoryski's 'Copernicana'."

196 See D. Knox, "Ficino and Copernicus," pp. 403–5, and accompanying notes; A. Goddu, *Copernicus and the Aristotelian Tradition*, pp. 225–9.

197 *De revolutionibus* I, 10 (Rosen, p. 18; Zekl, p. 124): "With regard to Venus and Mercury, however, differences of opinion are found. [...] Hence Venus and Mercury are located above the sun by some authorities, like Plato's *Timaeus* [...]."

198 *De revolutionibus* V, intr. (Rosen, p. 227; Gansiniec, p. 233): "In Plato's *Timaeus* these five planets are each named according to its aspect, Saturn is called 'Phaenon', as though you were to say 'bright' or 'visible', for it is invisible less than the others, and emerges sooner after being blotted out by the sun. Jupiter is called 'Phaeton' from its brilliance. Mars is called 'Pyrois' from its fiery splendor. Venus is sometimes called 'Phosphorus', sometimes 'Hesperus', that is, 'Morning Star' and 'Evening Star', according as it shines in the morning or evening. Finally, Mercury is called 'Stilbon', on account of its twinkling and shimmering." Copernicus' attribution, however, was wrong. Plato never used the names he listed here, but he may have been deceived by reading pseudo-Plutarch's

There are also several occasions where Copernicus mentions only Plato's name, as in Chapter 5 of Book I of *De revolutionibus*, where he evokes Plato as someone who did not hesitate to travel to Italy in order to visit Philolaus the Pythagorean, an extraordinary astronomer who believed in the earth's axial rotation and its other movements.[199] In a passage of Chapter 10 of Book I of *De revolutionibus*, following a passage on the order of planetary spheres, Copernicus speaks about Plato's followers who believed that planets reflected the sunlight. This may have been, according to Rosen, a reference to *Timaeus* 39b.[200] Another occasion was a two page long section of the autograph that was initially intended to be placed at the end of Book I. Copernicus made a mention of Plato's remark that only a few mastered the theory of the heavenly motions.[201]

Theoretical correspondence between Copernicus and Plato. Copernicus must have been familiar with Plato's works through Ficino's Latin translation and I believe he even possessed and read Plato in Greek, perhaps even in manuscript form, before Aldo Manuzio published the Greek edition of Plato's dialogues in 1513.[202] But the question that needs to be answered is whether his familiarity with and study of Ficino's or Greek Plato in any way instrumental in the development of his heliocentric hypothesis?[203]

My answer is definitively yes, but not just for the reasons that have been adduced thus far in the literature on Copernicus. The vast majority of scholars see Plato's influence on Copernicus only in his invocation of the sun as the ruler of

---

Placita philosophorum II, 15, 889b in Greek, where it was maintained that Plato used these planetary names. See also, D. Knox, "Ficino and Copernicus," p. 404, n. 23; A. Goddu, *Copernicus and the Aristotelian Tradition*, p. 226.

199 *De revolutionibus* I, 5 (Rosen, p. 12; Zekl, pp. 100–3). See supra, p. 209. Goddu and Knox count as indirect citation also *Preface* and *De revolutionibus* I, 5, where support for axial rotation by several philosophers is mentioned, but Plato's name does not appear. Copernicus knew of Cicero's interpretation of *Timaeus* 40b in his *Academica priora* II, 39, 123, and even transcribed it in his copy of Pliny the Elder. See infra, p. 364, and n. 362.

200 See his *Commentary*, p. 355.

201 See *De revolutionibus* (Rosen, p. 25; Zekl, p. 148): "But only a keen mind and persevering study could understand these subjects. They were therefore unfamiliar to most philosophers at that time, and Plato does not conceal the fact that there were then only a few who mastered the theory of the heavenly motions." It seems to me that Copernicus refers to Plato's *Laws* 817e–818d.

202 See also A. De Pace, *Niccolò Copernico e la fondazione del cosmo eliocentrico*, p. 103, n. 305.

203 D. Knox, "Ficino and Copernicus," p. 407, asks the same question.

the universe in Chapter 10 of Book I of *De revolutionibus*,[204] I see it more broadly in Plato's understanding of the universe's order on the one hand and the status, tasks and role of astronomy in discovering that order on the other. This theoretical, philosophical concordance between Plato and Copernicus is the seventh, last and most important piece of evidence of Copernicus' Platonism and its relevance for heliocentrism.

But before I go any further in defending my thesis, I would like to stress one thing: Copernicus was not a Platonist completely and without reservation. Plato's reflections on astronomy were complicated and perhaps even contradictory. Several aspects of his doctrine regarding astronomy are still disputed by modern scholars.[205] I do not argue that Copernicus was aware of all the intricacies of Plato's thoughts on universe and astronomy. My thesis is simply that he adopted some of Plato's fundamental ideas concerning the order of the universe and the status and role of astronomy that had become the universal legacy of Platonism. I am not arguing that he was convinced by Plato's more specific solutions or suggestions; for instance, I do not believe that Copernicus envisioned the stars as living beings or, to put it differently, I cannot find any textual evidence to support that.

Copernicus summarized these basic ideas in a very condensed form in the *Introduction*, which is – together with some passages from *Preface* and Chapter 10 of Book I of *De revolutionibus* – in my opinion key to understanding Copernicus' entire project of astronomical renovation. The *Introduction* is much more than just

---

204 See E. Garin, "La rivoluzione copernicana e il mito solare"; T. Kuhn, *The Copernican Revolution*, p. 130; A. Koyré, *The Astronomical Revolution*, p. 66; F. A. Yates, *Giordano Bruno and the Hermetic Tradition*, pp. 151–5; B. Biliński, *Il pitagorismo*, pp. 39–43, 71–3, 96–103, 127–8; F. Schmeidler, *Kommentar zu De revolutionibus*, p. 88. I, again, agree with D. Knox, "Ficino and Copernicus," pp. 407–8, n. 31, and his, in my opinion accurate, discussion of the passage, on pp. 407–12.

205 For a thorough discussion of Plato's view on astronomy and related matters, see, to list just a few studies, A. Gregory, *Plato's Philosophy of Science*, passim; I. Bulmer-Thomas, "Plato's Astronomy"; F. M, Cornford, *Plato's Cosmology*; A. Gregory, "Astronomy and Observation in Plato's *Republic*"; N. S., Hetherington, "Plato and Eudoxus: Instrumentalists, Realists, or Prisoners of *Themata*?"; W. Knorr, "Plato and Eudoxus on the Planetary Motions"; A. P. D. Mourelatos, "Astronomy and Kinematics in Plato's Project of Rationalist Explanation"; A. P. D. Mourelatos, "Plato's 'Real Astronomy': *Republic* VII. 527d–531d"; I. Mueller, "Ascending to Problems: Astronomy and Harmonics in *Republic* VII"; G. Vlastos, "The Role of Observation in Plato's Conception of Astronomy." On how Plato's astronomy and cosmology relate to his ethics, see G. R. Carone, *Plato's Cosmology and its Ethical Dimension*.

"a typical piece of mathematical-humanist rhetoric," as Westman argues.[206] In a similar vein Nicholas Jardine dismisses Copernicus' claims for nobility, dignity and the edifying power of astronomy in the *Introduction*, as "commonplaces, whose classical sources include Plato, Aristotle and Ptolemy and which have many parallels in the period in prefaces and introductory orations by authors who show no inclination in practice to vaunt the prerogatives of astronomy over those of natural philosophy."[207] I agree with Jardine that Copernicus' claims are to a certain extent expressed also by some other authors, but this does not mean that they lack internal philosophical coherence. As I establish below, there is a very profound Platonist philosophical background that links all the ideas which Copernicus expresses into a coherent philosophy. In short, the *Introduction* reveals Copernicus' deep connection and debt to Plato and the role Platonism played in formulation of his heliocentric hypothesis.

The *Introduction* can be divided into two parts. In the first part, astronomy, "this divine rather than human science," is praised and its goals are described, the second part is dedicated to its "perplexities."

(i) Beauty and divinity of the heaven: heaven as the visible god. Copernicus hails astronomy as the art that studies "the most beautiful objects, most deserving to be known," which is concerned with the divine revolutions of a "visible god":

> Among the many various literary and artistic pursuits which invigorate men's minds, the strongest affection and utmost zeal should, I think, promote the studies concerned with the most beautiful objects, most deserving to be known. This is the nature of the discipline which deals with the universe's divine revolutions, the stars' motions, sizes, distances, risings and settings, as well as the causes of the other phenomena in the sky, and which, in short, explains its whole appearance. What indeed is more beautiful than heaven, which of course contains all things of beauty? This is proclaimed by its very names, *caelum* and *mundus*, the latter denoting purity

---

[206] R. S. Westman, "The Astronomer's Role in the Sixteenth Century," p. 109.

[207] N. Jardine, "The Copernican Orbs," p. 187. He refers in note to P. L. Rose, *The Italian Renaissance of Mathematics*, but does not provide any further references. Westman and Jardine are also evoked by A. De Pace, *Niccolò Copernico e la fondazione del cosmo eliocentrico*, p. 24, and n. 57.

and ornament, the former a carving. On account of heaven's transcendent perfection most philosophers have called it a visible god.[208]

It seems, at least at first sight, that Copernicus does not say anything extraordinary here: the divinization of the heavens has been known since Antiquity. Aristotle speaks about "divine bodies,"[209] Ptolemy in Chapter 1 of Book I of *Almagest*, tells us that he was drawn to "the theory concerning divine and heavenly things,"[210] and in Chapter 2 of Book XIII, describes the heavenly constructions as "divine."[211] Heaven was called the most beautiful of all things by several authors. In explaining the terms *caelum* and *mundus* by their etymologies, Copernicus relies, as is well known, on Pliny the Elder's *Historia nauralium*: "The Greeks have designated the world by a word that means 'ornament', and we [Romans] have given it the name of *mundus*, because of its perfect finish and glance. As for our word *caelum*, it undoubtedly has the signification 'engraved', as is explained by Marcus Varro."[212] The most probable source for Copernicus' reference to "most philosophers" who called the heaven a "visible god" is Cicero's *De natura deorum*, which lists several philosophers – Alcmaeon of Croton (ca. fifth century BC), Xenophanes, Aristotle, Heraclides of Pontus, Theophrastus (ca. 371–ca. 287 BC), Cleanthes (ca. 330–ca. 230 BC), Chrysippus (ca. 280–207 BC) – who believed that *mundum deum esse*.[213]

But it is in Plato that all of this florid expressionism, including the reference to the beauty of heaven, finds its ultimate theoretical consistency and philosophical justification. In the *Timaeus*, Plato explains that the craftsman made the world by looking "at what is always changeless" and used it as his model. Since he reproduced its form and character, "then, of necessity, all that he so completes is beautiful."[214] This thought was repeated in the works of several Platonists, Ficino, for example, in his comment on the passage in *Compendium in Timaeum* states:

---

[208] *Introduction* (Rosen, p. 7; Zekl, p. 80).

[209] *De caelo* II, 12, 292b32 (Leggatt, p. 149).

[210] *Almagest* I, 1 (Toomer, p. 36; Heiberg, pars 1, p. 6).

[211] See also M.-P. Lerner and A.-Ph. Segonds, "Éditer et traduire Copernic," p. 394, n. 2.

[212] Pliny the Elder, *Historia naturalium* II, 3, 8–9 (Rackham, vol. 1, pp. 174–5). Varro, *De lingua latina* V, 18, derives etymology of *caelum* from *caelare*. As we know now, these etymologies are false. See also, M.-P. Lerner and A.-P. Segonds, "Éditer et traduire Copernic," p. 395, n. 3.

[213] See *De natura deorum* I, 25–39.

[214] *Timaeus* 28a–b (Zeyl, pp. 1224–5). Cornford, *Plato's Cosmology*, p. 22, translates *kalos* as "good."

"And because beauty is the glory of goodness, Plato adds that the transient world was brought forth by almighty God in the fullness of beauty as an image of the eternal model."[215]

Copernicus might have obtained the reference to the "visible God" from Macrobius' *In somnium Scipionis,* where the whole universe was deemed "the temple of God,"[216] but Cicero in *De natura deorum* reveals that the origin of this idea is to be found in Plato's *Timaeus* and *Laws.*[217] Heaven is first called a "blessed god,"[218] and a perceptible, or visible god at the end of the *Timaeus*:

> Here at last let us say that our discourse concerning the universe has come to its end. For having received in full its complement of living creatures, mortal and immortal, this world has thus become a visible living creature embracing all that are visible and an image of the intelligible, a perceptible god, supreme in greatness and excellence, in beauty and perfection, this Heaven single in its kind and one.[219]

It is questionable which section in the *Laws* Cicero had in mind,[220] but there is more than just a reference to the universe as "visible god" that is of interest as far as Copernicus' reliance on Plato's conception of the universe and astronomy is concerned.

Copernicus goes on in the *Introduction* to explain that the value of the arts depends on their subject matter; that astronomy is called the summit of the liberal arts; and that it is most worthy of a free man. Astronomy provides man with pleasure, and because it treats the things established in the finest order that are directed by divine management, it also stimulates man to the best and admiration

---

215 *Compendium in Timaeu*, 15 (Farndell, p. 15; *Opera omnia*, p. 1444). See also ibid., 16 (Farndell, p. 25; *Opera omnia*, p. 1444). Compare with Alcinous, *The Handbook of Paltonism* (Dillon, pp. 20–1). See also Bessarion, *In calumniatorem Platonis* II, 6 (Mohler, p. 111): "Certe, si pulcher hic mundus est, et ipsius mundi opifex bonus, sempiternum exemplar respexisse conditorem manifestum est."

216 See *In somnium Scipionis* I, 14, 2 (Stahl, p. 142; Neri, pp. 346–8): "His designation of the whole universe as the temple of God was appropriate, too, and was for the edification of those who think that there is no other god except the sky itself and the celestial bodies we are able to see [...]."

217 *De natura deorum* I, 30 (Rackham, pp. 34–5): "Yet both in the *Timaeus* and the *Laws* he says that the world, the sky, the stars, the earth and our souls are gods."

218 See *Timaeus* 34b (Cornford, p. 58).

219 *Timaeus* 92c (Cornford, p. 359).

220 Perhaps *Laws* 821a–b.

of the Maker of everything. Through the contemplation of the heaven one is led to the highest Good. There are benefits of astronomy for the community as well, as Plato states in the *Laws*. Everyone who wants to access higher learning must study astronomy and no one lacking the knowledge of the sun and the moon and other heavenly bodies can ever become godlike:

> If then the value of the arts is judged by the subject matter which they treat, that art will be by far the foremost which is labeled astronomy by some, astrology by others, but by many of the ancients, the consummation of mathematics [or knowledge] (*mathematices consumationem*). Unquestionably the summit of the liberal arts and most worthy of a free man, it is supported by almost all the branches of mathematics. Arithmetic, geometry, optics, surveying (*geodesia*), mechanics and whatever others there are all contribute to it. // Although all the good arts serve to draw man's mind away from vices and lead it toward better things, this function can be more fully performed by this art, which also provides extraordinary intellectual pleasure. For when a man is occupied with things which he sees established in the finest order and directed by divine management, will not the unremitting contemplation of them and a certain familiarity with them stimulate him to the best and to admiration for the Maker of everything, in whom are all happiness and every good? For would not the godly Psalmist in vain declare that he was made glad through the work of the Lord and rejoiced in the works of His hands, were we not drawn to the contemplation of the highest good by this means, as though by a chariot? The great benefit and adornment which this art confers on the commonwealth (not to mention the countless advantages to individuals) are most excellently observed by Plato. In the *Laws*, Book VII, he thinks that it should be cultivated chiefly because by dividing time into groups of days as months and years, it would keep the state alert and attentive to the festivals and sacrifices. Whoever denies its necessity for the teacher of any branch of higher learning is thinking foolishly, according to Plato. In his opinion it is highly unlikely that anyone lacking the requisite knowledge of the sun, moon, and other heavenly bodies can become and be called divine.[221]

Copernicus himself points out his source for the last part of the quoted passage. As recognized by many scholars, Copernicus freely summarized Ficino's

[221] *Introduction* (Rosen, p. 7; Zekl, pp. 80–2).

translation of that passage, 809c–d, from the *Laws*.[222] Moreover, the next two sentences of the *Introduction* regarding the necessity to possess the knowledge of astronomy for the teacher of any branch of higher learning, are also references to the *Laws*, as is the last one in which he argues that no one can be called divine without the knowledge of the movement of the sun, the moon and other heavenly bodies. Further on, in Book VII of the *Laws*, at 817e-818d, Plato explains what Bessarion summarized in Chapter 12 of Book IV *In calumniatorem*. There are, besides music, three related disciplines that befit a free man: "computation and the study of the numbers; measurement of lines, surfaces and solids; the mutual relationship of the heavenly bodies as they revolve in their courses."[223] This is not a study for the multitude of men, but for a chosen few. A man, claims Plato,

> will fall a long way short of such godlike standards, if he can't recognize one, two, three, or odd and even numbers in general, or hasn't the faintest notion how to count, or can't reckon up the days and nights, and is ignorant of the revolutions of the sun and the moon and the other heavenly bodies. It's downright stupid to expect that anyone who wants to make the slightest progress in the highest branches of knowledge can afford to ignore any of these subjects.[224]

Apart from that, Copernicus, as we have seen, also maintains the following in the *Introduction*:

* that the dignity of the arts depends of their subject matter;
* that astronomy is called the summit of the liberal arts (or of the whole knowledge);
* that it is supported by almost all branches of mathematics;
* that astronomy is most worthy of a free man;
* that all *artes* lead to the better things, but astronomy does it better;

[222] See *Laws* 809 c–d (Saunders, pp. 1476–7), where Athenian explains that youth "must acquire such knowledge about the heavenly bodies in their courses – sun, moon and stars – as will help them with the arrangements that every state is forced to make in this respect. [...] We mean that the days must be grouped into months, and the months into years, in such a way that the seasons, along with their various sacrifices and festivals, may each receive proper recognition by being duly observed into their natural sequence. The result will be to keep the state active and alert to render the gods due honor, and to make men better informed on these matters."

[223] See also *Laws* 819a–b.

[224] *Laws* 819c–d (Saunders, p. 1485).

- ∗ that astronomy treats the things established in the finest order and directed by divine management;
- ∗ that the contemplation of these thing stimulates man to the best and to admiration for the Maker of everything, in whom are all happiness and every good;
- ∗ that through the contemplation of the heaven, as a product of God, one is led to the highest Good.

Similar motives and ideas related to fundamental issues of Copernicus' project are also to be found in Plato's dialogues. However, other allusions to Platonic thought were, unlike the last section of the first part of the *Introduction*, not directly from Plato, but were more likely from different Platonist and Neo-Platonist sources.[225]

(ii) Astronomy as the consummation of mathematics (or knowledge). Michel-Pierre Lerner and Alain Segonds suggest, for instance, that Copernicus' expression *mathematices consumatio* derives from Giorgio Valla's paraphrase of Proclus' *Commentary on the First Book of Euclid's Elements.* According to these authors, the expression does not mean the "the consummation of mathematics" but rather "the consummation of knowledge (*l'achevement supreme du savoire*)." Giorgio Valla in *De expetendis et fugiendis rebus* paraphrases Proclus and explains that mathematical sciences are the consummation of all human knowledge and decisively support all other sciences, from theology, physical science, political philosophy and moral philosophy:

> From what we have said it is clear that mathematical science makes a contribution of the greatest importance to philosophy (*ad philosophiam mathematica praecipuam utilemque praebeat consumationem*) and to its particular branches [...]. For theology, first of all, mathematics prepares our intellectual apprehension. [...] Mathematics also makes contributions of the very greatest value to physical science. It reveals the orderliness of the ratios according to which the universe is constructed and the proportion that binds together in the cosmos, making, as the *Timaeus* somewhere says, divergent and warring factors into friends and sympathetic companions.[226]

[225] The suspects are Ficino, Bessarion, Pliny the Elder, Cicero, Plutarch and pseudo-Plutarch, Alcinous, Giorgio Valla, Proclus, Eusebius of Caesarea, Apuleius and perhaps some others.

[226] *A Commentary on the First Book of Euclid's Elements*, Prologue, Part One, VIII, 22 (Morrow, pp. 18–9). Reference to *Timaeus* is to 32b–c. Here is Giorgio Valla's Latin

Copernicus' enumeration of the mathematical sciences that support astronomy (arithmetic, geometry, optics, geodesy, and mechanics) could also have been derived from Proclus' *Commentary on the First Book of Euclid's Elements*, where Geminus' division of mathematical sciences is discussed. While Geminus included arithmetic, geometry, mechanics, astronomy, optics, geodesy, canonics and calculation,[227] neither Proclus nor Valla claim that *astronomy* is the consummation of human knowledge (*consumatio mathematices*) or the summit of the liberal arts (*ingenarum artium caput*). But Copernicus does. At this point Copernicus' seems to import Plato's conception of astronomy from the *Epinomis*, to which we shall come shortly, in which astronomy is considered the highest science that leads to the understanding of the divine. The idea of astronomy's high status among the sciences was, as is well known, also shared by Ptolemy and Regiomontanus.[228]

In the introductory chapter of the *Almagest* Ptolemy makes clear that astronomy is, as a part of mathematics (itself a part of theoretical philosophy), concerned with "divine and heavenly things"[229] devoted to "the investigation of the eternally unchanging."[230] Because of that, "it too can be eternal and unchanging (which is a proper attribute of knowledge) in its own domain, which is neither unclear nor disorderly."[231] Astronomy is also relevant for the other two parts of theoretical philosophy, that is, theology and physics, and for "virtuous conduct in practical actions and character."[232] "[T]his science above all things, could men see clearly: from the constancy, order, symmetry (*symmetria*) and calm which are associated

---

paraphrasis, *De expetendis et fugiendis rebus* I, 19, "De pertinentibus ad scientiam mathematicam": "Caeterum quod ad philosophiam mathematica praecipuam utilemque praebeat consumationem, vel ex hoc nobis esse manifestum potest [...]. Ad theologiam sane capessandam, intellectiuas dirigit coniectationes [...] Ad naturam quoque insipiciendam rerum maximopere mathematica conducit agnitio, rationum seriem qua universum contructum est patefaciendo. Proportionem aperit qua mundi cuncta reuinciuntur, ut Timaeus, ut Plato docet."

[227] *A Commentary on the First Book of Euclid's Elements*, Part One, XIII, 38–42 (Morrow, pp. 31–5) See also M.-P. Lerner and A.-Ph. Segonds, "Éditer et traduire Copernic," p. 397, n. 8.

[228] See also A. De Pace, *Niccolò Copernico e la fondazione del cosmo eliocentrico*, p. 308,

[229] *Almagest* I, 1 (Toomer, p. 36; Heiberg, pars 1, p. 6).

[230] Ibid.

[231] Ibid. (Toomer, p. 36; Heiberg, pars 1, p. 7)

[232] Ibid. (Toomer, p. 37; Heiberg, pars 1, p. 7).

with the divine, it makes its followers lovers of this divine beauty, accustoming them and reforming their natures, as it were, to a similar spiritual state."[233]

In a similar vein, Regiomontanus praises astronomy in his Padua oration.[234] After he discusses history of geometry and arithmetic, mechanics and kinematics, he comes to astronomy, "the pearl of mathematics, which outshines not only the other mixed sciences but also geometry and arithmetic."[235] For Regiomontanus the highest form of mathematics is the study of astronomy. Through astronomy man is brought closer to God, just as through the other arts man is separated from the animals.[236]

(iii) Plato's axiom; orderly motion of the heavens and divine management. Let us return to the *Laws* and Plato's ideas relevant for understanding not only the *Introduction* but the whole Copernican project. In the passage *Laws* 897c Plato states that "the whole course and movement of the heavens and all that is in them reflect the motion and revolution and calculation of reason, and operate in a corresponding fashion, then clearly we have to admit that it is the best kind of soul that cares for the universe and directs it along the best path." Next is the passage at 898a–b, in which Plato, according to Andrew Gregory, expresses thoughts that might have been at the origin of the postulate of the uniform circular movement of the heavenly bodies. That motion must be regular and ordered according to some reason. In this passage, explains Gregory, Plato speaks of regular and uniform motion,

> and then of motion which is according to *hena logon kai taxin mian ampho*, one reason and order. *Laws* 898b contrast motion which is not regular or uniform and also motion which is *med'en kosmoi med en taxei mede en tini logoi*, 'not arranged or ordered according to some reason'. On the evidence of the *Laws*, it would seem then that Plato asks for regular, circular and ordered motion.[237]

233 Ibid.

234 See P. L. Rose, *The Italian Renaissance of Mathematics*, pp. 95–7, and especially N. Swerdlow, "Science and Humanism in the Renaissance: Regiomontanus's Oration on the Dignity and Utility of the Mathematical Sciences."

235 Quoted from P. L. Rose, *The Italian Renaissance of Mathematics*, p. 96.

236 See also N. Swerdlow, "Science and Humanism in the Renaissance: Regiomontanus's Oration on the Dignity and Utility of the Mathematical Sciences," pp. 151–3.

237 A. Gregory, *Plato's Philosophy of Science*, p. 96. See *Laws* 898a–b (Saunders, pp. 1554–5): "Of these two motions, that taking place in a single location necessarily implies

The idea that a systematic, orderly motion of the heavens depends on divine management is asserted by Plato in Book XII of *Laws*, 966d–968a. There he presents the context of the discussion, i.e. the belief in gods, and two arguments that support it. The first argument is the existence of a soul which is older and more divine than all other things, and "[a]nother argument is systematic motion of the heavenly bodies and the other objects under the control of reason, which is responsible for the order of the universe."[238] The universe is not governed by necessity but by the divine will. It is evident that the heavenly bodies are animated and therefore rational and that it was "reason that imposed regularity and order on the heavens." No mortal can ever get a firm grasp of God unless he understands that, in addition to the fact that reason is immortal and far older than any created thing; it is reason that governs and dominates the stars.[239]

(iv) Order of the cosmos made by demiurge; men will become godlike by studying the cosmos. Plato speaks about the order and arrangement of the universe made by the *demiourgos* also in the *Timaeus*.[240] In his desire to make all things good and perfect the demiurge brought all the discordant and unordered motions of the visible world into order; he judged the order to be in any way better.[241] A little bit further, he specifies that the universe was brought into being and concord by means of proportion.[242] Plato teaches that the cosmic soul and

continuous revolution round a central point, just like wheels being turned on a lathe, and this kind of motion bears the closest possible affinity and likeness to the cyclical movement of the reason. [...] Take reason on the one hand, and motion in a single location on the other. If we were to point out that in both cases the motion was determined by a single plan and procedure and that it was (a) regular, (b) uniform, (c) always at the same point in space, (d) around a fixed centre, (e) in the same position relative to other objects, and were to illustrate both by the example of a sphere being turned on a lathe, then no one could ever show us up for incompetent makers of verbal images. [...] Now consider the motion that is never uniform or regular or at the same point in space or around the same centre or in the same relative position or in a single location, and is neither planned nor organized nor systematic. Won't that motion be associated with every kind of unreason?"

238 *Laws* 967a (Saunders, p. 1614).

239 See also *Phaedo* 97b–98b.

240 See more on this and related subjects in G. R. Carone, *Plato's Cosmology and its Ethical Dimension*, pp. 54–78.

241 See *Timaeus* 30a.

242 See ibid. 32b–c.

the universe that it animates are gods and have mathematical structure.[243] The world soul, which is described according to harmonic intervals, rules over the motions of the divine stars and planets.

The theme of order – with the additional motive that the orderly universe is a gift given to us as a model for our souls, for humans to apprehend and reproduce in ourselves – is developed in *Timaeus* 47a–c, where the basic subject is the purpose of sight. But Plato makes it clear that for the achievement of happiness, which is the ultimate goal of human reasoning, one has to study the order of the universe and in doing so rely more on reasoning – that is, calculations – than on sight. To attain happiness human beings should fulfill their capacity of reason:

> [...] the god invented and gave as vision in order that we might observe the circuits of intelligence in the heaven and profit by them for the revolutions of our own thought, which are akin to them, though ours be troubled and they are unperturbed; and that, by learning to know them and acquiring the power to compute them rightly according to nature, we might reproduce the perfectly unerring revolutions of the god and reduce to settled order of the wandering motions in ourselves.[244]

We should learn unerring revolutions of intelligence in heaven, that is, astronomy, in order to correct the erring of our own thoughts, which were acquired by birth.

That man, endowed with divine reason, should and must become godlike by reproducing in his own nature, as Cornford says, "the beauty and the harmony revealed in the cosmos, which is itself a god, a living creature with soul in body [...],"[245] is a subject elaborated upon in more detail in at the end of the *Timaeus*.[246] There Plato encourages us to follow and learn the reasoning and revolutions of the universe, which, according to Gabriela Roxana Carone, means "not just the physical visible motions of the corporeal universe but the intellections in which the motions of its soul consists. This is done by thinking, which is assimilated to the object of thought."[247] Such learning of astronomy, she continues, explaining

[243] See ibid. 35b et seq.

[244] Ibid. 47a–c (Cornford, pp. 157–158).

[245] F. M. Cornford, *Plato's Cosmology*, p. 34. See also A. De Pace, *Niccolò Copernico e la fondazione del cosmo eliocentrico*, pp. 19–20, n. 38.

[246] See *Timaeus* 90b–d.

[247] B. R. Carone, *Plato's Cosmology and its Ethical Dimension*, p. 71.

Plato, "is based on the apprehension of mathematical relationships (such as those governing the relative distances between the heavenly bodies [...]."[248]

Besides, in this passage Plato also talks about man's happiness in thinking about immortal and divine realities and about the fulfillment of the best life in this way. These immortal and divine realities are the harmonies and revolutions of the universe. There is, according to Plato, a most sovereign form of soul in man given to humankind as a guiding genius. This soul, a divine part of man, should lift him or her from earth towards celestial affinity. If man wants to attain immortality, claims Plato, he needs to maintain his guardian genius in good estate by nourishment and movements proper to it:

> The motions akin to the divine part in us are the thoughts and revolutions of the universe; these, therefore, every man should follow, and correcting those circuits in the head that were deranged at birth, by leaning to know the harmonies and revolutions of the world, he should bring the intelligent part, according to its pristine nature, into the likeness of that which intelligence discerns, and they win the fulfillment of the best life set by the gods before mankind both for his present time and for the time to come.[249]

(v) The role of astronomy in reaching the Good; real objects of astronomy and their relationships are accessible to reason and thought alone. However, the *Laws* and the *Timaeus* are not the only works relevant for our discussion here. As far as knowledge of and access to the Good through mathematical disciplines, including astronomy, is concerned, one should consult also Book VII of the *Republic*, in which Plato explains the Good, in John Cooper's words, "as a perfect example of rational order, conceived in explicitly mathematical terms: a complex, ordered whole, whose orderliness is due to the mathematical relationships holding among its parts."[250] There Plato prescribes a curriculum for the guardians of the state that is in its nature mathematical and that should prepare the way for dialectic, which itself culminates in the knowledge of the Good.[251] The guardians should learn arithmetic and geometry, harmony (= music) and astronomy, which should lead them to cognition of the Good.[252]

---

248 Ibid. For the last point, she refers to 35b et seq., and 36d.

249 *Timaeus* 90c–d (Cornford, p. 354).

250 J. M. Cooper, "The Psychology of Justice in Plato," p. 155.

251 See *Republic* 524d et seq.

252 See also A. Gregory, *Plato's Philosophy of Science*, p. 93.

Here Plato develops one additional and maybe the most important idea that was also followed by Copernicus. As in the *Timaeus*, where he suggests that one should go beyond visible motions of the corporeal universe, Plato makes the same demand also in the *Republic VII,* 528e–530c. Astronomy must be learned differently from the way in which it is learned at present. We should consider the decorations in the heaven as the best and most exact visible things. But we should at the same time admit that these motions fall short of the true ones – "those motions which the real speed and the real slowness in [their] true numbers and in all [their] true figures move relatively to each other and carry along whatever is in them."[253] And these, of course, are accessible only "to reason and thought, but not eyesight." The decorations in the heavens are just models, an excellent starting point to discover real movements of the stars, but by no means the real things. It is just like as if someone would come upon some thoroughly well-drawn and perfected diagrams of some skilled craftsman or artist, such as Daedalus, for instance. He or she would consider them beautifully crafted, but would "think it laughable to scrutinize them zealously, expecting to find in them true equality or duplicity or any other relation of *symmetria*."[254] The true astronomer will feel the same when he looks at the motions of the stars. He finds the tracing beautiful but it would be absurd for him to seek to obtain the truth "of the relation of *symmetria* of night to day, of these to months, and of the [periods of the other] stars to these and to one another from the visible appearances."[255] This is extremely important: the Real Astronomer, according to Alexander Mourelatos, "does not dismiss questions concerning the *symmetria* of celestial periods."[256] On the contrary, Plato demands that the True or Real Astronomer discovers the true *symmetriai* – that is, "commensurable proportions" – of celestial periods, which exist beyond visible motions: the Real Astronomer "realizes that the *aletheia* concerning these *symmetriai* cannot be elicited from the observed periods of the celestial bodies."[257]

---

253 *Republic* 529d (Vlastos, p. 2). I quote Vlasotos' translation from "The Role of Observation in Plato's Conception of Astronomy," but I amend it with Mourelatos's clarifications from "Plato's 'Real Astronomy': *Republic* 527d–531d."

254 *Republic* 529e–530a (Mourelatos, p. 3; Grube, p. 1146). See also A. Gregory, *Plato's Philosophy of Science*, pp. 65–6, and especially excellent analysis of A. Mourelatos, "Plato's 'Real Astronomy': *Republic* 527d–531d." I agree with his interpretation of the passage.

255 Ibid. 530a (Mourelatos, p. 3).

256 A. Mourelatos, "Plato's 'Real Astronomy': *Republic* 527d–531d," p. 39.

257 Ibid. See also his discussion of the meaning of *symmetria* on pp. 39–41. *Symmetriai*, p.

(vi) One natural bond. Elaboration of some of the most relevant topics regarding astronomy is nicely and coherently provided in the *Epinomis*, an addition or appendix to the *Laws*. The *Epinomis* was very likely not written by Plato himself, but reflects the views of his school near the time of his death. This possibility, however, was ignored in the sixteenth century, and in Copernicus' time the *Epinomis* was believed to be genuinely Plato's.

In the *Epinomis*, the nature of wisdom and the studies necessary to achieve it are explained. After dismissing several disciplines that are useful only for the necessities of human life, the author turns to those that make a person genuinely wise and a good citizen. He starts by discussing god's gift to man, that is, the gift of heaven itself. This gift is the knowledge of number. It is a divine thing, since where there is number there is order. The ability to count distinguishes humans from animals. We learn to count by observing daily changes in the face of the moon, for example. Plato, or whoever wrote the dialogue, then reminds the reader that the soul is older than the body, and that the two together form a single structure called the "living being" or "animal." There are five solids and five corresponding forms of body (fire, water, air, earth, and aether), and five corresponding kinds of animals or living beings. Two kinds are of special importance: those composed predominantly of earthly matter and those composed predominantly of fire.[258]

At *Epinomis* 982c–d is given a cosmological justification for the order and disorder in the universe. The earthly element dominates the make-up of humans, animals and plants, all of which live on the earth. All these things are the outcome of all kinds of bodies, yet for the most part they are of earth and of solid nature. Fire dominates in gods, stars and planets. They are second in generation (after the invisible soul) and visible. They also have some earth and air and slight portions of other elements. They are divine, imperishable and immortal. Comparison shows that earthly things are in disorder, whereas those comprised mostly of fire move with perfect order. What moves in disorder is unintelligent and what has an orderly path in the heaven has intelligence. It always proceeds in its course uniformly and without variation.

It is by necessity that a soul acquires mind: the soul governs as the ruler not as the ruled. The soul's decisions which are in accordance with the best intelligence

---

40, "are not any old proportions, but specifically such as would appeal to us *a priori* as maximally coherent yet simple, and structurally elegant."

[258] The other three kinds of living things that are composed mostly of air and water are discussed in passage 984b–985e; there is this hierarchy from above: fire, air, water, then earth.

or mind are unalterable. The stars and their whole circuit do not alter their paths by wandering, but remain true to their original intention. The stars have minds, the writer concludes, and souls. This opinion that the stars have souls distinguishes true philosophers from those who believe that, because they always do the same things uniformly and that the heavenly motions happen by necessity, the stars must not have souls. What is more, man could uplift himself and could understand that "whatever operates uniformly, without variation, and through the same causes is for that very reason to be regarded as intelligent."[259] Such man could understand that this is the nature of the stars, the most beautiful thing to see, that "moving through their march and dance, the finest and most magnificent dance there is, they bring to pass what all living things need."[260]

A little bit further, the author again characterizes astronomy as divine, which means that heavenly beings are either themselves gods or likenesses of gods, made by gods themselves.[261] The author then, at *Epinomis* 989d–990a, affirms that astronomy is the summit of knowledge. In the words of Cooper, "though certain preliminary studies are also needed, wisdom is constituted solely by the knowledge of astronomy – of the single, mathematically unified system of the constant movements of the heavenly bodies [...]."[262] Proper reverence for gods is learned solely through astronomy. The astronomer is the wisest person, and in particular the seasoned astronomer. The wise astronomer is not the one who practices astronomy by "observing risings and settings of stars, but the one who has observed seven of the eight circuits, each of them completing its own path in a way no one can easily contemplate who is not endowed with an extraordinary nature."[263] In the words of Alfred Taylor, the author of the *Epinomis* argues that the "study we need to lead us to true piety, the greatest of the virtues, is thus astronomy, knowledge of the true orbits and periods of the heavenly bodies, pursued in the spirit of pure science, not in that of Hesiod's farmer's calendar."[264]

How shall we learn astronomy? The first to be studied is the moon, then the sun, and then the sun's followers, Venus and Mercury. Studies of paths of other planets should be conducted by youth. Novice astronomers should study different mathematical disciplines: first theory of numbers, not corporal numbers but

---

[259] *Epinomis* 982e (McKirahan, p. 1625).

[260] Ibid.

[261] See ibid. 983e–984a.

[262] J. M. Cooper, introductory paragraph to the *Epinomis*, in Plato, *Complete works*, p. 1617.

[263] *Epinomis* 990a–b (McKirahan, p. 1631).

[264] A. E. Taylor, *Plato*, p. 500.

numbers in themselves, and then the theory of geometry and stereometry, both of which are theories of different kinds of numbers. Plato concludes the discussion of stereometry with an assertion that "the sequence based on both of these means [e. g. ratios 3:2 and 4:3] has been granted to the human race by the blessed choir of the Muses and has bestowed upon us the use of concord and symmetry to promote play in the form of rhythm and harmony."[265]

And finally the goal of learning all of these mathematical sciences is discussed.[266] To ascertain the point of learning mathematical sciences author refers to "the divine element in generated world, which consists of the finest and most divine sort of visible things God has permitted humans to observe."[267] They can only be learned with the help of mathematical sciences. Plato continues by explaining that this is an important matter. Anyone who comprehends each of these studies correctly, through the right method, will be greatly benefited. To those who learn in a proper manner, every diagram, every system of numbers and harmony, and the ordered motions of stars, will reveal themselves to be, as Gregory puts it, "a part of a unitary and coherent (*homologian*) whole, as there is *a single bond between these things* [i. e. the mathematical objects or phenomena], and we must keep this in mind while investigating."[268] Plato writes:

> In all these studies, though, the following point must be kept in mind: anyone who comprehends each of them through the right method is greatly benefited in doing so; otherwise, it is better to call on God for help. The right method is this – I must say this much at least. To the person who learns in the right way it will be revealed that every diagram and complex system of numbers, and every structure of harmony and the uniform pattern of the revolution of the stars are a single thing, applying to all this phenomena. And it will be revealed to anyone who learns correctly, as we say, fixing his eyes on unity. To one who studies these subjects in this way, there will be revealed a single natural bond that links them all.[269]

This is an extremely important metaphysical and epistemological point. All mathematical objects and phenomena, studied by mathematical sciences,

[265] *Epinomis* 991b (McKirahan, p. 1632).

[266] See ibid. 991b–d.

[267] Ibid. 991b (McKirahan, p. 1632).

[268] A. Gregory, *Plato's Philosophy of Science*, p. 89. Emphasis and text in square brackets added.

[269] *Epinomis* 991d–992a (McKirahan, p. 1632).

are linked together with one natural bond; they form a coherent whole, bound together by one single principle.

In sum: the citations, paraphrases and explanation of Plato's dialogues above provide a sufficient basis for the conclusion that the first part of the *Introduction* reveals Copernicus as a Platonist in his fundamental philosophical orientation. This does not mean – let me repeat this once more – that Copernicus drew all his ideas and formulations directly from Plato. Ideas expressed in *Introduction, Preface* and Chapter 10 of Book I of *De revolutionibus* are ultimately the usual Copernican rag-tag of different Platonist and Pythagorean sources. I am certain that one could find some additional *loci* in Plato and different Platonists and doxographers to support our thesis of Copernicus' Platonism. Some additional sources will be evoked in the next section, when we shall see how Copernicus' Platonism sheds light on his astronomical reform.

Let me conclude this section with the most important Platonist idea, one which Copernicus expresses only indirectly in *Introduction*, but which comes forward more distinctly in the *Preface* and Chapter 10 of Book I of *De revolutionibus*. This fundamental Platonist idea, critical for understanding Copernicus' project, was very pertinently summarized by Konrad Gaiser: "The goodness (*arete*) of a thing is shown by its permanence, beauty and form. These qualities depend on order (*taxis, cosmos*), that is, on a well proportioned arrangement of parts within the whole. The basis of order is then unity, and thence unity or oneness is the cause of all good, or the good in itself."[270] Against this background Copernicus' *Introduction* and consequently his attempt at the restoration of astronomy takes on a completely new dimension. The *Introduction* is not vain and empty Renaissance rhetoric, but – a little bit condensed and adjusted to his goals – exposition of a completely coherent Platonist philosophy, regarding some fundamental philosophical issues, such as man's way to the highest Good through the contemplation of universe's order, divine management of that order, and the role of astronomy in that endeavor.

Let us read the *Introduction* against this backdrop once again. Astronomy deals with revolutions of the universe, which are in their nature "divine." Every art must be judged by its subject matter, and astronomy has as its subject matter the most beautiful of all things, heaven, which because of its transcendent perfection is called by philosophers a "visible God." Since the value of an art is relative to its subject matter, astronomy is possibly the "consummation of knowledge," but it is certainly the summit of the liberal arts, supported by almost all mathematical

270 K. Gaiser, "Plato's Enigmatic Lecture 'On the Good'," p. 12.

disciplines. It is a study most appropriate to a free man. Astronomy leads men towards "better things," and does it more fully then all other arts. It also provides intellectual pleasure; astronomy enables men to see things established in the finest order and as directed by divine management. This contemplation and familiarity with the ordered cosmos, directed by divine management, stimulates men to the best and to the admiration of the Maker of everything, in whom is all happiness and every good. We are, as is attested by the Psalmist, led through the work of god (the cosmos) to "the contemplation of the highest good (*ad summi boni contemplationem*)." In combination with Copernicus' statement from the *Preface* that he was at the time of his confrontation with the problems of astronomy – maybe already during his studies in Italy and even later, after his return to Warmia – concerned about the fact that the "movements of the world machine, created for our sake by the best and most systematic Artisan of all, were not understood with greater certainty by philosophers," we have a rather clear answer to the question as to why Plato was important to Copernicus and why Copernicus – and no one else – was so annoyed by the problems of astronomy. It was because he shared the Platonist conception of the divine management of the heavens and the Platonist conception of the lofty status of astronomy among the arts.

4. Copernicus' Platonism, the problems of existing astronomy and his solutions

Now we are in a position to answer our further questions: Why were all the astronomical issues which Copernicus lists in his *Commentariolus*, *Introduction* and *Preface* as critical stimuli for the reformation of astronomy, so important to him? How were they related? What is the common denominator that links the problems of the equant and the *forma mundi*? It is Copernicus' Platonism. In other words, it is his conception of cosmic order, the purpose of that order for humankind's access to the highest Good, and the role of astronomy in that pursuit.

To fulfill the task of bringing humankind to the supreme Good, the supreme Artisan created a perfectly ordered, divine cosmos, a visible god which contains all things of beauty. Since it is divine and beautiful, and since it is intelligently created and ordained, it cannot be irregular; it cannot lack structure; it cannot be a monster. *Mundus, caelum* was created by the superior Artisan and is still regulated by him, which means that it must be perfectly ordered. In accordance to this idea and according to the explicitly formulated "Plato's axiom" the movements of the stars cannot be but absolutely uniform and circular. On the other hand the revolutions of the stars must also as a whole be organized in certain, definite structure of *symmetria*, that is, as a well proportioned, harmonious whole. But the universe is not divinely ordered for its own sake. There is a purpose in this order. In understanding and contemplating this order humankind has access to the highest Good. The divinely ordered cosmos, supreme beauty, visible god, was made in the best order *for us*, to enable us to discover through its beauty and order its Maker, the supreme Good, and admire the invisible god.

This general point has several sub-points. First, if the existing astronomy does not reflect the order the divine Artisan bestowed upon the universe, it is necessarily wrong and must be corrected. Therefore an astronomical system which cannot determine the exact length of the tropical year, which uses an equant that contravenes the principle of the uniform circular movement in the heavens, and which fails to exhibit certainty in establishing *symmetria* of the whole does not yield a suitable explanation of the actual state of matters. Without an astronomy that reflects postulated divine order in the heavens, we do not have access to the supreme Good and cannot have access to the invisible God.

On the other hand, the general Platonist metaphysical backdrop of Copernicus' heliocentric reform also implies that humankind is necessarily endowed with the capacity to discover the universe's divine order – otherwise God's creation would be in vain. Compared with "the fundamental epistemological resignation of scholasticism,"[271] this is a completely new situation. "Copernicus," affirms Fernard Hallyn, "conceptualized the relation between God and man as a profound affinity, not a radical break."[272] Traditionally, historians have argued that this epistemological breakthrough was enabled by Renaissance humanism and the regained value of man it inspired.[273] However I believe there is a more specific answer: the rejuvenation of Platonic philosophy. As we have seen, there is, according to Plato, a divine nature in man that enables him to discover the divine order of the heavens made for us, and to be led by the contemplation of it to the supreme Good. Marsilio Ficino expresses that view very clearly in *Theologia platonica*, where he almost identifies man with God. According to Ficino, not just anybody "can take the work of a clever artisan that has been skillfully constructed and discern by what rational principle or manner it has been constructed: this can only be done by someone who has acquired the same skill in the art."[274] In order to discern how

---

[271] F. Hallyn, *The Poetic Structure of the World*, p. 39 et seq.

[272] Ibid., p. 55.

[273] See, for example, A. Goddu, "Reflections on the Origin of Copernicus's Cosmology," p. 45.

[274] *Theologia platonica* XIII, 3 (Allen and Hankins, vol. 4, pp. 176–7). See also Ficino's letter

Archimedes constructed a model of bronze spheres and gave them the motions just like those in the heaven, one has to have a similar genius. A person with a like genius could build a similar model, provided that he has the needed materials:

> Since this person sees the order of the heavens, whence they are moved, whither they proceed, whith what measures, and to what they give rise, who will deny both that he has almost the same genius so to speak, as the author of the heavens, and that he is capable in a way of making the heavens, should he ever obtain the instrument and the celestial material, since in fact he is making them now, albeit from a different material, yet in the same order?[275]

Now let us take a closer look at the problems of existing astronomy.

The Calendar. In the light of Copernicus' Platonist convictions it is easy to understand why he points to the problem of the length of tropical year and, therefore, the calendar. Plato praises astronomy in the *Laws* for benefiting the commonwealth with the calendar and thus keeping the state attentive to the religious festivals and sacrifices. If existing astronomy cannot fulfill this task, there must be something wrong with it, something that must be corrected. Naturally, the then-ongoing discussions of calendar reform played a part in Copernicus' concerns, without forgetting that his philosophical background was, as we have seen, Plato. The exact determination of the length of the year is, however, a relatively minor issue in comparison to the other two critical moments of Ptolemaic astronomy: equant and *forma mundi*. These, too, assume a different meaning when seen from the Copernicus' Platonist perspective.

The equant and Plato's axiom. Ever since Noel Swerdlow's important studies (mostly written in 1970s),[276] the equant has been seen as problematic for mechanical reasons: the equant does not raise any difficulty when considered as a constraint on the motion of a point in a two-dimensional mathematical construction used for the computation of planetary longitudes.[277] It becomes

---

to Bernardo Bembo, in *Epistolarum liber* I (Members of the Language Department of the School of Economic Science, vol. 1, letter 123, p. 190; *Opera omnia*, p. 670), where Ficino copies the passage from *Theologia platonica*.

275 Ibid.

276 For the list of his articles on the subject, see supra, p. 61, n. 67.

277 See supra, pp. 56–65.

problematic for the postulated uniformity of the motion of a sphere only when taken as a three-dimensional physical entity. A planet-carrying sphere can move only in a uniform motion around a diameter that passes through its centre. If it moves on an axis that does not pass through the centre of the sphere, its motion cannot be uniform. The only motion permitted to the sphere (which is "perfectly spherical") is uniform rotation about its own axis. Any motion around a straight line other than its diameter (as in the case of the equant) cannot be uniform. According to this interpretation, Copernicus' objection to Ptolemaic astronomy, voiced in the *Commentariolus*, as well as in the *Introduction*, in the *Preface*, and elsewhere in *De revolutionibus*, stems from his belief that a planet is moved by a three-dimensional, material sphere to which it is attached. Accordingly, Swerdlow claims that the principle of uniform, circular movement "should not be understood as a philosophical, much less metaphysical, principle about the motion proper to the substance of the heavens,"[278] since "[s]peculations about such things do not belong to the domain of mathematical astronomy."[279]

I agree with Swerdlow that Copernicus believed in the existence of real, three-dimensional orbs, both total and partial, and I also agree with him that Copernicus' criticism has (almost) nothing to do with the substance of the heavens, but I also doubt that it has anything to do with celestial "mechanics." On the contrary, his critical attitude towards Ptolemaic astronomy originates from within his Platonist philosophical and metaphysical convictions. Copernicus finds the equant problematic, first and foremost, for a metaphysical reason: because it contravenes "Plato's axiom." Speculations about such things certainly do not belong to the domain of mathematical astronomy, which explains why no Latin astronomer before Copernicus was bothered by the equant. As I emphasized above, Peurbach, Regiomontanus and Brudzewo, all of them extremely capable astronomers and all of them also supporters of real, three-dimensional spheres as carriers of the planets and stars, did not find the equant problematic. Copernicus was the first in the Latin West to address this question with such vigor because he was not just an astronomer, but a philosopher as well. He was, as Galileo would later label him, an "astronomer-philosopher."[280] More precisely: Copernicus was a

---

[278] "The Derivation and First Draft of Copernicus' Planetary Theory: A Translation of the *Commentariolus* with Commentary," p. 435.

[279] Ibid.

[280] See *Dialogo sopra i due massimi sistemi del mondo* (Drake, p. 341; Besomi and Helbing, p. 373). See also M. Clavelin, "Le copernicianisme et la mutation de la philosophie naturelle," pp. 356–7.

Platonist astronomer-philosopher. He took Plato's ideas about the divine creation and management of the universe seriously and since the universe is a visible god, its movements must be divine, absolutely uniform and circular. "Plato's axiom" must be respected. This is the ultimate reason why the equant for Copernicus is not acceptable and Ptolomey's theory not *rationi satis concinna*.[281]

As is well known, Ptolemy's violation of "Plato's axiom" was the subject of continuous improvement efforts by Islamic astronomers.[282] Beginning in the first half of the thirteenth century, Islamic astronomers, many of whom worked at the Marāgha observatory, devised new techniques that did not violate the accepted principles. One of the most important concepts, if not *the* most important one, that needed to be amended was Ptolemy's concept of the equant. Of these astronomers at least three deserve special mention because of their connection with Copernicus: Mu'ayyad al-Dīn al-Urdī (d. 1266), Nasīr al-Dīn al-Tūsī (1201–1274), and Ibn al-Shātir of Damascus (1304–1375/6). While al-Urdī discovered so called al-Urdī's lemma, al-Tūsī invented the famous geometric technique called the "Tūsī device" (also "the Tūsī couple," or "Tūsī's reciprocation mechanism"). Together with Urdī's lemma this model represents one of the most significant results of the Marāgha astronomical tradition. The "Tūsī couple" is a mathematical device

[281] A. Goddu, "Reflections on the Origin of Copernicus's Cosmology," p. 41, also believes that Copernicus "may have interpreted the axiom as having physical causes and consequences," but he underlines that the "violation of the axiom, however interpreted, is sufficient to explain his rejection of the equant." Goddu, however, does not explain that it is "Plato's axiom" that Copernicus set to defend. See also W. Hartner, "Copernicus, the Man, the Work, and Its History," p. 416, n. 7: "I accept with pleasure a casual remark made in a private conversation by Professor Noel Swerdlow that leaving aside philosophical principles it would seem sufficient to state that a sphere revolving with variable speed about its axis is mechanically inconceivable. Nevertheless, philosophical reasons are adduced by Arab astronomers as well as by Copernicus (De rev., I 4)." I am inclined to believe that some retroactive effect took place after Swerdlow's "mechanical" interpretation became prevalent. Historians of Arabic astronomy perhaps reinterpreted Arab astronomy in the light of Swerdlow's account. It does not mean that Arabic astronomers and Copernicus did not believe in the solid spheres (both total and particular); it does not mean that Swerdlow's reasoning is in principle wrong; but it means that "mechanical" reasoning, if ever taken into account, was secondary to primarily metaphysical concerns. But for more affirmative claim one would have to reexamine the sources.

[282] On Islamic astronomy, see G. Saliba, "Arabic Planetary Theories after Eleventh Century AD"; *A History of Arabic Astronomy* and *Islamic Science and the Making of the European Renaissance*. There is a nice summary of the list of violations in F. J. Ragep, "Copernicus and his Islamic Predecessors," p. 66.

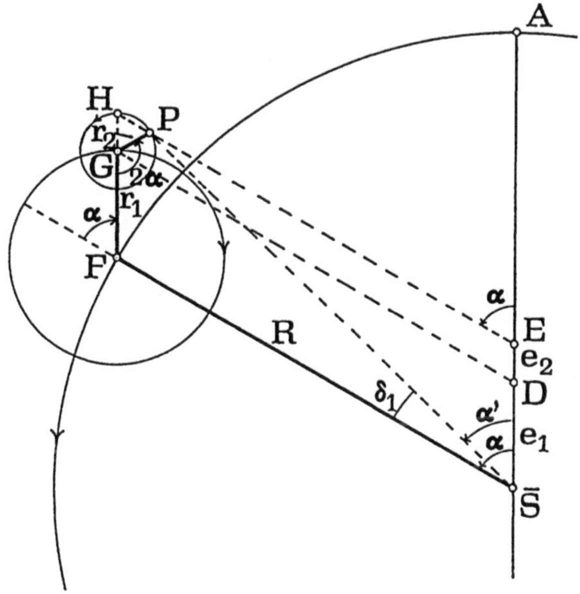

Figure 21.
Ibn al-Shātir's model. Reconstruction of N. Swerdlow, "The Derivation and First Draft of Copernicus' Planetary Theory: A Translation of the Commentariolus with Commentary," p. 468.

that starts from the combined movement of two spheres or circles and generates straight line motion. Ibn al-Shātir invented yet another important model which employed an additional epicycle in order to avoid the difficulties of the equant.

In the *Commentariolus* Copernicus used exactly the same replacement of the equant by two epicycles as found in Ibn al-Shātir (see *Figure 21*).

In all the planetary models, described in the *Commentariolus*, the combination of a deferent and two epicycles substitutes the movement of the deferent relative to the equant. Copernicus also uses Ibn al-Shātir's model for Mercury, which includes the Tūsī couple. And there is one more similarity. Copernicus' lunar theory is identical to that of Ibn al-Shātir. Besides, Copernicus uses the Tūsī couple throughout the *Commentariolus*.[283] Copernicus' use of Ibn al-Shātir's model suggests that he very probably encountered the problem of the equant and its solution at practically the same time. Besides his Platonism which sharpened his attitude towards the equant, this might have been additional stimulus to evaluate the equant critically. Copernicus took the opportunity and used this ready-made solution to the equant problem very hastily, without any critical examination.

283 Copernicus uses Urdī 's lemma in *De revolutionibus*.

As Swerdlow claims, he did not understand "the fundamental properties of his model for the first anomaly."[284]

There are different speculations as to how and when Ibn al-Shātir's model, the Tūsī couple, and other models of Islamic origin were transmitted to the Latin West. Some scholars defend Copernicus' independent (re)discovery, but since the Tūsī couple was used not only by Copernicus in the *Commentariolus* and later also in *De revolutionibus*, but also by Fracastoro in one of the final chapters of the *Homocentrica*, and by Amico in *De motibus*, all of whom studied at the University of Padua, it seems most plausible that Copernicus found these models in some manuscript in Padua or in nearby Venice.[285] This, in turn, suggests that

---

284 "The Derivation and First Draft of Copernicus' Planetary Theory," p. 469.

285 On this issue, see N. Swerdlow, "The Derivation and First Draft of Copernicus' Planetary Theory," p. 424; Swerdlow and Neugebauer, *Mathematical Astronomy in Copernicus's De revolutionibus*, p. 48. For a discussion of Copernicus' use of Arabic models and transmission problems, see O. Neugebauer, *The Exact Sciences in Antiquity*, pp. 197, and 203–4; E. S. Kennedy, "Late Medieval Planetary Theory"; E. S. Kennedy and V. Roberts, "The Planetary Theory of Ibn Ash Shâtir"; W. Hartner, "Copernicus, the Man, the Work, and Its History"; W. Hartner, "Ptolemäische Astronomie im Islam zur Zeit des Regiomontanus"; W. Hartner, "Astronomy from Antiquity to Copernicus"; G. Saliba, *A History of Arabic Astronomy: Planetary Theories during the Golden Age of Islam*, pp. 243–305; G. Saliba, *Islamic Science and the Making of the European Renaissance*, pp. 193–255; F. J. Ragep, "Copernicus and his Islamic Predecessors: Some Historical Remarks"; M. Di Bono, "Copernicus, Amico, Fracastoro and Tūsī's device: Observations on the use and transmission of a model"; M. Di Bono, *Le sfere omocentriche di Giovan Battista Amico nell' astronomia del Cinquecento*"; N. Guessoum, "Copernicus and Ibn al-Shatir: Does the Copernican Revolution have Arabic Roots?"; I. N. Veselovsky, "Copernicus and Naşīr al-Din al-Tūsī"; A. Goddu, *Copernicus and the Aristotelian Tradition*, pp. 262–9, and 476–86; H. Hugonnard-Roche, "Influence de l'astronomie arabe en Occident medieval," in particular the section "La révolution copernicinne et l'astronomie arabe," pp. 323–8. For most recent ideas on the subject, see Y. T., Langermann, "A Compendium of Renaissance Science: Ta'alumot hokmah by Moses Galeano," and R. Morrison, "An Astronomical Treatise by Mūsā Jalīnūs alias Moses Galeano." There is a specific problem regarding the relationship between Amico and Fracastoro and their use of the Tūsī device. Since the same device was also used by Amico, there is, according to E. Peruzzi, "Note e richerce sugli *Homocentrica* di Girolamo Fracastoro" (summarized in M. di Bono, "Teme e fonti dell' *Homocentrica* di Fracastoro," pp. 121–2, and in "Copernicus, Amico, Fracastoro and the Tūsī device," p. 148), a possibility that Fracastoro had seen *De motibus* and copied the model from there. See M. Di Bono, "Copernicus, Amico, Fracastoro and Tūsī's device: Observations on the use and transmission of a model," p. 133; M. Di Bono, *Le sfere omocentriche di Giovan Battista Amico nell' astronomia del Cinquecento*, pp. 70–1"; N. Swerdlow, "Amico's Homocentric Spheres."

even if they had not have known each other directly, they at least moved in the same extended circle of people and that there was some kind of a philosophical and astronomical discussion going on in Padua which stimulated the use of new Islamic material for various purposes. Since Fracastoro relates homocentric revival to the *Timaeus,*[286] I think it is reasonable to assume that these discussions were incited by Nicholas Leonicus Tomaeus' activities, his partial translation of the *Timaeus,* and the more general revival of Platonism.

The *forma mundi* problem and distance-period relationship. Basically the same reasoning applies to the *forma mundi* problem. Why is the lack of certain *symmetria* of the parts of Ptolemaic universe and its monstrosity problematic for Copernicus? As I already stressed, Muslim astronomers never even attempted a new arrangement of the universe, despite their concern that Ptolemy's astronomy violated uniform circular motion, or Plato's axiom.[287] What is more, with only minor exceptions Islamic astronomers did not find Ptolemy's arrangement of the universe to be the least bit problematic. In contrast, however, Copernicus objected to Ptolemaic astronomy because one was unable to deduce from the eccentrics the principal consideration, that is, *forma mundi* and certain *symmetria* of its parts. From the parts, which are very nicely depicted in themselves, one can only construct a monstrous picture of the universe. This is due to combining parts which do not respond or correspond to each other, or which do not form a *symmetria.* Just like the objection to the equant, this objection to Ptolemaic astronomy clearly does not stem from any astronomical concerns but from Copernicus' Platonist considerations. In order to understand the specifics of Copernicus' objection, as well as his solution, two closely related issues need to be explained: (1) Copernicus' stress on certainty in establishment of the order of the universe; and (2) the meaning of *symmetria* and *harmonia.*

(i) Certainty. Let us first focus on the fact that one cannot deduce from Ptolemaic astronomy the true form of the world. In other words, there is no certainty in regard of the ordering of celestial spheres. Copernicus carries throughout Book I of *De revolutionibus* illustrations of the lack of certainty in existing astronomy

---

286 See supra, p. 291, and n. 121.

287 N. Swerdlow, "The Derivation and First Draft of Copernicus' Planetary Theory," pp. 471–80, suggests that Copernicus arrived to his heliocentrism *via* first anomaly.

and the philosophical imperative for certainty in explanation of celestial appearances and of cosmic order.

Copernicus complains in the *Preface* about the *uncertainty* of the mathematical traditions concerning the derivation of the motions of the universe's spheres; he is annoyed by the fact that the movements of the world machine are not understood with greater *certainty* by philosophers.[288] In Chapter 5 of Book I of *De revolutionibus*, he argues again that it is impossible to find *certain* explanation of the celestial phenomena without the answer to the question whether the earth moves and what place in the universe it occupies. And in Chapter 10, he repeats this in a slightly different form: "Either the earth is not the centre to which the order of the planets and orbs is referred, or there really is no *certain* explanation of the order (*aut certe rationem ordinis non esse*) nor any apparent reason why the highest place belongs to Saturn rather than to Jupiter or any other planet."[289] Finally he concludes with the claim that the heliocentric arrangement of the planets manifests "a marvelous *symmetria* of the universe, and *certain* harmonious linkage (*certum harmoniae nexum*) between the motion of the orbs and their size."[290]

We have already seen that Copernicus' complaints about uncertainty have their justification in Ptolemy. Ptolemy's ordering of the planetary spheres in the *Almagest* (Saturn, Jupiter, Mars, the sun, Venus, Mercury, the moon, and the earth) is not based on any objective criterion; Ptolemy himself confesses that there could be other orderings.[291] We can infer from Chapter 10 of Book I of *De revolutionibus* that Copernicus read arguments for different orderings of the planetary spheres primarily in Regiomontanus' *Epytoma Almagesti*, Pico della Mirandola's *Disputationes* and Proclus' *Hypotyposis*. Regiomontanus recognized the issue was controversial and therefore uncertain, and Pico built on that to vigorously attack both astronomy and astrology. But nobody before Copernicus proposed any solution to the controversy, even less so one as radical. This is completely understandable since there is no astronomical need for certainty in ordering the heavenly spheres. Astronomers could predict planetary positions without knowing their respective arrangement and order.

---

[288] The same complaint is repeated in the *Introduction* with regard to the determination of the year (Rosen, p. 9; Zekl, p. 82): "[…] it is well known […] how different the opinions concerning it have always been, so that many have abandoned all hope that *certain* determination of it (*certam eius rationem*) could be found." Emphasis added.

[289] *De revolutionibus* I, 10 (Rosen, p. 20; Zekl, p. 130). Emphasis added.

[290] Ibid. (Rosen, p. 22, Zekl, p. 136). Emphasis added.

[291] See, supra, p. 70.

Uncertainty only becomes a problem from the astrological point of view, as proved by Pico della Mirandola, and from Platonist philosophical point of view, as is the case with Copernicus. This was very clearly recognized already by Galileo Galilei in his *Dialogo sopra i due massimi sistemi del mondo, tolemaico e copernicano.*[292] Galileo explains that the task of "pure astronomers is to give reasons just for the appearances of celestial bodies, and to fit to these and to the motions of the stars such a structure and arrangement of circles that the resulting calculated motions correspond with those same appearances."[293] Pure astronomers "are not much worried about admitting anomalies which might in fact be troublesome in other respects."[294] According to Galileo, Copernicus himself

> rectified astronomical science upon the old Ptolemaic assumptions, and corrected the motions of the planets in such a way that the computations corresponded much better with the appearances, and vice versa. But this was still taking them separately, planet by planet. He goes on to say that when he wanted to put together the whole fabric from all individual constructions, there resulted a monstrous chimera composed of mutually disproportionate members, incompatible as a whole. Thus however well the astronomer might be satisfied merely as a calculator, there was no satisfaction and peace for the astronomer-philosopher (*astronomo filosofo*).[295]

Galileo's account is obviously historically partly inaccurate – Copernicus never improved Ptolemaic astronomy by establishing better models for individual planets taking them separately. But this historical inaccuracy does not affect the truth of his more general theoretical point. The *forma mundi* problem is not a problem for "pure astronomers"; it is a problem for an "astronomer-philosopher," specifically a Platonist one. Given that the *machina mundi* was created for our sake by the best and most systematic Artisan of all, to enable humankind to ascend through the contemplation of his orderly work to the supreme Good, the uncertainty about the order of the universe is completely unacceptable. Copernicus' demand for certainty is then not motivated astronomically, or not primarily astronomically, but metaphysically. Without certainty in the explanation of the celestial appearances and in the structure of the universe, one cannot have access to the supreme Good.

---

[292] *Dialogo* (Drake, p. 341; Besomi and Helbing, p. 372).

[293] Ibid.

[294] Ibid.

[295] Ibid.

(ii) *Symmetria* and *harmonia*. But uncertainty is only one side of the coin. On the other side is the specific kind of universe Copernicus is expecting to find. In the *Preface* Copernicus attacks Ptolemaic astronomy for its inability to discover *forma mundi* and certain *symmetria* of its parts. He uses the same term in Chapter 10 of Book I of *De revolutionibus* when he explains that heliocentrism not only achieves marvelous *symmetria* in the universe but also provides a harmonious linkage of the universe's parts. The universe that Copernicus is expecting to find in astronomy has certain, definite structure, and it is a kind of *symmetria*, which in other words means that its parts are harmoniously linked together. Since Ptolemaic astronomy cannot establish certain *symmetria* of the cosmos, it is, as Copernicus explains in the *Preface*, a monster.

Let us leave aside the specific reasons why he uses the term "monster" and concentrate on another term he uses: *symmetria*.[296] His major objection to the Ptolemaic astronomy has been, in a way justifiably, characterized as an aesthetic one.[297] Copernicus tells us that the Ptolemaic universe resembles a picture of a man, composed of various, very well depicted pieces, but not corresponding or belonging (*respondere* is the term Copernicus uses) to each other. Copernicus' description of the Ptolemaic universe as a monster, whose parts, elaborately depicted as they may be, are completely incongruous, is traditionally explained by reference to *De arte poetica* (or *Ars Poetica*), in which Horace compares bad poetry to a painting of some kind of a creature possessing "neither head nor tail."

But as has been recently pointed out by Giora Hon and Bernard Goldstein, Horace speaks neither of a human body, nor about *symmetria* – or *harmonia*, for that matter.[298] De Pace, on the other hand draws attention to the fact that Copernicus' formulation most resembles that of Eusebius' *Praeparatio evangelica*, but although he does speak of a human body and his description really resembles that of Copernicus, there is, again, no sign of *symmetria* and *harmonia*.[299]

---

[296] See infra, p. 369, n. 368.

[297] See, for instance, O. Gingerich, "'Crisis' versus Aesthetic in the Copernican Revolution."

[298] See G. Hon and B. Goldstein, "Symmetry in Copernicus and Galileo," p. 273. Compare with Horace, *The Art of Poetry: An Epistle to the Pisos* (Fairclough, pp. 450–1): "If a painter chose to join a human head to the neck of a horse, and to spread feathers of many a hue over limbs picked up now here now there, so that what at the top is a lovely woman ends below in a black and ugly fish, could you, my friends, if favored with a private view, refrain from laughing? Believe me, dear Pisos, quite like such pictures would be a book, whose idle fancies shall be shaped like a sickman's dreams, so that neither head nor foot can be assigned to a single shape."

[299] See *Praeparatio evangelica* XV, 7 (Gifford): "[…] just like some of the more modern

According to very insightful analysis of Hon and Goldstein Copernicus' use of the term *symmetria* could derive from his reading of Marcus Vitruvius' *De architectura*, first published in Venice in 1511. Vitruvius (ca. 80–after ca. 15 BC) believes that architecture consists of ordering (*ordination*, called *taxis* in Greek), of design or arrangement (*dispositio*, called *diathesis* in Greek), eurythmy (*eurythmia*), symmetry (*symmetria*), décor (*decor*) and distribution or allocation (*distributio*, called *oikonomia* in Greek).[300] Two of these concepts, *eurythmya* and *symmetria*, deserve special attention.

Vitruvius defines *eurythmia* as "an attractive appearance and coherent aspect in the composition of the elements."[301] "*Eurythmia* is "achieved when the elements of the work are proportionate in high to width, length to breadth, and every element corresponds (*respondent*) in its dimensions to the total symmetries of the whole (*summam omnia respondent suae symmetriae*)."[302] In words of Hon and Goldstein, *eurythmia* "is associated with the overall coherence of an entity and related proportions of the parts to the whole,"[303] where *respondere* implies correspondence, a mutual relation between the parts and the whole.

Symmetria, on the other hand, is defined as "the appropriate arrangement of the elements of the work itself (*ipsis operis memebris conveniens consensus*), a correspondence (*responsus*), in any given part, of the separate parts to the entire figure as a whole. Just as in the human body there is a symmetric quality of eurhythmies (*symmetria est eurythmiae qualitas*) expressed in terms of the cubit, foot, palm, digit, and other small units, so it is in perfect works [of architecture]."[304] Here again *responsus* is associated with the relationship of the parts of the structure of the whole. Vitruvius' example of the human body is also very telling and he explains the symmetry by an example of *symmetria* in temples. *Symmetria*, explain Hon and Goldstein, "expresses the relation among the members of the

sculptors. For they too, when they have copied the head of one statue, and the breast of another, and the waist of another, sometimes put together things which do not suit each other, and persuade themselves that they have made something original: and indeed the whole, which any one would blame as being unsymmetrical (*inconcinnum*), is their own; but the contributions which are brought together in it, and have some beauty, are not theirs." See also A. De Pace, *Niccolò Copernico e la fondazione del cosmo eliocentrico*, pp. 26–7.

300 See Vitruvius, *De architectura* I, 2, 1 (Rowland, p. 24; Maufras, vol. 1, p. 44).

301 Ibid. I, 2, 3 (Rowland, p. 25; Maufras, vol.1, p. 44).

302 Ibid.

303 G. Hon and B. R. Goldstein, "Symmetry in Copernicus and Galileo," p. 273.

304 Vitruvius, *De architectura* I, 2, 4 (Rowland, p. 25; Maufras, vol. 1, pp. 44–6).

body and the body itself."[305] *Symmetria* is therefore extremely important when an architect aims to build a beautiful building:

> The planning of temples depends upon symmetria and architects must diligently apprehend its principle (*ratio*). It arises from proportion (*proportio*), which is called *analogia*. Proportion consists in taking a fixed module, in each case, both for the part of a building and for the whole, by which the principle of symmetries is put into practice. For without *symmetria* and proportion no temple can be designed according to reason; that is, it must have an exact proportion worked out after the fashion of a finely-shaped human being.[306]

According to Vitruvius *symmetria* requires a fixed module, or, as Goldstein and Hon put it, "a fundamental unit, in relation to which a proportion is fixed."[307] Each proportion in a building must also have a common relationship in such a way that all the parts or elements can be bound into a whole. In sum, in Vitruvian aesthetics *symmetria* is the quality of the whole (the human body, a building, a machine) when its parts are joined together so that they correspond to each other and the whole and form a beautiful and well-proportioned unity.[308]

One can find similar thoughts in Alberti's *De re aedificatoria*, first published in 1485, although he does not use the term *symmetria*. According to Leon Battista Alberti (1404–1472) the fundamental and absolute rule of nature is that of *concinnitas*.[309] Beauty is therefore defined as "a form of sympathy and consonance of the parts within a body, according to definite number, outline, and position, as dictated by *concinnitas*."[310] Beauty is "that reasoned harmony

---

305 G. Hon and B. R. Goldstein, "Symmetry in Copernicus and Galileo," p. 274.

306 Vitruvius, *De architectura* III, 1, 1 (Rowland, p. 47; Maufras, vol. 1, pp. 240–2).

307 G. Hon and B. R. Goldstein, "Symmetry in Copernicus and Galileo," p. 275.

308 Compare with the formulation of G. Hon and B. R. Goldstein, ibid., p. 276: *Symmetria* is "an attribute of a whole object, namely, a human body, a building, or a machine, when the parts are so joined that, in effect, they 'answer' (*respondere*) one another as well as the whole in such a way that they form a beautiful and well coordinated entity [...]." On the meaning of the term *symmetria*, see also the intervention of G. Simon, in *Avant, avec, après Copernic*, pp. 136–7. See also F. Hallyn, *The Poetic Structure of the World*, p. 96, who refers to the *Canon* of Polycleitus as passed on by Galen as Copernicus' possible source.

309 Copernicus uses the term *concinnitas* in the *Commentariolus*, see supra, p. 268, claiming that Ptolemy's *speculatio* was because of the equant *rationi non satis concinna*.

310 *De re aedificatoria* IX, 5 (Rykwert, Leach and Tavernor, p. 303).

(*concinnitas*) of all the parts within a body, so that nothing may be added, taken away, or altered, but for the worse."[311]

The idea of beauty expressing itself in well-proportioned arrangement of the parts within the whole, which rules heavens, was defended also by Marsilio Ficino. Ficino explains in *Theologia Platonica* that the universal order can be understood only if more or less valuable parts are distinguished among themselves, since "the order of the whole consists in their intervals and proportions."[312] Things which are, when seen in isolation, commonly regarded to be without value and not beautiful "most aptly accord both with themselves and with everything else in the whole order and in the order of the whole. Orpheus sang of all this in his 'Hymn to Apollo': You possess the limits of the whole world. The beginning and the end are in your care. Through you everything flourishes. You tune the whole sphere with the sound of your lyre."[313] Here Ficino does not use the term *symmetria*, but he links heavenly order with proportions and intervals and with Apollo's tuning of the entire heavenly sphere, thus implying musical harmony of the heavens.[314]

In one of his letters to Pellegrino Agli (1440–1469), Ficino related that Plato "thinks that the harmony which we make with musical instruments and voices is the image of divine harmony, and that the symmetry (*aptissima compositio*) and comeliness (*convenientia*) that arise from the perfect union of the parts and members of the body are image of divine beauty."[315] He repeats the idea, stating

---

311 Ibid. VI, 2 (Rykwert, Leach and Tavernor, p. 156). Compare this with Copernicus' *Preface* (Rosen, p. 5; Zekl, p. 74), where he explains that by heliocentric arrangement of planetary spheres heaven itself is "so linked together that in no portion of it can anything be shifted without disrupting the remaining parts and the universe as a whole." For more on Vitruvius and Alberti, see J. Hendrix, *Platonic Architectonics*, pp. 99–148.

312 *Theologia platonica* II, 9, 7 (Allen and Hankins, vol. 1, pp. 156–7.)

313 Ibid.

314 Compare also with Nicolaus Cusanus, *De docta ignorantia* II, 13 (Hopkins, p. 100; Senger, p. 112: "Who would not admire this Artisan, who with regard to the spheres, the stars, and the regions of the stars used such skill that there is – though without complete precision—both a harmony (*concordantia*) of all things and a diversity of all things? [This Artisan] considered in advance the sizes, the placing, and the motion of the stars in the one world; and He ordained the distances of the stars in such way that unless each region were as it is, it could neither exist nor exist in such a place and with such an order – nor could the universe exist. [...] And He established the interrelationship of parts so proportionally that in each thing the motion of the parts is oriented toward the whole."

315 *Epistolarum liber* I (Members of the Language Department of the School of Economic

that the soul receives "the sweetest harmonies and numbers through the ears,"[316] and by these echoes is "reminded and aroused to the divine music which may be heard by the more subtle and penetrating sense of mind."[317] There are two kinds of divine music. One exists in the mind of God, and the second one is "in the motions and the order of the heavens, by which the heavenly globes and their orbs make a marvelous harmony (*concentum*)."[318] As a matter of fact, God rejoices in harmony so much that He seems to have created the world especially for this reason: "that all its individual parts should sing harmoniously to themselves and to the whole universe."[319] This Universe "should resound as fully as it can with the intelligence and goodness of its author."[320] God, the author of the world, "arranged the spheres and regulated their movements in relation to one another, as the Pythagoreans and Platonist teach, that they make a harmony and melody beyond compare."[321]

Regardless of whether Eusebius, Vitruvius, Alberti or Ficino were actually immediate sources for Copernicus, his use and pairing of the concepts of *symmetria* and *harmonia* clearly reveal that he was inspired by Pythagorean and Platonist philosophical conceptions.[322] Plato himself in his *Republic*, as we saw, uses term *symmetria* ("commensurable proportion"), and demands that the True or Real Astronomer seek the truth of the relation of *symmetria* of night to day, of these to months, and of the periods of the other stars to these and to one

Science, vol. 1, letter 7, p. 44; *Opera omnia*, p. 613). Compare also with *Epistolarum liber* I (Members of the Language Department of the School of Economic Science, vol. 1, letter 47, p. 91; *Opera omnia*, pp. 631–2): "The beauty of the body lies not in the shadow of matter, but in the light and grace of form; not in dark mass, but in clear proportion; not in sluggish and senseless weight, but in harmonious number and measure. But we come to that light; that grace, proportion, number and measure only through thinking, seeing and hearing."

316 *Epistolarum liber* I (Members of the Language Department of the School of Economic Science, vol. 1, letter 7, p. 45; *Opera omnia*, p. 614).

317 Ibid.

318 Ibid.

319 *Epistolarum liber* VI (Members of the Language Department of the School of Economic Science, vol, 5, letter 21, p. 38; *Opera omnia*, p. 822).

320 Ibid.

321 Ibid.

322 G. Hon and B. Goldstein, "Symmetry in Copernicus and Galileo," pp. 277–9, examined several possible sources (Plato, Aristotle, Pliny the Elder, Pseudo-Aristotle) for an association between *symmetria* and *harmonia*, but could not find any. They therefore concluded that this combination of harmony and symmetry is unique to Copernicus.

another. The True Astronomer should therefore seek, among other things, the commensurable proportion of the periods of stars, including planets, and should do this by going beyond the observable periods of celestial bodies. As we already pointed out by quoting Gaiser,[323] Platonist doxographers summarized as the most fundamental Platonist doctrine exactly the ideas that found echo in above mentioned authors. The goodness of the universe is displayed by its permanence, beauty and form. All of these concepts depend on order. For a Platonist, order means a well-proportioned arrangement of parts within the whole. This order amounts, we may add, to *symmetria* and *harmonia*. Keeping that in mind, we can see that Copernicus' objections to the problems of the equant and the *forma mundi* are evidently motivated by his Platonist philosophy. For him, as for Plato, the universe is, in light of Chapter 10, Book I of *De revolutionibus*,[324] a "symmetric temple" whose constituent parts (planetary spheres) are arranged so as to form most beautiful and good whole. Copernicus thus, along with Pythagoreans and Platonists, believes that the entire universe is governed by mathematical harmony and that its parts are arranged by *symmetria* or commensurable proportion. When put together they must be parts of one and the same whole; they must correspond, or respond, to each other and to the whole. By failing to reproduce the postulated *symmetria* of the parts to the whole, by failing to subsume the universe under mathematical *symmetria* and *harmonia*, Ptolemaic astronomy produced a monster instead of the true *mundi formam*.

That this Pythagorean and Platonist idea was clearly on Copernicus' mind is also evident from Rheticus' account in the *Narratio prima*. Rheticus speaks of the cosmic order, mostly in terms of harmony but also once of symmetry. According to Rheticus, Copernicus must do more than Ptolemy, as he "must arrange in a certain and consistent scheme or harmony the series and order of all the motions and appearances (*seriem et ordinem omnium motuum et apparentium [...] in certam sibique motuo consentientem rationem seu harmonium colligere*)."[325] The uncertainty of astronomy is due to the fact that its masters fashioned their theories

---

[323] See supra, p. 337, for a quote from Gaiser.

[324] See *De revolutionibus* I, 10 (Rosen, p. 22; Zekl, p. 136): "At rest, in the middle of everything is the sun. For in this most beautiful temple, who would place this lamp in another and better position than that from which it can light up the whole thing in the same time?"

[325] *Narratio prima* (Rosen, pp. 131–2; Hugonnard-Roche and Verdet, p. 52). See also ibid. (Rosen, p. 132; Hugonnard-Roche and Verdet, p. 53): "Nevertheless, the observations of all scholars and heaven itself and mathematical reasoning convince us that Ptolemy's hypotheses and those commonly accepted do not suffice to establish the perpetual and

and explanations with "too little regard for the rule which reminds us that the order and the motions of the heavenly orbs agree in an absolute system."[326] In establishing "the harmony of the motions" they should "imitate the musicians who, when one string has either tightened or loosened, with great care and skill regulate and adjust the tones of all the other strings, until all together produce the desired harmony, and no dissonance is heard in any. [...] And since this one rule was neglected, we should have had to face at some time, if we intend to speak the truth, the collapse of all astronomy." [327]

This neglect of the harmonious arrangement of the heavens means that Ptolemaic astronomers committed one fundamental mistake: they disregarded the unity and internal coherence of the cosmos. According to Platonism, the entire cosmos is governed by a single principle that organizes the whole. For a Platonist the fundament of order is unity, therefore, unity or oneness is the cause of all good. As is nicely explained in the *Epinomis*, the uniform pattern of the revolutions of the stars, the agreement of the revolution of the stars depends on one thing: there is "a single natural bond" that links all the subjects in question. The supreme Artisan created a harmonious universe, a universe of *symmetriai*, in which one bond connects all the stars and their revolution into well-proportioned whole. It is this requirement for unity in an explanation of the cosmos which in turn should result in certain *symmetria* of the universe which the Ptolemaic astronomy was in no way able to satisfy. The question that remains to be answered is: What is the natural bond that links revolutions of the stars together in a harmonious whole?

(iii) The distance-period principle. In his important study "Copernicus and the Origin of his Heliocentric System," Goldstein argued that for Copernicus the accepted natural bond is the distance-period relationship. Acording to this principle "the periods of the planets are longer as their orbs are further from the centre of motion."[328] I agree with Goldstein that Copernicus' acceptance of heliocentric hypothesis depended on this principle; I would like to add some additional reflections on why he accepted it. There are, in my opinion, two reasons why he accepted this principle.

---

consistent connection and harmony of celestial phenomena (*ad perpetuam sibique invicem consentientem colligantiam et harmoniam rerum coelestium*)."

[326] Ibid. (Rosen, pp. 138–40; Hugonnard-Roche and Verdet, p. 56).

[327] Ibid.

[328] B. Goldstein, "Copernicus and the Origin of his Heliocentric System," p. 220.

First, this relationship already works out to certain extent in Ptolemaic astronomy. It works for Saturn, Jupiter and Mars, but does not, however, work out for Mercury, Venus and the sun. The order of Mars, Jupiter, and Saturn follows their sidereal periods, while Mercury and Venus, just like the sun, have period of one year. Therefore, there is a problem of where to place Venus and Mercury in relation to the sun: Plato believed they were above the sun, Martianus Capella and others believed they circled around the sun, and Ptolemy believed they were below the sun.

The second reason for Copernicus' acceptance of this principle could be again associated with his Platonism. Goldstein considers it very likely that Copernicus "took the distance–period relationship from Aristotle, *De caelo* II, 10, 291a29–b11, as interpreted by his commentators."[329] The distance–period relationship was actually a commonplace. In Cracow Johannes Versoris discussed it in his *Quaestiones de caelo et mundo* (ca. 1485), where he also noted that the principle was not respected by Ptolemy.[330] In Bologna Achillini mentioned it in *De orbibus*, and quoted in that context Averroes's comment 44 on Aristotle's *Metaphysics* XII: "It is also evident that the rank of these movers relative to the first mover must follow the order of the spheres in space because their precedence in position and magnitude determines their hierarchy in nobility. But in regard to their velocity of motion, we find them in the opposite situation; the closest to the earth is the fastest."[331] According to Averroes, Aristotle's views were reconcilable only with the opinion of those who believed that the sun is located below Mercury and Venus, and not above.

But use of the distance–period relationship was not confined to the Aristotelian tradition; it was mentioned and discussed by a broader circle of authors. Goldstein provides another possible source, Vitruvius' *De architectura*,[332] but there are other clear uses as well, like Macrobius' *In somnium Scipionis*. Macrobius relates that the Ancients believed planets "err" because they move with their own course from west to east, contrary to the direction of the greatest sphere, that is, the sphere of

---

329 Ibid., p. 226.

330 See A. Goddu, "Reflections on the Origin of Copernicus's Cosmology," p. 50, n. 36, and *Copernicus and the Aristotelian Tradition*, pp. 124–5.

331 See *De orbibus*, fols. 34vb and 35ra. For Averroes, see *Metaphysica* XII, comm. 43 (Genequand, p. 172). See B. Goldstein, "Copernicus and the Origin of his Heliocentric System," p. 225; R. S. Westman, *The Copernican Question*, p. 61.

332 B. Goldstein, "Copernicus and the Origin of his Heliocentric System," mentions and quotes (see pp. 220, 223–6, 230, 231) Vitruvius, *De architectura* IX, I, 14–15 (Rawland, p. 111; Panckouke, vol. 2, pp. 322–34).

the fixed stars. "Moreover, they all have similar movements and travel at the same rate of speed, and yet they do not all complete their circles and orbs in the same amount of time. He [i.e. Cicero] described their speed as amazing, for although it is the same for all and none can accelerate or retard its speed, the time required for their revolution varies."[333] He explains how it is possible:

> In addition, this diagram will show us with equal facility why some planets course through the same zodiac and the same constellations in a longer period of time and others in a shorter period. Whenever many concentric orbs are placed one within another, the outermost one is the greatest and the innermost the smallest, and likewise the one nearest the outermost orb is greater than those beneath it, and the one next to the innermost circle is smaller than those above it. The time required for each planet's revolution is determined, therefore, by its position among these seven spheres. The planets that course through the greater spaces complete their circuits in longer periods of time, those in the shorter spaces in less time. It is a fact that no planet moves more swiftly or more slowly than the others; since they all move at the same rate of speed, the difference traversed is alone responsible for the difference in time consumed. For example, omitting the intermediate planets to avoid repetition, Saturn in thirty years passes through the same signs that the moon traverses in twenty-eight days. The only explanation for this difference is the difference in distance covered by their spheres, one being the greatest and the other the smallest; thus the time consumed in each planet's revolution is in proportion to the amount of space traversed.[334]

Those are already several *prisci philsophi*, as Copernicus names them in Chapter 10 of Book I of *De revolutioniubus*, who "wanted to arrange the planets in accordance with the duration of the revolutions,"[335] assuming "that of objects moving equally fast, those farther away seem to travel more slowly,"[336] but the list is still not complete. One has to look further in the history of philosophy, beyond Aristotle, to Plato's *Timaeus*, to find one of the oldest if not the oldest formulation of that principle. Plato, explaining how god created and ordered the planets, maintains in very similar words to those used by Copernicus in the *Commentariolus*,

---

[333] *In somnium Scipionis* I, 14, 26–27 (Stahl, p. 148; Neri, p. 358).

[334] Ibid. I, 21, 5–6 (Stahl, p. 176; Neri, p. 412). Compare also to Martianus Capella, *De nuptiis Philologiae et Mercurii* VIII, 861 (Stahl, p. 335; Ramelli, p. 614).

[335] *De revolutionibus* I, 10 (Rosen, p. 18; Zekl, p. 124).

[336] Ibid.

that some planets are "moving in greater circles, some in lesser; those in the lesser circles moving faster, those in the greater more slowly."[337] The problem is that the arrangement of the planets Plato uses in the both the *Republic* and the *Timaeus* (the earth, the moon, the sun, Venus, Mercury, Mars, Jupiter, Saturn, the fixed stars), contradicts his own principle. According to Plato, Venus and Mercury have the same period as the sun: one year.[338] This means that Plato himself, just like Ptolemy, is not equal to the task he has set himself. One bond, the distance–period relationship, which was used by Artisan when creating the cosmos, should govern all the planets, not only Saturn, Jupiter and Mars.[339] Copernicus, as Plato's Real Astronomer, must discover, if Mourelatos' words may be used, "what formula, rule,

337 *Timaeus* 38e–39a (Cornford, p. 112). Compare with the *Commentariolus* (Swerdlow, p. 440; Zekl, p. 8): "Also, one [planet] exceeds another in rapidity of revolution in the same order in which they traverse the larger or smaller perimeters of [their] circles (*Eodem quoque ordine alius alium revolutionis velocitate superat; secundum quod maiora minorave circulorum spatia emetiuntur*)." Plato's statement was paraphrased by Ficino, *Compendium in Timaeum* (Farndell, p. 113; *Opera omnia*, p. 1468): "Those of them which are borne by a smaller sphere return to the same point more quickly, while those borne by a larger sphere return more slowly." Compare also with Chalcidius, *In Platonis Timaeum Commentarius* CXIV (Moreschini, pp. 328–9): "Planetum uero quidam citius, alii tardius cursus suos peragunt. Merito; circulorum enim minorum compendia aduersum maiorum anfractus comparata modicis temporum impendiis curriculorum spatia complent, ut luna quae iuxta cubicum numerum uiginti et septem diebus circulum suum lustrat, cum Saturnium sidus triginta prope annis redeat ad pristinum exordium."

338 See *Republic* 624b–621d, where so called Myth of Er is told, and *Timaeus* 38d (Cornford, p. 105). "Having made a body for each of them, the god set them in the circuits in which the revolution of the Different was moving – in seven circuits seven bodies: the moon in the circle nearest the earth; the sun in the second above the earth; the Morning Star (Venus) and the one called sacred to Hermes (Mercury) in circles revolving so as, in point of speed, to run their race with the sun [...]. As for the remainder, where he enshrined them and for what reason – if one should explain all these, the account, through only by the way, would be a heavier task than that for the sake of which it was given." On this, see for instance, F. Cornford, *Plato's Cosmology*, pp. 72–114; W. Knorr, "Plato and Eudoxus on the Planetary Motions," pp. 313–4; A. Gregory, *Plato's Philosophy of Science*, pp. 125–36.

339 According to A. Mourelatos, "Plato's 'Real Astronomy': *Republic* 527d–531d," pp. 56–7, Real Astronomer of Plato's *Republic* 527d–531d should discover the system of *symmetriai* which "cannot fail to have some relation to whatever principles the divine craftsman employs in fashioning the cosmos," and suggests that the divine demiurge put into the heaven a system of *symmetriai* that is "essentially similar to the maximally coherent *symmetriai* [...] the divine craftsman of the *Timaeus* put into elementary bodies, or of

or process of mathematical-geometrical construction would establish a system of *symmetriai* that articulates in a coherent and perspicuous way"[340] the numbers established by observations. Is this possible at all?

(iv) Copernicus' solution to the *forma mundi* problem: a non-traditional application of the distance–period relationship. Many authors noticed the special bond between Mercury, Venus and the sun – that they all have a period of one year. One author thought that this special relationship, the bond of the planets with the sun, should be transformed into a non-traditional cosmological order. Roman encyclopedist Martianus Capella in his *De Nuptiis Philologiae et Mercurii*, published in 1499 in Venice and again in 1500 in Modena,[341] argued that it was not the earth that was at the centre of the movement of Mercury and Venus but the sun (see *Figure 22*). Here are some statements he used to support his position: "Three of these [planets], together with the sun and the moon, circle around the orb of the earth, but Venus and Mercury do not go around the earth."[342] Venus and Mercury are said to have their centre of movement in the sun: "Now Venus and Mercury, although they have daily risings and settings, do not travel around the earth at all; rather they encircle the sun in wider revolutions. The centre of their circles is set in the sun. As a result they are sometimes above the sun; more often they are beneath it in a closer proximity to the earth."[343] These two planets travel around the sun as it makes its celestial revolution and their circles "are epicycles; that is, they do not encompass the globe of the earth within their orbits, but describe an orbit to one side, in some way."[344] And specifically for Venus: "Located on its own circle, Venus goes around the sun, varying its course; sometimes it passes ahead of the sun, sometimes it follows after it, and does not catch it; again at times it is borne above the sun and at times beneath it."[345]

---

that matrix of harmonically related intervals the same divine craftsman cut into the fabric of the world soul."

340 Ibid. p. 55.

341 On these editions, see B. Eastwood, "Kepler as Historian of Science," pp. 370–1, and n. 16 and p. 18; A. De Pace, *Niccolò Copernico e la fondazione del cosmo eliocentrico*, p. 371, n. 223. See also Eastwood's discussion on Capellan astronomy in *Ordering the Heavens*, pp. 179–311. There is bilingual, Italian and Latin, edition of all medieval commentaries on Capella, in I. Ramelli, ed., *Tutti commenti a Marziano Capella*.

342 *De nuptiis Philologiae et Mercurii* VIII, 854 (Stahl, p. 332; Ramelli, pp. 608).

343 Ibid. VIII, 857 (Stahl, p. 333; Ramelli, pp. 609).

344 Ibid. VIII, 879 (Stahl, p. 341; Ramelli, pp. 626).

345 Ibid. VIII, 882 (Stahl, p. 342; Ramelli, p. 630).

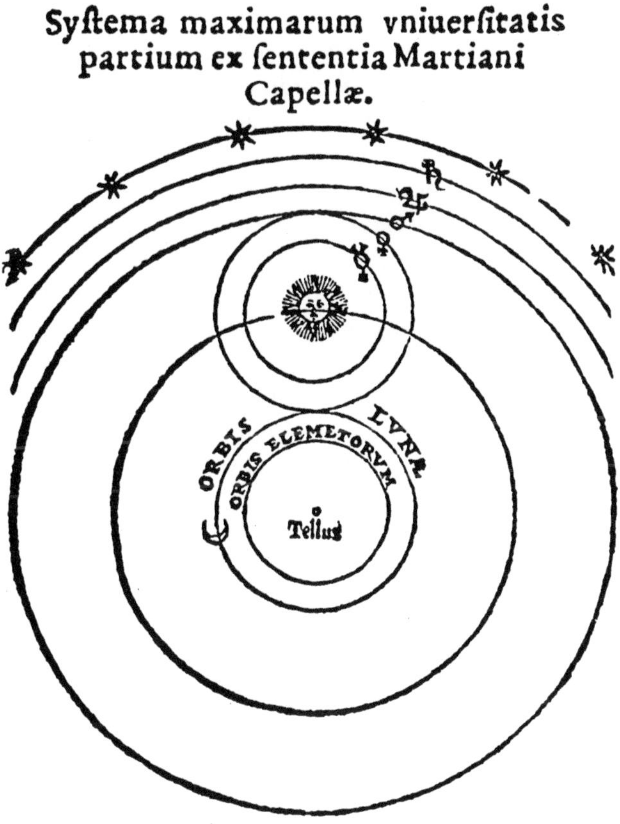

Figure 22.
The Capellan system. From Valentin Naibod (or Naboth), Primarum de coelo et terra institutionum quotidianarumque mundi revolutionum, libri tres *(Venice, 1573).*

Some other authors beside Martianus Capella – the usual suspects are Pliny the Elder, Vitruvius, Plato, Chalcidius, and Macrobius, – allegedly also reported or supported the idea of a geostatic and partly heliocentric system, in which the earth was situated at the centre of the universe with Saturn, Jupiter, and Mars circling around it, and Mercury and Venus circling around the sun, which in turn circles around the earth.[346] The Capellan system has one enormous advantage: it resolves the problem of a special relationship of Mercury and Venus with the

---

[346] See Pliny the Elder, *Historia naturalium* II, 14, 72–4; Vitruvius, *De architectura* IX, 1, 6–12; Plato, *Timaeus* 38c–d; Chalcidius, *In Timaeum* 110–111; Macrobius, *In somnium Scipionis* I, 18. For a discussion of these sources, see B. Eastwood, "Kepler as Historian of Science," pp. 370–1; A. De Pace, *Niccolò Copernico e la fondazione del cosmo eliocentrico,*

sun, a relationship which expresses itself in common one year period.[347] The Capellan solution observes the distance–period relationship and arranges the above mentioned planets accordingly, but with one difference: the reference point of the revolutions is not the earth but the sun.

Once Copernicus came across this arrangement, he was, as he reports in Chapter 10 of Book I of *De revolutionibus*, practically one step away from the solution to the *forma mundi* problem.[348] But why did he not settle for the Capellan solution?[349] He would gain a lot. There seems to be no astronomical reasons against it and on the other hand it would save him from the need to affirm the absurd notion of the movement of the earth.[350] The answer is, again, his Platonism. As has been pointed out several times already, the universe is a homogenous entity for Platonists. For Plato and the Platonists, parts of the universe must be subsumed in a unitary, coherent, homogenous whole, and there must be a single bond linking them together. As Ficino explains it in one of his letters, "Pythagoras teaches

p. 372, n. 224; B. Goldstein, "Copernicus and the Origin of his Heliocentric System," pp. 229–30.

347 Moreover, some astronomers noticed a special relationship of inner and outer planets with the sun. R. Westman, *The Copernican Question*, pp. 50–1, cites some passages from Peurbach's *Theoricae planetarum novae*. Westman draws attention also to Francesco Capuano da Manfredonia, *Theoricae novae planetrum Georgii Purbachij astronomi celebrissimi [...] expositio*, 1495, who echoed the original text referring the mean motion of the Mercury and Venus "as if mirror and measure" of the sun's motion. Compare with Swerdlow's criticism of Westman in "Copernicus and Astrology," p. 366. See also Swerdlow's "Copernicus and Astrology: Appendix," pp. 11–6, for his translation of pertinent sections and explanation of the relations of geocentric and heliocentric models.

348 One could argue that this was already at the time of the publication of his translation of *Letters* of Theophylactus Simocatta from *Epistolae diversorum philosophorum, oratorum, rhetorum* (Venice, 1499). The Latin *Epistolae* were dedicated by Aldo Manuzio to Antonio Urceo Codro. Copernicus published his Latin translation of Simocatta's *Letters*, which were not included in the Aldine edition, in 1509. I am inclined to think that Corvinus referred to this description in his eulogy. Copernicus could tell him that he made a new cosmological model by drawing on the Capellan system that was familiar to Corvinus. See also M. Clutton-Brock, "Copernicus's Path to His Cosmology: An Attempted Reconstruction," p. 203.

349 This question was posed to me by Peter Barker.

350 It is worth noticing that the "Calvinist Copernicans" were very much impressed by Capellan system and at certain point it became a standard view among Dutch scholars. See R. Vermij, *The Calvinist Copernicans*, pp. 32–42.

us that unity looks towards the good, but diversity towards the evil."[351] Capellan system lacks unity and homogeneity. There are actually two systems, not one, which account for the order of the planetary spheres in the universe and this is, in my opinion, the reason Copernicus did not choose this explanation. There is no *symmetria* and no one bond that links the parts in well-proportioned, unitary, coherent whole. There is no unity.

Nevertheless, as Copernicus himself admits, the Capellan system is a step in the right direction. It indicates that the problem is not the distance–period relationship, but the reference point of that principle. In fact, the distance–period relationship works very well for all of the planets, but the centre of their motion must be transformed. If that does not work, there is no principle governing the order of the planetary orbs, there is no certainty, and there is no order in the universe, which, of course, is unacceptable from the Platonist perspective. In order to save the distance–period relationship, a bond that unites all the planets and stars into coherent system, Copernicus has to change the centre of the motion: all six planets move around the sun. In order to maintain a modified form of the distance–period relationship he has to find the periods of the planets around the sun, rather than around the earth. Since the order of the periods for Saturn, Jupiter and Mars is not under question, it is only necessary "to demonstrate that the heliocentric periods of Venus and Mercury were less than a year, and that the period of Mercury was less than the period of Venus."[352]

And so he does – in his Uppsala notes fol. 15^{v}, usually referred to as "U." That note, consisting of results of calculations, gives the following ratios (from the upper part of U):[353]

10,000 : 6583 for Mars,
10,000 : 1917 for Jupiter,
10,000 : 1083 for Saturn,
376[0] : [10,000] for Mercury, and
[7200 : 10,0000 for Venus].
The bottom part of U has the respective ratios:
38 : 25 for Mars,

---

351 *Epistolarum liber* III (Members of the Language Department of the School of Economic Science, vol. 2, letter 30, p. 39; *Opera omnia*, p 635).

352 B. Goldstein, "Copernicus and the Origin of his Heliocentric System," p. 221.

353 Here I am closely following A. Goddu, "Reflections on the Origin of Copernicus's Cosmology," p. 39.

130 : 25 for Jupiter,
230 $^{5/6}$: 25 for Saturn,
9 $^{2/3}$: for Mercury, and
18 : 25 for Venus.[354]

The calculation of ratios gives the following result: 1.52 for Mars, 5.2 for Jupiter, 9. 23 for Saturn, 0.376 for Mercury, and 0.72 for Venus, which is, as Goddu puts it, "very close to 1.52, 5.2, 9.5, 0.387, and 0.72, the estimates of the orbital radii with the earth's orbital radius set at 1."[355]

In this way, Copernicus can explain the orderly principle governing the Universe, which is the order through which the distance of each planet from the sun agrees with its period of revolution.[356] His calculations prove that the distance–period relationship works for all the planets. They also prove that there is one bond (the sidereal period of each planet) that connects them all in a certain *symmetria*. Or, as Rheticus puts it in the *Narratio prima*: "the admirable symmetry and interconnection of the motions and orbs (*orbium symmetria et nexus*), as maintained by the assumption of the foregoing hypotheses, are not unworthy

354 Compare with the *Commentariolus* (Swerdlow, p. 465; Zekl, p. 20): "But the orb of Saturn completes a revolution in the 30th year, the orb of Jupiter in the 12th, and the orb of Mars in the 23rd month, just as if the size of the orbs slowed down these revolutions. For where the semidiameter of the great orb is given 25 parts, the semidiameter of the sphere of Mars will receive 38 parts, that of Jupiter 130 5/12 parts, and that of Saturn 230 5/6 parts. I call the *semidiameter* the distance from the centre of the orb to the centre of the first epicycle."

355 A. Goddu, "Reflections on the Origin of Copernicus's Cosmology," p. 39. I cannot but refer to A. Mourelatos's abstract definition of Plato's *symmetria* from his "Plato's 'Real Astronomy': *Republic* 527d–531d," p. 40: "Two quantities (numbers, lines, figures, solids) F and G are in a relation of *symmetria* if and only if G can be mathematically constructed through a simple operation from distinct and easily determinable parameters of F in such a way that determination of these parameters also determines common sub-multiples of F and G." Compare also with his speculation (inspired by Kepler), p. 57, about what a Real Astronomer might do to establish the system of *symmetria*. The Real Astronomer "might fill in the diagram of the heavens with imaginary circles and radii. Then assuming, as he might, some intelligible intermediate connection between comparable periods and radii [Mourelatos refers in the n. 40 to *Republic* 616e–617b, and *Timaeus* 39a], he may be able to reduce the system of *symmetriai* he seeks to one involving the radii, or some parameters of the radii, or perhaps some geometrical construction based on the radii."

356 See also A. Koyré, *The Astronomical Revolution*, p. 49; D. Shapere, "Copernicanism as a Scientific Revolution," p. 102; T. S. Kuhn, *The Copernican Revolution*, pp. 174–6.

of God's workmanship (*opifice Deo*) and not unsuited to these divine bodies."[357] Thus the six movable planetary orbs achieve "celestial harmony," "[f]or they are all so arranged that no immense interval is left between one and another; and each, geometrically defined, so maintains its position that if you should try to move any one at all from its place, you would thereby disrupt the entire system."[358]

This *symmetria* and *harmoniae nexus*, of course, presupposes the movement of the earth around the sun. And seen from this perspective, it is the movement of the earth that answers Plato's demand for one bond that links all the parts of the universe in well-proportioned, harmonious whole. Or, to borrow again Rheticus' words, this time from his other Copernican *Treatise on Holy Scripture and the Motion of the Earth*:

> I do not see how the wisdom and infinite power of God may so clearly be grasped by the uncomprehending human mind in any [other] part of nature, than it is in accepting the motion of the earth, where it appears that God desired to establish one particular bond of all visible things, something which Plato saw was necessary and urged men to investigate, even though he did not perceive what it really was.[359]

(v) Confirmation of the earth's movement by the authority of ancient philosophers and writing of the *Commentariolus*. With this solution at hand, all Copernicus needs to do is to find some ancient philosophical authority or authorities that held as possible the movement of the earth around the sun as well as around its axis. At that moment, when he had done the basic math, he started reading, as he says in the *Preface*, all philosophers available to him

[357] *Narratio prima* (Rosen, p. 145; Hugonnard-Roche and Verdet, p. 59). See also ibid. (Rosen, p. 139; Hugonnard-Roche and Verdet, p. 56): "Under the commonly accepted principles of astronomy, it could be seen that all the celestial phenomena conform to the mean motion of the sun and that the entire harmony of the celestial motions is established and preserved under its control."

[358] Ibid. (Rosen, p. 147; Hugonnard-Roche and Verdet, p. 60). Compare with ibid. (Rosen, pp. 164–5; Hugonnard-Roche and Verdet, p. 69): "But if anyone desires to look either to the principal end of astronomy and the order and harmony of the system of the spheres (*systematis orbium rationem ac consensum*) or to ease and elegance and a complete explanation of the causes of the phenomena, by the assumption of no other hypotheses will he demonstrate the apparent motions of the remaining planets more neatly and correctly."

[359] *Treatise on Holy Scripture and the Motion of the Earth*, 20 (Hooykaas, Latin, p. 49; English, p. 75).

in order to see whether anyone had different ideas about the motions of the worlds' spheres than those exposed by astronomers in schools. It was only after Copernicus settled on a heliocentric system resulting from the distance–period relationship that is, after he obtained the correct heliocentric periods, that he started investigating physical possibility of the mobility of the earth.

In the *Commentariolus* Copernicus cites the Pythagoreans as authorities confirming movement of the earth, but he is more specific in his later *Preface* to *De revolutionibus*. There he cites two specific sources in which he found reports of ancient philosophers defending the earth's mobility.

The first one was Cicero, whose work probably came to Copernicus' attention in Padua through Nicholas Leonicus Tomaeus.[360] Cicero tells us in *Quaestiones academicae* that Theophrastus maintained that, according to the last-generation Pythagorean Hicetas of Syracuse, heavens stood still while the earth rotated around its axis and that some believed that such was also Plato's opinion.[361] Copernicus, who himself read *Quaestiones academicae* published in 1498 in Milan, copied this entire passage in his copy of the 1487 edition of Pliny's *Historia naturalium* II, 12:

> According to Cicero, *Quaestiones Academicae*, Hicetas of Syracuse, as Theophrastus has it, thought that the heavens, sun, moon, planets and all higher things stood still and that the earth alone in the cosmos moved, which turns around its own axis with the supreme speed and by its turning produces everything what would happen if the earth were standing still and the heavens moved. Some think that Plato too said this in the *Timaeus*, though more obscurely.[362]

As we have seen, the name of Nicetas, as it is properly spelled, can also be found in Chapter 5 of Book I of *De revolutionibus*.

---

360 See supra, p. 290, and B. Biliński, "Intorno ai problemi del Copernico padovano," p. 105; L. A. Birkenmajer, *Mikolaj Kopernik*, p. 567.

361 See *Quaestiones academicae* (or *Academica Priora*) II, 39, 123. For quote from the *Preface*, see supra, p. 75.

362 "Apud Ciceronem libro secundo Academicorum quaestionum: Nicetus Syracosius, ut ait Theophrastus, celum, Solem, Lunam, stellas, supera denique omnia stare censet neque praeter terram rem ullam in mundo moveri, quo cum circa axem se summa celeritate convertat et torqueat eadem effici omnia, que si stante terra celum moveretur. Atque hoc etiam Platonem in Timeo dicere quidam arbitrantur, sed paulo obscurius." Quotation is taken from B. Biliński, "Intorno ai problemi del Copernico padovano," p. 104. It is partly translated in P. L. R. Rose, *The Italian Renaissance of Mathematics*, p. 125.

Later, Copernicus states in the *Preface*, he found in Plutarch that Philolaus the Pythagorean argued that the earth moved and that Heraclides of Pontus and Ecphantus the Pythagorean made the earth move like a wheel. And he transcribed Plutarch's Greek text: "Some think that the earth remains at rest. But Philolaus the Pythagorean believes that, like the sun and moon, it revolves around the fire in an oblique circle. Heraclides of Pontus, and Ecphantus the Pythagorean make the earth move, not in a progressive motion, but like a wheel in a rotation from west to east about its own centre."[363] Since Copernicus cites this report in the *Preface* in Greek, he must have copied it from *Plutarchi Opusculi LXXXXII*, published in Venice in 1509, a copy of which existed in Warmia. But this report can also be found in Giorgio Valla, *De expetendis et fugiendis rebus opus*, published by Aldo Manuzio in 1501 in Venice, where Copernicus probably first encountered it.[364]

With the movement of the earth corroborated by the philosophical authorities of Antiquity, mostly by the Pythagoreans, but also, according to some, by Plato, Copernicus could set out to write the *Commentariolus*. Since it displays "a good deal of carelessness and incomplete understanding on Copernicus's part,"[365]

363 *Preface* (Rosen, p. 5; Zekl, pp. 72–3).

364 Valla's ten volumes, divided into forty-nine books, are a kind of encyclopedia combining different disciplines, including arithmetic, geometry, music, mechanics, optics, medicine and astronomy. What is of concern to Copernicus is found in *Physica*, in Books XX and XXI. Chapter 45 of Book XXI contains, in the anonymous summary of pseudo-Plutarch's *Placita philosophorum*, which is in fact a report on philosophical opinions of Aëtius, a section titled *De motu terrae. Placita philosophorum* III, 13, contains a Latin report of Philolaus, Heraclides of Pontus and Ecphantus: "De motu terrae - Quidam ac fere omnes stare terram volunt. Philolaus autem Pythagoreus circa ignem circumferri per obliquam circulum simili modo quo Sol et Luna. Heraclides Ponticus et Ecphantus Pythagoreus terram quidem movent, nec tamen transitive, at rotae modo revolutam ab occasu in ortum circa suum centrum." On Giorgio Valla, see B. Biliński, B., *Il Pitagorismo di Niccolò Copernico*, pp. 64–5; L. A. Birkenmajer, *Stromata Copernicana*, p. 154 et seq. There is another ancient source Copernicus cited in the manuscript of *De revolutionibus*, Aristarchus of Samos (310–ca. 230 BC), but eventually decided to delete the passage from the final version. Aristarchus is also mentioned in G. Valla, *De expetendis et fugiendis rebus opus*, in *Placita philosophorum* II, 24. Valla mistranslated Aristarchus, stating that he "locates the sun beyond the fixed stars." Copernicus may have also read in Plutarch's *Concerning the Face Which Appears in the Orb of the Moon* about the ancient philosopher who thought that the Greeks should indict Aristarchus of Samos on the charge of impiety because he sent in motion the hearth of the universe. He tried to save the phenomena by assuming that the heavens are stationary and that the earth revolves in an oblique circle, while at the same time rotating about its axis. See E. Rosen, *Commentary*, pp. 360–1.

365 N. Swerdlow, "The Derivation and First Draft of Copernicus' Planetary Theory," p. 429, lists as problematic: the precession, the lunar latitude theory, the latitude theory

Swerdlow believes – and I with him – that it was written in haste, very probably "shortly after Copernicus developed his new theories and before he correctly understood all the necessary details of his models. Thus the derivation of the heliocentric theory, the writing of U, and the writing of the *Commentariolus* probably came very close together, possibly within a few months or even weeks."[366] However, rather than publishing it, Copernicus decided to circulate it among his friends. In front of the larger public he preferred, in the manner of the Pythagoreans, to keep quiet about the movement of the earth. Later he started writing *De revolutionibus*, which contains all the fundamental ideas of the *Commentariolus*, with the only difference that he wrote it on the imitation of Ptolemy's *Almagest* and changed some of his planetary models from the *Commentariolus*, so that his models include not only Ibn al-Shātir's models and Tūsī couple but also al-Urdī's lemma. After several decades his friends finally persuaded him to publish it. When everything was ready, he replaced the original *Introduction* with the *Dedication to the Pope*, partly also because he needed to dissuade Pope Paul III from granting his patronage to a homocentric competitor Fracastoro, and mostly because he needed some advance defence of the movement of the earth.

---

of Venus and Mercury, and the description of the variation of the radius of Mercury's orbit.

[366] Ibid.

Chapter THIRTEEN

The Platonist astronomer-philosopher and the Scientific Revolution

What happened to astronomy with the appearance of Copernicus as Platonist astronomer-philosopher? What did his project bring into the astronomical landscape compared to those of his immediate predecessors and contemporaries? Did he "only" trigger the Scientific Revolution or are there any significant epistemological and philosophical changes already present in his work?

A brief comparison of Copernicus' astronomy with that of his immediate predecessors and contemporaries reveals that he retained several common major features while at the same time changing some radically.

With the spread of Peurbach's *Theoricae novae,* three-dimensional orbs became the accepted causes of the movements of celestial bodies. What Regiomontanus praised as a novelty, the majority of subsequent astronomers, including Copernicus, held as an indisputable premise. Peurbach's and Regiomontanus' *Epytoma Almagesti* provided astronomers with high quality Ptolemaic astronomy and Copernicus took full advantage of it. There was, also during Copernicus' lifetime, a constant criticism of Ptolemaic astronomy and awareness of the possibility of homocentric alternative.

The first major change brought into astronomy by Copernicus' Platonist reform is very obvious when his work is compared with various Averroist projects. His change of orientation is plain: there is a radical shift in the philosophical questions that should occupy astronomer's minds. The second aspect of Copernicus' novelty concerns the obvious changes which Copernicus implemented into the framework of Ptolemaic astronomy. These have radical, even revolutionary, epistemological and philosophical implications. In short, Copernicus' Platonism constitutes the foundation of Copernicus' genuine Copernican Revolution and inspired his major contribution to the Scientific Revolution.

1. Copernicus in comparison with Averroists

The main concern for defenders of the homocentric astronomy was, as we have seen in the section on astronomy before Copernicus, thoroughly philosophical. From Roger Bacon to Regiomontanus, philosophers and astronomers were concerned about bringing astronomy in compliance with the principles of Aristotelian natural philosophy. They believed that the duty of every astronomer was also a philosophical one. Every astronomer must, to use Regiomontanus' word, "philosophize." Some of them just attacked Ptolemaic astronomy or repeated the Averroists' arguments against it (its violation of some of the Aristotelian natural philosophical principles, as by Brudzewo and Achillini); others tried to establish homocentric planetary models (Regiomontanus); still others, like Copernicus' contemporaries, Amico and Fracastoro, actually provided their own versions of homocentric astronomical systems aimed to obtain accurate predictions of planetary positions while remaining in accord with the Aristotelian philosophy.

Fracastoro, for example, claims in the dedicatory preface to his *Homocentrica* that the causes of the celestial motions can be explained either through homocentric or eccentric orbs.[367] Each option has its difficulties and advantages. Those who use homocentric spheres are incapable of reproducing the observed motions of the celestial bodies and thus "demonstrating the appearances" or "saving the appearances." Eccentrics and epicycles, on the other hand, seem to demonstrate

---

367 See, Fracastoro, *Homocentrica*, "Sanctissimo Pavlo Pontifici Maximo." See also Amico, *De motibus corporum coelestium*, "Amplissimo Cardinali Nicolao Rodulpho" (Di Bono, pp. 134–5), who also claims that there is a dispute between astronomers and philosophers, in regard of the question, which are the causes that explain the variety of celestial motions. Whereas astronomers attributed all these appearances to eccentrics and epicycles, philosophers denied that such orbs could exist in the nature. But so far philosophers were unable to explain the celestial appearances by combination of homocentric orbs. Compare with M. Granada and D. Tessicini, "Copernicus and Fracastoro: The Dedicatory Letters to Pope Paul III, the History of Astronomy, and the Quest for Patronage," pp. 458–64.

appearances, but draw from physically unacceptable hypotheses and make unsuitable assumptions on celestial bodies. Every genuine philosopher should regard eccentrics and epicycles as monstrous because they are ontologically and physically impossible in the realm of divine and most perfect heavenly bodies. There can be only one centre of motion (i.e. the centre of the world, the earth). Therefore, "[u]ntil now, no philosopher could find how to set those monstrous spheres among those divine and most perfect bodies."[368] For this reason, after the triumph of Ptolemy's astronomy and in the absence of any noteworthy proposal to supplant it, the monstrous astronomy is all that has remained. Since all homocentric astronomy, which complies with the demand of Aristotelian physics for one centre of the movements in the heavens and with the nature of celestial substance, has so far not been able to "demonstrate the appearances," he will provide just such one. His *Homocentrica* will remain faithful to principles of Aristotelian philosophy of nature and at the same time "demonstrate the appearances." The task Fracastoro sets for himself is obviously not only astronomical but philosophical at the same time. He is an astronomer-philosopher, but an Aristotelian astronomer-philosopher.

Copernicus is also an astronomer-philosopher. He too strives to establish an astronomical system that would not only provide precise calculations for the positions of celestial objects but also comply with philosophical principles. He agrees with the demand for philosophically feasible astronomy and even with the Fracastoro's demand to avoid monstrosity in the heaven. However, apart from this general program, and his use of the Tūsī couple, Copernicus has nothing else in common with Fracastoro and other Averroists. First of all, according to Copernicus no variation of homocentric astronomy can "save the appearances." In his view the Ptolemaic astronomy of eccentrics and epicycles is the only astronomy that can accomplish fundamental astronomical task to give accurate prediction of the celestial phenomena, but it needs reform or renovation. The reform of Ptolemaic astronomy is not needed because of eccentrics and epicycles, nor is Ptolemy's system monstrous, as Fracastoro claims, because of these models. Ptolemy's universe is monstrous, Copernicus argues, because it is unable to satisfy Plato's ideal of well-ordered universe, governed by unity. He sees, therefore, the

368 Fracastoro, *Homocentrica*, "Sanctissimo Pavlo Pontifici Maximo." Copernicus' knowledge of Fracastoro's formulations is, according to very persuasive analysis of M. Granada and D. Tessicini, "Copernicus and Fracastoro: The Dedicatory Letters to Pope Paul III, the History of Astronomy, and the Quest for Patronage," the reason why Copernicus also characterized Ptolemy's universe as monstrous.

philosophical problems of Ptolemaic astronomy in completely different, Platonist terms. Copernicus is Platonist astronomer-philosopher; he therefore implements the program of philosophical rather than exclusively mathematical astronomy. But unlike Regiomontanus, Amico, Fracastoro and other Averroists on the one hand, and traditional Ptolemaists on the other, Copernicus is convinced that this task can only be accomplished by the heliocentric version of the Ptolemaic astronomy of eccentrics and epicycles. For only a heliocentric version of Ptolemaic astronomy is able to "save the appearances" and be in accord with Platonist philosophy. In sum: while defenders of the homocentric astronomy are concerned by questions of how to reconcile astronomy with Aristotelian philosophical principles, Copernicus strives to bring astronomy in accord with Platonist philosophy.

There is yet another, more fundamental and crucial distinction between Copernicus and homocentric astronomers-philosophers. Whereas Fracastoro, for instance, regarded homocentric astronomy as a means to bring inappropriate geometric models in compliance with Aristotelian physical principles, Copernicus chose a contrary alternative: he retained models that provided more accurate computations of the celestial positions and found them appropriate physical and philosophical principles. Or, to formulate it more appropriately: according to Fracastoro, the selection of mathematical models depends on physical assumptions, whereas for Copernicus physical assumptions (movement or rest of the earth) depend not on mathematical models but on mathematical reasoning (distance–period relationship). Because of that, Copernicus has to take one step further: on the basis of his Platonist mathematical considerations and results, and in order to fulfill the demands of Platonist astronomy, he has to change the basic assumption of the Aristotelian philosophy of nature and force Aristotle's immovable earth into "unnatural" motion and throw it off its "natural" place in the centre of the universe. This gesture brings us to the threshold of the question of the Copernican and Scientific Revolution.

2. Copernicus' revolutionary reform of Ptolemaic astronomy

According to standard accounts of the Copernican and consequently Scientific Revolution, the radical consequences of Copernicus' work were fully grasped only by the members of later generations. The Copernican Revolution was not a single great event (i.e. the publication of *De revolutionibus*), but rather a lengthy process, consisting of many small steps. A comparison between Copernicus' version of Copernicanism with those of Kepler and Galileo reveals three conceptually completely different systems. It was only the "modern" Copernicanism of Galileo and Kepler that finally brought about a veritable conceptual revolution. The difference between the Copernicanism of Kepler and Galileo and that of Copernicus is according to these accounts greater than the difference between Copernicus' heliocentric astronomy and geocentric astronomy. The works of Kepler and Galileo introduced huge epistemological transformations into Copernicanism. Kepler supplanted the concept of "path" with that of "orbit," rejecting the ancient postulate of circular motion and replacing it by the elliptical motion.[369] This replacement was based on the requirement for accurate observations of motions of the celestial bodies and the elimination of solid orbs, both established by Tycho Brahe (1546–1601). Now that the planets freely moved through space, they could move in non-circular motion, which in turn raised the question of their dynamics. The other radical epistemological shift introducing new empirical astronomical facts and physics came with Galileo's telescopic discovery of "things never seen before": the irregularity of the moon's surface, the satellites of Jupiter, the "handles" of Saturn, the phases of Venus and sunspots, to name just the most important ones.[370] In short, the true incommensurability, the revolutionary transformation of the conceptual structure and introduction of new epistemological themes in relation to the Ptolemaic astronomical system was introduced by Kepler's astronomy, which was incommensurable with both that of Ptolemy and Copernicus, and by Galileo's empirical discoveries.

In standard accounts, Copernicus' work and that of his immediate followers is therefore seen as conceptually consistent with Ptolemaic astronomy. While Copernicus' methodology and epistemology remained consistent with Ptolemy's, his work set in motion a train of events that led to a decisive epistemological shift, without itself representing such a shift at all.[371] In other words, Copernicus' *De revolutionibus* seems, according to Kuhn, in every respect except the earth's motion "more closely akin to the works of ancient and medieval astronomers than to the

[369] See, for example, P. Barker, "Incommensurability and Conceptual Change during the Copernican Revolution"; B. Goldstein and G. Hon, "Kepler's Move from Orbs to Orbits: Documenting a Revolutionary Scientific Concept"; P. Barker, and B. Goldstein, "Distance and Velocity in Kepler's Astronomy"; P. Barker and B. Goldstein, "The Theological Foundations of Kepler's Astronomy."

[370] On this see, for instance, N. Swerdlow, "Galileo's Discoveries with the Telescope."

[371] See K. Tredwell and P. Barker, "Copernicus' First Friends: Physical Copernicanism from 1543 to 1610," p. 162.

writings of the succeeding generations […]."[372] However, rather than stopping here, Kuhn continues: "[the succeeding generations] who based their work upon Copernicus' and who made explicit *the radical consequences that even its author had not seen in his work*."[373] The succeeding generations not only based their work on Copernicus' model but derived the radical consequences, which were already present in his work, but of which he was not aware. Copernicus himself, as Kepler maintained in *Astronomia nova*,[374] inadvertently generated a great many premises and a great many riches – *Copernicus divinitiarum suarum ipse ignarus* – which the succeeding generations recognized and refined. Copernicus established many relatively minimal conceptual changes and "incommensurabilities" with the traditional Aristotelian physics and cosmology that ultimately culminated in the Scientific Revolution.

I would go even further and defend several stronger, closely related theses. First, several of Copernicus' minimal "incommensurabilities" with the traditional Aristotelian physics and cosmology introduced to account for the movement of the earth, such as his theory of gravity, are Platonist. Second, these conceptual changes and "incommensurabilities" are not very coherent; sometimes they are even contradictory. His physics and cosmology from the Book I of *De revolutionibus* are far from being thoroughly elaborated. Third: his inconsistencies and lack of adequate physics and cosmology are result of the fact that Copernicus' fundamental preoccupation is not natural philosophy, but Pythagorean-Platonist idea of a mathematically ordered universe. Fourth and most importantly: his quest for a Pythagorean-Platonist, well-proportioned universe brought about a major epistemological and philosophical shift. In its most fundamental aspect, Copernicus' work is not minimally incommensurate with traditional astronomy and philosophy but radically different from it, both epistemologically and philosophically. These constitute a genuine Copernican Revolution or, to put it slightly differently, a major element of the Scientific Revolution. It was Copernicus who made the epistemological break, rather than merely prepared the ground for it. This break opened a new horizon of thought essential for the entire Scientific Revolution. This radical epistemological and philosophical break with the Aristotelian tradition is operational in Copernicus' argumentation in favor of the earth's motion, which ultimately stemmed from his Platonist astronomical-philosophical orientation.

---

[372] *The Copernican Revolution*, p. 135.

[373] Ibid. Emphasis added.

[374] See *Astronomia nova* XIV (Donahue, p. 232).

Minimal incommensurabilities. In order to justify the earth's motions, arising from the program of Platonist astronomy, an astronomy which furnished certain *symmetria* of the world, Copernicus has to make more or less significant – and more or less coherent – conceptual changes in various fields. Several of these changes and transformations of traditional Aristotelian philosophy of nature are in their nature essentially Platonist.

Let us start with his technical astronomy. In his technical astronomy, as we have already noted, Copernicus owes a great debt to Islamic astronomers. He uses their improved Ptolemaic models to replace the equant and for other purposes. Indeed, these models represent transformations of Ptolemy's work rather than their radical change to it. Besides, it seems that he did not always understand very well the models he used. He also was not aware that his models that were supposed to get rid of the equant retain a hidden equant point.[375]

Copernicus introduces many philosophical changes in the organization and function of the universe.[376] He enlarges the universe or, more accurately, increases the distance between Saturn and the sphere of the fixed stars. Understanding the immensity of the heavens, the result of the absence of detectable annual parallax, is the first step towards the conception of an infinite universe. Although Copernicus still considers the planetary orbs as real, three-dimensional bodies, in his system the planetary orbs are separated by vast empty spaces, which was something utterly inconceivable for the traditionally *plenum* universe.[377] His introduction into the "world machine" vast intercalated spaces that played no role in planetary motion could be seen as the beginning of a process that led not only to the abolition of real three-dimensional celestial orbs, but to the abolition of the sphere of the fixed stars itself. Copernicus still considers uniform circular motion as a privileged motion of the celestial bodies, that is, celestial spheres carrying planets and stars. He suggests that the cause of circular motion of the spheres is not their inner nature or the substance they are composed of, but their figure or shape, which somehow expresses and actualizes itself by this circular motion. If such had been the case, he would have had to – in order to be consistent – ascribe circular movement to every sphere in the universe. He, of course, does not do that.

---

[375] See J. Evans, "The Division of the Martian Eccentricity from Hipparchus to Kepler: A History of the Approximations to Kepler Motion," pp. 1012–14, and *The History and Practice of Ancient Astronomy*, pp. 420–2. Compare also with O. Neugebauer, "On the Planetary Theory of Copernicus," pp. 92–6.

[376] See also *Le monde des sphères*, vol. 2, pp. 67–73.

[377] See also M.-P. Lerner, *Le monde des sphères*, vol. 1, pp. 131–8, and *Figure 7*, supra, p. 63.

In his system the sphere of the fixed stars is necessary immobile and there is no indications that he thought that the sun and planets rotate as does the earth. He seems to hold that the mere spherical shape does not cause the circular movement, but only those that by doing so establish Platonist order of the universe. He also changes traditional Aristotelian cosmology with Platonist elements. His *terraquea sphaera*, which includes the adjacent air, forms a spherical whole of one and the same nature. It also seems that in his cosmology there is no place for the sublunary element of fire (there is only earthly fire), while there seems to be a place for celestial fire. This fire is the substance of the regions above the air.

Copernicus, furthermore, supplants the Aristotelian concept of heavy and light with the Platonist concept of gravity. But, as we showed previously, Copernicus' doctrine of elemental motion and gravity is inconsistent and contradictory: Copernicus considers rectilinear motion, which is due to gravity, as both natural and non-natural. What is more, when he explains that the rectilinear motion of falling bodies returning to their whole (gravity) produces impetus, which is the cause of acceleration and makes the rectilinear motion nonuniform and therefore non-natural, he should, provided that he thoroughly examined his own arguments, draw the conclusion that the earth does not rotate.

Copernicus' relatively minor, mostly Platonist, philosophical changes or "minimal incommensurabilities" with the Aristotelian philosophy of nature were, despite their inconsistencies, productive premises for the modern Copernicans, who used them more or less skillfully and intensely in their respective new epistemological paradigms. But this, of course, does not mean that Copernicus provided a coherent heliocentric and geokinetic physics and cosmology. The inconsistencies in Copernicus' physics and cosmology reveal that physics and cosmology were not his priority or guiding principle of his thought. They are due to the necessity of the movement of the earth, which is due to, as Rheticus explains, mathematical reason. A well-proportioned universe governed by one principle entails movement of the earth around its axis and around the sun, so Copernicus had to provide at least some physical justification for that. His guiding principle was search for well-proportioned universe, governed by unity, complying with Pythagorean-Platonist ideas. My major thesis is that exactly with that gesture – by subjecting himself to this ideal – Copernicus produced a major epistemological and philosophical shift.

Major incommensurabilities: the hierarchy of sciences; the criterion of truth and appearance. Let me recapitulate my basic argument so far. Copernicus is a Platonist astronomer-philosopher, or, in other words, he is Plato's Real Astronomer. His Platonism explains his objections to Ptolemaic astronomy. The most fundamental are its disregard for the principle of the uniform circular motion of the planets (the equant) and its inability to give an account of the *forma mundi*, that is, its lack of certainty in regard of universes' *symmetria*. Although the equant problem has by itself nothing to do with the *forma mundi* problem – Copernicus does not need the earth's motion to solve it – and although he considers it secondary to the problem of the "monstrosity" of the Ptolemaic astronomy, there is a natural connection that links it to the question of the arrangement of the universe. Both problems are philosophically, or better, metaphysically associated, but only from a Platonist perspective. For a Platonist the philosophical problem of neglecting the principle of the uniform circular movement carries the same metaphysical weight as the monstrosity of the universe's structure. The supreme Artisan did not create a universe with non-uniform motion of spheres or a universe which is not organized as *symmetria*. Heliocentrism and equant elimination are motivated by Copernicus' adherence to the Platonic doctrine of cosmic order and unity.

Copernicus believes he solves the equant problem with the use of Ibn al-Shātir's model and the *forma mundi* problem by strict application of the principle of a single natural bond, distance–period relationship, which links together all the parts of the universe. Since he builds on the idea of cosmic order that is mathematically determined, his heliocentrism and geokineticism follows from mathematical arguments. The beautiful, divine whole is composed of well-proportioned parts and governed by unity. Cosmic order depends on unity: there is one rule that governs in it and connects all its parts into harmonious arrangement within a whole. This requires one small modification of the traditional organization of the universe regarding the reference point or the centre of the motion of the planetary spheres: it is no longer the earth that lies stationary in the middle of the celestial ballet but the sun, while the earth circles around the sun as its centre and around its axis.

(i) Hierarchy of sciences. Copernicus' mathematical solution has obvious physical consequences: the earth must be capable of moving around the sun and around its axis. By advocating the reality of the earth's motions on mathematical grounds Copernicus violates the established Aristotelian hierarchical order of sciences: theology, philosophy, astronomy. This, of course, did not go unnoticed. One of his first critics, the Aristotelian theologian Tolosani, claims that

Copernicus, well versed in mathematical and astronomical sciences, is not only in contradiction to theology, but also in contradiction to philosophy of nature and dialectics. He contradicts many of the principles of the natural philosophy. These principles are organized in such a way that the inferior science (astronomy as a mathematical science) gets its principles from the superior one (natural philosophy). All the sciences are interconnected in a hierarchical order in such a manner that inferior sciences need superior ones. One cannot be a perfect astronomer unless he or she knows the physical sciences. Astronomy presupposes the natural celestial bodies and their natural movement.[378]

But Copernicus is not an Aristotelian; he is a Platonist. The counterintuitive consequences of movements of the earth, his contradiction to Aristotelian physics and his violation of Aristotelian hierarchy of sciences are protected by the Platonic doctrine that privileges mathematical arts and specifically astronomy as the highest art. For Copernicus mathematical art (astronomy) is more important than physics, which in turn means that even if astronomical arguments contradict physics, mathematics takes precedence. This explains the whole mystery of why Copernicus is not bothered by the conflict with Aristotelian physics and Aristotelian hierarchy of sciences.

(ii) Appearance and truth. On the other hand, Copernicus' Platonist reliance on mathematical argument in establishing movements of the earth produces yet another and fundamental epistemological shift, neglected by Kuhn and other scholars, which is extremely far-reaching. It concerns one of the most fundamental concepts of the entire modern science and philosophy, that is, the

---

[378] See *Opusculum quartum*, cap. 2, 340^r (Lerner, pp. 700–3): "Peritus est etiam in scientiis mathematicis et astronomicis, sed plurimum deficit in scientiis physicis ac dialecticis, nec non in sacrarum literarum imperitus apparet, cum nonnullis earum principiis contradicat, non absque infedilitatis periculo et sibi et lectoribus libri sui. Contra negantem autem prima scientiarum principia non est disserandum, quoniam ex primis principiis conclusiones rationis discursus deducuntur. Scientia quoque inferior a superiori principia comprobata recipit. Itaque omnes scientiae sibi invicem connectuntur, ita ut inferior superiore indigeat, et se invicem adiuvant. Non potest enim esse perfectus astronomus, nisi prius didicerit scientias physicas, cum astrologia presupponat naturalia corpora coelestia et naturales eorum motus. Nec homo potest esse perfectus astronomus et philosophus nisi per dialecticam sciat discernere inter verum et falsum in disputationibus, et habeat argumentorum notitiam: quod requiritur in medicinali arte, in philosophia, theologia et ceteris scientiis. Unde cum praedictus Copernicus non calleat scientiam physicam et dialecticam, non est mirum si ipse in hac opinione decipiatur, et falsum pro vero recipiat ex imperitia earum scientiarum."

concept of appearance and related criteria for distinguishing between appearance and reality. Or, to put it slightly differently, it has far-reaching consequences for the establishment of the criteria for determining the objective, scientific truth. Copernicus, again in complete accord with Plato's demands from the *Republic*,[379] argues for a well-proportioned universe which entails movement of the earth by mathematical arguments. This in other words means that he establishes the movement of the earth by going beyond appearances, not arguing from appearances.

Let us start with celestial appearances. Astronomical investigation centres on appearances or more accurately, celestial appearances or phenomena. The Greek expression *phainomenon*, from which the English *phenomenon* is derived, stems from the verb *phaino*. Used in different voices, forms and persons, it may signify, among others, "to cause to shine," "to bring to the light," "to discover," "to reveal," "to display," "to clarify," "to make evident," "to announce," "to come to view," "to appear as something," "to seem," "to be evident." The Latin verb *appareo*, and its English derivative noun "appearance," has similar meanings: "to appear," "to become visible," and "to be seen." "Phenomenon" or "appearance" is thus what we notice, see, what is visible or seen. In the case of the celestial appearances, an appearance, a phenomenon is what is seen in the heavens.

What appearances do we see in the heavens? Following Copernicus' description in Chapter 4 of the Book I of *De revolutionibus*, we can distinguish two classes of celestial appearances. The first class of celestial appearances is the daily revolution of the fixed stars and of the whole universe around us. The next class of the appearances is the movement of the planets. We can see that some stars or, to be more precise, "wandering stars" or planets, also have a different motion relative to the sphere of the fixed stars, that is, their own, proper motion. In addition to performing a daily rotation from east to west together with the entire heavens, these stars also move with their own motion proceeding from west to east. However, in this case, planets do not move uniformly, but sometimes proceed more slowly and at other times more rapidly. What is more, the five planets possess a retrograde motion that causes them to move backwards, come to a halt and resume their forward motion again. They sometimes also come closer to the earth and at other times recede farther away from it. Copernicus', or Ptolemy's for that matter, description of what is seen in the heavens stems – though not exclusively – from the point of view of *sensus communis*, which observes these appearances from a motionless earth at the centre of the universe. The celestial appearances as they

---

379 See supra, pp. 332–3.

are seen from the geocentric and geostatic perspective represent the observer's "direct impression."

From such a perspective, the daily motion of the heavens is observable, direct sensory evidence based on the assumption that the earth is motionless and situated in the centre of the universe – or at least somewhere in its relative proximity. This commonsensical, natural, and direct experience of human sense of sight as well as spatial and motion perception is also preserved in everyday language (the stars rise and set, just like the sun, etc.). At the same time, the viewpoint from the earth as the centre of the whole bears an association to the relationship between an individual and the universe. A man, being part of the whole, is an immovable pole around which the whole moves. This *sensus communis*, the universal sensory experience or commonsensical perception of the universe, is preserved in different versions of Aristotelian-Ptolemaic cosmology and astronomy, the essential feature of which is that they put an appearance – that which is seen in the heavens – on a par with the truth of the universe.

But, of course, they only do so to a certain degree. As we saw earlier, with regard to the celestial phenomena, astronomy takes primary interest in that which, according to Regiomontanus, contradicts the "primordial and intrinsic equality" of the celestial spheres, that is, the "phenomenon of inequality." The task of an astronomer is to save the "appearance of inequality," that which does not pertain to the nature of celestial bodies but to "human observers." Celestial anomalies are the result of our view, not their nature; they are such for us rather than in themselves. Appearances of irregularity are an illusion, not the truth. Therefore, it is the duty of astronomy to account for the apparent irregularities by uniform circular motion of the celestial bodies, which depends on various combinations of the celestial orbs.

Both Aristotelian (or Averroist) homocentric astronomy and Ptolemy's astronomy of eccentrics and epicycles try to "save the appearances" and explain the appearance of anomalies in the motions of bodies by their true motions. Both seek the truth, the true motion, which is hidden behind the apparent motion and which – as we know – must be uniform and circular. In view of their philosophy, it can rightly be said that for both Aristotelians and Ptolemaists, as Olaf Pedersen puts it, "things are not what they seem" and that the true structure of the world can only be revealed through the "eyes of our mind."[380] He is absolutely correct to note that for Ptolemy – if we focus on him, although the same also applies to homocentric astronomy – "the real structure of the universe is hidden from our

---

[380] See O. Pedersen, *Survey of the Almagest*, p. 33 et seq.

senses" and that it "can only be revealed through the eyes of our mind."[381] Ptolemy must "distinguish the observable phenomena from their hidden causes."[382] His task is clear: "by means of a suitable intellectual effort he must try to prove beyond doubt that even the most confused and disorderly celestial phenomena can be explained in terms of invariable, orderly laws [that is, by means of uniform and circular motion]."[383] All this is true. Notwithstanding, the Aristotelian-Ptolemaic tradition, despite its obvious efforts to find the truth behind the apparent irregularities, despite distinguishing between what is seen and what needs to be known, leaves the foundation, the firm ground on which the "saving of appearances" stands, unshaken.

According to Ptolemy and his astronomy, celestial anomalies, the irregular motions of the planets or, the "phenomena of inequality" as Regiomontanus put it, pertain to human observers. The apparent approaching and withdrawing of a planet, its apparent irregular motion, as well as its apparent retrogradation – all these phenomena depend on the position of the human observer. Ptolemy holds it most important that the observer, although standing on a motionless earth in the middle of the universe, is never situated exactly at the centre of the movement of the particular celestial body but slightly off to it, like in the case of an eccentric or an epicycle. Copernicus sees himself as the heir of this tradition by regarding anomalies, irregular celestial phenomena, as the result of the observer's off-centredness. In order to preserve the "intrinsic equality" of the motion of the celestial orbs, he, too, has to shift the observer away from the centre. But in contrast to Ptolemaic astronomy, Copernicus makes a small but crucial step further. In order to accommodate the observer's view of the appearances, Ptolemy shifts the observer away from the centre of a planet's motion, but the "eccentric" observer still remained stationary. Copernicus, on the other hand, launches the observer in continual, perpetual motion around his or her axis and around the centre. And this makes very essence of the Copernican Revolution.[384] Ptolemy's differentiation between appearance and the truth is still founded on the unchallenged assumption of a motionless earth on which the observer of the celestial phenomena is situated. The consequence of insisting on this fixed foundation is that in both the Aristotelian and Ptolemaic traditions, the truth of at least one class of celestial phenomena is derived from their immediate appearance.

---

381 Ibid.

382 Ibid.

383 Ibid. Text in square brackets added.

384 See also J.-J. Szczeciniarz, *Copernic et la révolution copernicienne*, p. 160.

Because we see that the entire heavens move around us it must be so. In other terms: Ptolemy leaves unchanged and unchallenged the basic framework (the earth–heavens relationship), on the basis of which he differentiates between the appearance and the truth of the celestial phenomena. So, the identification of this appearance (celestial motion) and its truth remains unchanged as well. The heavens move, because we see them move, whereas the earth stands still at the centre of the world.

At this point the Aristotelian-Ptolemaic tradition directly equates what is seen in the heaven and the objective truth. But that is not all. Even in its attempt to provide a philosophical justification for a stationary earth occupying the centre of the universe, this tradition mostly draws on what is seen, on appearances, In other words: in determining the status of the earth in relation to the heavens (is it the earth that moves or the heavens), the Aristotelian-Ptolemaic distinction between truth and appearance relies on direct or indirect sensory evidence, that is, experiences or appearances. We have seen in the formulations of Buridan and Oresme lists of several appearances which are supposed to prove that the earth does not rotate. But besides the appearance of moving heavens, which is direct evidence for a stationary earth, and several more indirect appearances (a thrown stone or arrow), also the philosophical arguments for a motionless earth and moving heavens, which at first glimpse seem to follow from an abstract conceptual analysis that has nothing to do with sensory experience or appearances, are derived from the unchallenged direct sensory evidence or appearances. This is indirect sensory evidence.

Aristotle's philosophical arguments for the stability of earth in the middle of the universe from *De caelo*, as we have seen, make relatively little reference to phenomena and observations. The organization, structure and function of the universe are depicted on the basis of far more abstract arguments derived from a conceptual analysis. Rather than reasoning on the basis of sensory experience or appearances via induction, as he demands from a "scientist" in the *Analytica posteriora*, Aristotle arrives at a surprising number of conclusions following from his "first principles," which seem to be established in advance, without alluding to visible phenomena. His fundamental argument for the stationary earth in the centre of the universe – which consequently implies that the entire heavens revolve around it – rests on his doctrine of natural movements and natural places of simple bodies, as well as on the theory of weight and lightness on the one hand and observational experiments on the other. However, regardless of the fact that, in *De caelo*, Aristotle is more inclined to provide evidence for the building blocks of his cosmos on the basis of logic, and despite his evident distrust of observations,[385] all basic theories of his physical cosmology – the theory of natural movements, the theory of natural places of simple bodies or elements of the sublunary (earth, water, air, and fire) and superlunary region (aether), as well as the theory of weight and lightness – arise from a vision of the world that, for Aristotle, had to be "immediate evidence (*l'evidence immediat*)."[386] As Paul Moraux puts it, Aristotle's reasoning is based on a "direct vision of a spherical and geocentric universe."[387]

Aristotle opens his treatise by using simple lines and figures to determine simple motions, from which he then derives simple bodies. He characterizes circular motion as revolution around a centre and rectilinear motion as a downward or upward movement, the former leading towards a centre and the latter away from it. Although it may appear that these definitions were developed *in abstracto*, and that Aristotle's thinking and reasoning are deductive, this is not the case. Aristotle identifies the upward movement with the movement of fire and air, and the downward movement with the movement of water and earth. Despite its apparently deductive character, Aristotle's reasoning shows that he starts from "a vision of a spherical universe in which the motionless earth occupied the centre, with its own centre being again taken as the point of reference in defining simple motions."[388]

Let me once again summarize the general outline of Aristotle's arguments in *De caelo* for the stationary earth at the centre of the world.[389] Certain philosophers assigned motion to the earth, Aristotle tells us. Some believed that it rotates on its axis and some even belied it to be a planet circling some centre. But Aristotle finds it absurd to ascribe any movement to the earth, because any circular motion of the earth would be forced. This is demonstrated by the motion of individual parts of the earth which always move in a straight line towards the centre of the universe. Circular motion, not being the earth's proper motion, can only be constrained and unnatural. Given that individual portions of earth naturally move in a straight line, the earth, if it did move at all, should move in the same natural motion. Furthermore, since portions of earth have the same nature as

[385] See P. Moraux, "Introduction," in Aristote, *Du ciel*, p. CXV.

[386] Ibid.

[387] Ibid.

[388] Ibid., p. CXVI.

[389] See *De caelo* II, 14, 296a24–227a8. See also supra, pp. 86–91, for more detailed examination of his arguments.

the earth and they naturally move in a straight line towards the centre of the universe, the earth should, then, also move in a straight line towards the centre. Being then forced and unnatural, the circular motion of the earth cannot be eternal, and this, in turn, contradicts the order of the universe, which is eternal. Aristotle recognizes the most crucial proof for a motionless earth at the centre of the universe in his theory on natural movements and natural places of simple bodies. If earth as a whole (as well as its parts) naturally moves towards the centre of the whole, it is impossible that any portion of earth should move away from the centre except by constraint:

> [f]or a single body has a single locomotion, but not contrary locomotions, and locomotion from the centre is contrary to that to the centre. If, therefore, it is impossible for any part to move from the centre, it is evident, that it is in fact still more impossible for the earth as a whole; for the place to which the part is such as to move is also the place to which the whole is such as to move.[390]

The earth's immobility is further supported by the "contributions of mathematicians to astronomy," as well as by the observable fact that objects thrown into the air fall back to the place whence they were thrown, which could not happen if the earth moved. Since "the earth is not just seen (*phainetai*) to remain at the centre but also to move to the centre,"[391] Aristotle concludes that earth always has a tendency to move towards its natural place (the centre of the universe) and that once it reaches it, it will come to rest. Aristotle also supposes that what holds for any part of earth also holds for earth as a whole. Hence, if the earth were removed to where the moon is now, various fragments of earth would each move not towards it but to the place in which it now is.[392] In much the same way as one can observe that earth always moves downwards, it is possible to notice that water, too, moves downwards, while fire and air always move upwards.

The Aristotelian spherical universe therefore also rests on circular demonstration. The silent assumption – a spherical universe with the motionless earth at the centre – which serves as the starting point for the apparently abstract deduction of individual building blocks of his universe (the theory of natural movements, the theory of natural places of simple bodies, the theory of heaviness and lightness), is demonstrated by these very building blocks, which follow from

---

[390] *De caelo* II, 14, 269b31–36 (Leggatt, p. 163).

[391] Ibid. II, 13, 295b19 (Leggatt, p. 159) et seq.

[392] See ibid. IV, 3, 310b3 et seq.

it. Aristotle's physical theory in support of the geocentric and geostatic universe is itself derived from this silent assumption, by which it then interprets a number of observable phenomena.

Aristotle, and subsequently the entire Aristotelian-Ptolemaic tradition, translates directly observed appearances into theory. However, he does so by drawing on this silent assumption, which gives them meaning in the first place. On a more abstract level, the conception of a static, immovable earth around which the heavens and the celestial bodies in the heavens revolve, is based, *directly* and *indirectly*, on sensory experience or appearances. The geostatic and geocentric conception follows from a non-explicit assumption, which, in simplified terms, equates direct vision and knowledge.[393] With regard to the status of the earth, knowledge depends on what is seen, which in turn depends on the viewpoint from which we see what we see and therefore know what we see. In the Aristotelian-Ptolemaic tradition, this is the viewpoint taken by the "natural," "sensory" "commonsensical" consciousness placed on a motionless earth at the centre of the universe. A stationary observer, a person – or the whole of humankind, – situated on a stationary earth is like a pole around which the entire universe revolves. This viewpoint is geocentric and thus anthropocentric. The criterion for distinguishing between an appearance and the truth with regard to the "celestial motion" (and "the immobility of the earth") is, despite insisting on distinguishing between the apparent anomalies of motion and their reality in uniform circular motion, grounded on a vision of the world that is, for Aristotle, direct evidence.

(iii) Copernicus' denies the relevance of the appearances and relies on purely mathematical reasoning. For Copernicus, the "direct evidence" or appearances almost completely vanishes from the argumentation *pro et contra* earth's rest or movement.

Copernicus himself tells us that in the *Commentariolus*, in the passage following the seven postulates. He announces that he will explain briefly how the uniformity of motions may be preserved and that he will leave out the mathematical demonstrations (which will be dealt with in a larger book), but he will nevertheless give some mathematical data (semidiameters of the orbs) in order to provide mathematical justification for the heliocentric arrangement of the circles: "Nevertheless, the lengths of the semidiameters of the orbs will be set

393 Of course, by saying that, I do not imply that Aristotle simply equated vision and knowledge throughout his work. For more on his characterization of *phainomena*, see, for instance, G. E. R. Lloyd, *Magic, Reason and Experience*, pp. 202–6.

down here in the explanation of their circles, from which anyone not ignorant of mathematics will easily understand how very precisely such an arrangement of circles (*circulorum compositio*) agrees with computations and observations."[394] And he continues that there will be a considerable mathematical evidence for the movement of the earth: "In the same way, in case anyone believes that we have asserted the movement of the earth (*mobilitatem telluris*) for no good reason along with the Pythagoreans, he will also receive considerable evidence [for this] in the explanation of the circles (*in circulorum declaratione*)."[395]

In brief, Copernicus promises, first, to provide mathematical demonstrations for the heliocentric arrangement of the circles, and, second, to provide evidence for the movement of the earth *in circulorum declaratione*, that is, again, in a mathematical way. And then he more or less explicitly dismisses the appearances as having any relevance in the argument *pro et contra* the earth's movement. According to him natural philosophers argue for the immobility of the earth principally from the appearances. But Copernicus explains that its immobility is also due to an appearance:

> And in fact, [the evidence] by which natural philosophers (*physiologi*) attempt so very hard to confirm the immobility of the earth (*stabilitatem eius*) depends for the most part upon appearances. All [their evidence] falls apart here in the first place, since we explain [or consider] it [i. e. the immobility of the earth] as being also due to an appearance.[396]

This passage has been translated in many different ways, the major problem being the very end of the sentence: "cum etiam propter apparentiam versemus eandem." What does the verb *versare* signify here: "to move," "to turn around"

---

394 *Commentariolus* (Swerdlow, p. 439; Zekl, p. 8).

395 Ibid.

396 Ibid.: "Etenim quibus Physiologi stabilitatem eius astruere potissime conantur, apparentiis plerumque innituntur; quae omnia hic in primis corruunt, cum etiam propter apparentiam versemus eandem." Swerdlow's translation does not make much sense to me: "And in fact, [the evidence] by which natural philosophers attempt so very hard to confirm the immobility of the earth depends for the most part upon appearances. All [their evidence] falls apart here in the first place since we overthrow the immobility of the earth also by means of an appearance." Similary also H. Hugonnard-Roche and J.-P. Verdet, in *Introductions à l'astronomie de Copernic*, p. 74: "Et, de fait, les arguments par lesqueles les physicienes s'efforcent per-desssus tout de prouver l'immobilité de la terre s'appuient généralement sur les apparences: ce sont tous ces arguments qui, les premiers, s'effondrent ici, dès lors que nous bousculons l'immobilité de la terre en raison même des apparences."

or "to think of," "to consider," and "to explain"? And what is *eandem* referring to: to the earth's stability (*stabilitas*) or to the earth if *versemus* means "to move," "to turn around"? As far as I was able to check in Copernicus' work, Copernicus never uses the verb *versare* to denote earth's movement. He also does not make the earth to move because of the appearances but *in spite* of them. Therefore I believe that *versemus* here means "to think of," "to consider," and "to explain" and that *eandem* refers to *stabilitas*.[397] But even if we take *versemus* to mean "to move" and *eandem* to refer to "the earth," as De Pace believes to be the case, the basic message Copernicus is delivering here is still the same: the appearances or phenomena which we experience can bring a decisive proof neither for the mobility nor for the immobility of the earth.[398]

In the Book I of *De revolutionibus* the irrelevance of the appearances for the argumentation in favour of the rest or of the movement of the earth is shown discoursively. Copernicus shows that all the appearances or experiences which are supposed to support earth's stability are easily refuted. With the help of optical relativity he, first, demonstrates that nothing changes in celestial appearances regardless of whether it is earth that rotates or heaven. Then he shows that all of the earthly appearances or experiences understood as arguments in favor of the earth's stability are also easily explained under the assumption of its rotation. And finally, he provides some alternative natural philosophical justification for its movement around its axis. But his natural-philosophical reasons are, as we showed, very weak and not very consistent. The reason behind this is obvious:

---

397 Similarly also E. Rosen. He first, in *Three Copernican Treatises*, p. 59, translated it as follows: "For the principal arguments by which the natural philosophers attempt to establish the immobility of the earth rest for the most part on the appearances; it is particularly such arguments that collapse here, since I treat the earth's immobility as due to an appearance." Later, in *Minor works*, p. 82, he changed slightly the last bit: "All these arguments are the first to collapse here, since I undermine the earth's immobility as likewise due to an appearance."

398 See A. De Pace (*Niccolò Copernico e la fondazione del cosmo eliocentrico*, pp. 32–3): "Infatti, gli argumenti con cui *physiologi* si sforzano di provare la sua [della Terra] stabilità poggiano sopratutto sulle apparenze, ma tutti questi argumenti qui per primi vengono a cadere, poiché anche a causa dell'apparenza [sotinteso quindi 'oltre ad altri motivi'] noi faciamo girare la medesima [cioè la Terra]." See also her discussion of different translations, p. 33, n. 80. H. G. Zekl (*Das neue Weltbild*, p. 9) also interprets the text in this way: "Denn die Beweise, auf denen die Naturforscher in besonderem Masse ihre Unbeweglichkeit errichten wollen, stützen sich allgemeinst auf die Erscheinungen; das alles bricht hier schon dadurch zusammen, dass wir ja eben der Erscheinungen wegen die Erde in Bewegung versetzen."

rotation of the earth, its first movement, is for Copernicus only a consequence of its second movement, its revolution around the sun. The earth must move around its axis as well because it moves around the sun. And it moves around the sun not because this would be supported by any appearance (earthly or celestial), but for mathematical reasons, because only with the movement of the earth around the sun the harmonious, well-proportioned universe is achieved.

In the homogeneous universe where there is no dividing line between the sublunary and superlunary spheres and where the earth is just one of many planets, everything is governed by the universal law of the relativity of motion perception. In such a universe one cannot determine whether or not the earth moves solely on the basis of what one sees in the heaven, on celestial appearances. Nor can one determine whether or not the earth moves on the basis of what can be seen on earth and in the air. In order to establish which motion is an appearance and which reality, it is necessary to transcend sensory experience and appearances. For Copernicus, direct or indirect sensory evidence cannot govern knowledge; on the contrary, his *savoire* (knowledge) determines and rectifies his *voire* (what he sees). He does not submit to the dictates of the body but shows that what we see is ultimately dependent on what we know. And what we know is dependent on mathematical reasoning. For Copernicus the Aristotelian-Ptolemaic geocentric and geostatic world is not natural, it is not supported by neutral, sensory experience. In his opinion, reason (mind) defines vision and not vice versa. In Aristotelian-Ptolemaic cosmology, it is through various intermediary stages – and with already mentioned exceptions – that we ultimately know what we directly observe.

In other words, Copernicus' conviction that the earth moves does not stem from sensory evidence, or "the eye of the body," but from a look at the universe through "the eye of the mind." The harmonious universe, created by the maker of the world according to an eternal model, says Ficino, "may be understood by reason and wisdom alone."[399] True astronomy must reach, as was explained by Plato in the *Timaeus* and the *Republic*, beyond appearances, beyond corporeal

---

399 See *Epistolarum liber* I (Members of the Language Department of the School of Economic Science, vol. 1, letter 43, p. 86; *Opera omnia*, pp. 429–30): "Let us consider again whether the maker of the world followed model that is always one and always the same, or whether we should say he followed one that has been created. If the world is beautiful and its creator is good, he would have preferred to follow an eternal model; if otherwise, which would be blasphemous to admit, he would followed a finite model in place of the eternal. Since in truth the world is the most beautiful of all things that are begotten and its author the best of all causes, there is no doubt that he followed a model without beginning or end. So the creation is brought forth according to that

motions of the visible universe, and discover the true order and relations of the motions, which are accessible to reason and thought, but not eyesight. Or, as Rheticus puts it in the *Narratio prima*:

> [t]he remarkable symmetry and interconnection of the motions and orbs [...] are not unworthy of God's workmanship and not unsuited to these divine bodies. These relations, I should say, can be conceived by the soul (on account of its affinity with the heavens) more quickly than they can be explained by any human utterance, just as in demonstrations they are usually impressed upon our souls, not so much by words as by the perfect and absolute ideas, if I may use the term of these most delightful objects. Nevertheless it is possible, in a general survey of the hypotheses, to see how the inexpressible harmony (*convenientia*) and agreement (*consensus*) of all things manifest themselves.[400]

The *symmetria* of the world can, because of its affinity with the heavens, be conceived only by the soul. Outside of the gods, only the human soul or human mind can understand harmony.[401] Or in the words of Galileo from *Dialogo sopra i due massimi sistemi del mondo*:

> Nor can I ever sufficiently admire the outstanding acumen of those who have taken hold of this opinion and accepted it as true; they have through sheer force of intellect done such violence to their own senses as to prefer

---

model. Which may be understood by reason and wisdom alone, and which abides unchanging. Thus it must follow that this world is an image of the other."

[400] *Narratio prima* (Rosen, p. 145; Hugonnard-Roche and Verdet, p. 59). See also ibid. (Rosen, p. 138; Hugonnard-Roche and Verdet 56): "Under the commonly accepted principles of astronomy, it could be seen that all the celestial phenomena conform to the mean motion of the sun and that the entire harmony of the celestial motions is established and preserved under its control."

[401] See *Narratio prima*, "Encomium Prussiae" (Rosen, p. 196; Hugonnard-Roche and Verdet, pp. 86–7): "In the *Phaedo* [see 86b–c and 92a–95a] Socrates rejects the opinion of those who called the soul a 'harmony'. And he did so rightly if by harmony they understood nothing but a mixture of the elements in the body. But if they defined the soul as a harmony because in addition to the gods only the human mind understands harmony – just as it alone knows number, wherefore certain thinkers did not fear to call it number – and also because they knew that souls suffering from the deadliest diseases are sometimes healed by musical harmonies, then their opinion will not seem unfortunate, inasmuch as it is principally the soul of a heroic man that it called a harmony. Hence we might correctly call those states happy whose rulers have harmonious souls, that is, philosophical natures."

> what reason told them over that which sensible experience plainly showed them to the contrary. For the arguments against the whirling of the earth which we have already examined are very plausible, as we have seen; and the fact that the Ptolemaics and Aristotelians and all their disciples took them to be conclusive is indeed a strong argument of their effectiveness. But the experiences which overtly contradict the annual movement are indeed so much greater in their apparent force that, I repeat, there is no limit to my astonishment when I reflect that Aristarchus and Copernicus were able to make reason so conquer sense that, in defiance of the latter, the former became mistress of their belief.[402]

As a Platonist Astronomer, Copernicus prefers reason over the sense, which plainly shows that the earth is motionless. Copernicus completely disregards all kinds of appearances, all sensory experience and *sensus communis*, which speak for the stability of the earth and justifies his geokinetic and heliocentric system by pure mathematical reasoning. Mathematical reasoning which goes beyond appearances is the only way to establish harmony that the world as God's handiwork must necessarily have.[403] As we already said several times, the movement of the earth is needed to achieve certain *symmetria* of the universe and harmonious linkage of various parts of the universe, a task at which the Ptolemaic astronomy failed to do. Or, to put it differently: the criteria for the truth of the entire system and, of course, various motions of the earth as the essential elements of this system are determined by mathematical reasoning alone. Thus Copernicus reverses the perspective: rather than accommodating astronomy to the demands of natural philosophy of nature, he accommodates philosophy of nature to Platonist mathematical astronomy. With Copernicus, mathematical astronomy becomes a superior science – a science that determines the principles to be followed by philosophy of nature.

To put Copernicus' argument in slightly different terms: geocentrism is achieved through "direct human experience," while heliocentrism can only be achieved through abstract, mathematical reasoning. We know that the sun, the planets and the stars do not rise and set, but that the earth rotates on its axis. However, this knowledge is not and cannot possibly be obtained through experience. Nor can science proceed to such a conclusion through direct experience, but

---

[402] *Dialogo sopra i due massimi sistemi del mondo* (Drake, p. 328; Besomi and Helbing, p. 358).

[403] Compare with M. Clavelin, "Le copernicianisme et la mutation de la philosophie naturelle," p. 356.

by reason. The same holds for heliocentrism. The scholar of astronomical science must leave behind the natural position from which he observes phenomena, assume an "unnatural" position and from there observe and reflect what he sees from the natural position.[404] The distinction between an appearance and the reality of an appearance can only be made on the basis of a criterion derived from a purely mathematical reasoning and not on the basis of sensory evidence (appearance). While the geocentric universe is a creation of a "natural" man who by way of direct experience forms the whole together with the earth and the heavens, the heliocentric universe is a universe of Platonist astronomer-mathematician who has emerged through the separation of reason from direct live experience.

Copernicus sees and observes the same celestial and terrestrial appearances as Aristotle and Ptolemy, but understands them differently. Some observed motions of the celestial bodies are simply appearances, illusions, leaving the truth – the various motions of the earth – invisible to the natural eye. This especially applies to the apparent motion of the heavens, which the Aristotelian-Ptolemaic tradition sees as the actual state of matters. Copernicus, on the other hand, regards it as an image, an illusion, a projection of the earth's actual motion on its axis. The same also holds for the irregular apparent motions of the planets advancing, coming to a halt, retrograding and resuming forward motion again. This appearance is explained as an effect, as Alexandre Koyré puts it, "of perspective resulting from the motion of the observer himself."[405] Copernicus thus preserves the traditional disparity between the reality and an appearance in the Aristotelian-Ptolemaic cosmology, but elucidates it within a different conceptual framework.

First, he universalizes the issue of appearance and truth, questioning not only the appearance and the truth of the irregular celestial motions but the entire universe. Whereas the Aristotelian-Ptolemaic tradition restricted the issue of distinguishing between the appearance and the truth to the "heavens" disregarding the earth, Copernicus extends it to include the earth and exclusively investigates the relationship between the earth and the heavens. He eliminates the fixed foundation – a motionless earth – on which the traditional distinction between an appearance and the truth of the celestial phenomena stands or falls: beyond the irregular celestial motions, he finds their true, uniform and circular motion. By universalizing the principle of relative motion perception and applying it to set the earth into rotation and later into revolution, Copernicus eliminates the dividing line between the earth and the heavens, which was left unchallenged

[404] Compare with N. Jardine, *The Birth of History and Philosophy of Science*, p. 293.
[405] A. Koyré, *The Astronomical Revolution*, p. 49.

in the Aristotelian-Ptolemaic tradition, and brings them to an equal level. There is no longer any predetermined relationship between rest and movement; both elements in the relationship, the moving heavens and the motionless earth are subjected to the same criteria for distinguishing between appearance and reality.

Secondly, Copernicus establishes appearance from reality in the universe by different criteria. "Direct evidence," which lies at the foundation of peripatetic cosmology and astronomy, is replaced by mathematical reasoning in which the decisive role is played by the "eye of the mind": every astronomical appearance is subject to the eye of the mind rather than sensory perception. Hence, the viewpoint that enables one to understand the truth of an appearance is completely theoretically constructed. Whereas in the Aristotelian-Ptolemaic tradition, the objectivity of the celestial appearances crucially rests on "direct evidence," Copernicus establishes it through the double negation of direct evidence: first, by observing and reflecting the sensory perception of appearances experienced by the observer moving with the earth on its axis, and second, by observing and reflecting the sensory perception of appearances experienced by the observer moving with the earth around the sun. Such a concept of an appearance, characterized by Jean-Jaques Szczeciniarz as a "rational appearance," manifests itself in two aspects: "in the motion–rest or the observer–observed relationship, which Copernicus considers the basis of his reasoning, and in the establishment of a central, rational observer moved away from the centre of terrestrial perception, placed on the sun, and endowed with a rational eye."[406] It can therefore be said that by placing the "eye of the mind" on the sun, Copernicus is finally able grasp the objective order of the universe. Or, if we borrow Leibniz's adequate summary of Copernicus' achievement (disregarding the legend of Alfonso X of Castile):

> It is only with the eyes of the understanding that we can place ourselves in a point of view which the eyes of the body do not and can not occupy. For example, if we consider the course of the stars from where we stand on our earth's sphere, we obtain a wonderfully complicated structure, which astronomers, just in the last few thousand years, have been able to reduce to a few certain laws, and these laws are so difficult and confusing that King Alfonso of Castille, having let tables be drawn up of celestial motions to fill up the lack of accurate knowledge, is supposed to have said that if he had been God's counselor, the world would have been laid out better. // However, after it had been finally discovered that we must place our eye at the sun if we

406 J.-J. Szczeciniarz, *Copernic et la révolution copernicienne*, p. 395.

> want to view the celestial motions correctly, and that as a result everything comes out wonderfully beautiful, then we see that the supposed disorder and complication were the fault of our understanding and not of nature.[407]

It is necessary, as Copernicus himself maintains, to look at the matter, or the motions in the universe, "with both eyes." Quite literally: with the "eye of the body" and the "eye of the mind." In order to see the objective state of affairs in the universe, we must observe them from two points of view *simultaneously*. Somehow we have to leave behind the terrestrial viewpoint and place ourselves on the sun. It is only from this viewpoint that we can see the true operations of the system of the universe. However, by placing ourselves on the sun, we must not reject terrestrial observations. From the viewpoint of the sun we have to analyze what we see from the first viewpoint. The sun is merely a new foundation from which it is possible to systematize all phenomena, including appearances observed from the earth. The first observations, too, must be made from this central point. However, being completely separated from the empirical observer, this viewpoint is accessible only indirectly, through reason.[408]

What does this mean for Copernicus' role in the Scientific Revolution?

On the one hand, Copernicus' astronomy is merely the cause of radically new physics. The attempts to explain the motions of the earth, which became central to scientific thought in the seventeenth century, culminated in the theory of universal gravity through the magnetic hypotheses of Gilbert and Kepler, and the theory of inertia through Galileo's constant, perpetual motion. All this was, together with several other notions of the absolute space of the Copernican universe, later combined by Newton into the first successful system of mathematical physics. Copernicus contributed to the Scientific Revolution not only by spurring certain developments but, as I have clearly demonstrated, also by triggering a shift towards the horizon of modern scientific thought. By showing that the universe is not necessarily as it appears to our senses, he demonstrated that in order to discover the truth about the natural world, a scientist must very seriously reflect upon what sensory perception and appearances tell him. Following the example of Copernicus, who demonstrated that the earth does not necessarily stand still in the middle of the world as our senses tell us, natural philosophers soon began to assume that the bodies perceived by our senses of color, smell and taste, are

---

407 Leibniz, *Von dem Verhängnisse* (Wiener, pp. 572–3). For more on the legend of Alfonso X of Castile, see M. Franssen, "Did King Alfonso of Castile Really Want to Advise God against the Ptolemaic System? The Legend in History."

408 See also J.-J. Szczeciniarz, *Copernic et la révolution copernicienne*, pp. 201–2.

made up of unperceivable atoms that occur colorless, odorless and tasteless but, combined as they are, only give us such an impression. The departure from *sensus communis* coincides with the increasing role of mathematics as a language that enables us to recognize and explain the objective state of the world.[409] His thesis that the earth moves and the sun stands still in the middle of the world also ushered in a new reflection on the position of man in the universe and was connected with the destruction of the geocentrism and anthropocentrism of the traditional world view, though not in the customary sense of the terms. The merit – or crime – of Copernicus' work lies not in throwing man from the centre of a hierarchically ordered cosmos, thus delivering a blow to his narcissism, but in "dehumanizing him" and demonstrating that he is incapacitated by a complete and irredeemable split and consequently reduced from *anthropos* to subject: "a subject of science."

---

[409] See also J. Henry, *Moving Heaven and Earth*, pp. 142–3.

BIBLIOGRAPHY

I. Editions and translations of Nicholas Copernicus' works

Translation of Theophylactus' Letters

Theophylactus Simocatta, the Moralist, his Ethical, Rustic, and Love Letters in a Latin Translation, in N. Copernicus, *Minor Works*, tr. and comm. E. Rosen, with the assistance of E. Hilfstein (Baltimore & London: Johns Hopkins University Press, 1992), pp. 27–50.

Commentariolus

Commentariolus, in N. Swerdlow, "The Derivation and First Draft of Copernicus' Planetary Theory: A Translation of the *Commentariolus* with Commentary," *Proceedings of the American Philosophical Society* 67 (1973), pp. 423–512.

Commentariolus, in H. Hugonnard-Roche, and J.-P. Verdet, *Introductions à l'astronomie de Copernic. Le Commentariolus de Copernic. La Narratio prima de Rheticus* (Paris: A. Blanchard, 1975), pp. 68–91.

Nicolai Copernici De Hypothesibus motuum coelestium a se constitutis Commentariolus [Latin-German], ed. and tr. H. G. Zekl, in N. Copernicus, *Das neue Weltbild* (Hamburg: Felix Meiner, 1990), pp. 1–35.

Commentariolus, in N. Copernicus, *Minor Works*, tr. E. Rosen with the assistance of E. Hilfstein (Baltimore & London: The Johns Hopkins University Press, 1992), pp. 81–90.

Letter against Werner

Letter against Werner, in N. Copernicus, *Minor Works*, tr. and comm. E. Rosen with the assistance of E. Hilfstein (Baltimore & London: Johns Hopkins University Press, 1992), pp. 145–50.

Epistola Copernici contra Wernerum [Latin-German], ed. and tr. H. G. Zekl, in N. Copernicus, *Das neue Weltbild* (Hamburg: Felix Meiner, 1990), pp. 36–57.

De revolutionibus

De revolutionibus orbium coelestium Libri VI (Nuremberg, 1543).

De revolutionibus orbium coelestium Libri VI (Basel, 1566).

Des révolutions des orbes célestes. I. Livre, ed. and tr. A. Koyré (Paris: A. Blanchard, 1970).

De revolutionibus. Codicis propria auctoris manu scripti imago phototypa, ed. P. Czartoryski (Warsaw & Cracow: Polish Academy of Sciences, 1973).

De revolutionibus libri sex, ed. R. Gansiniec, comm. [Book I] A. Birkenmajer; comm. [Books II–VI] G. Dobrzycki (Warsaw & Cracow: Polish Academy of Sciences, 1975).

On the Revolutions of Heavenly Spheres, tr. A. M. Duncan (London: Newton Abbot, 1976).

De revolutionibus Orbium Caelestium Liber primus [Latin-German], ed. and tr. H. G. Zekl, in N. Copernicus, *Das neue Weltbild* (Hamburg: Felix Meiner, 1990), pp. 59–153.

On the Revolutions, tr. and comm. E. Rosen (Baltimore & London: The Johns Hopkins University Press, 1992).

On the Revolutions of Heavenly Spheres, tr. C. G. Wallis (New York: Prometheus Books, 1995).

De revolutionibus. Liber primus [Latin-Italian], in A. De Pace, *Niccolò Copernico e la fondazione del cosmo eliocentrico con testo, traduzione e commentario del Libro I de Le rivoluzioni celesti* (Milan: Bruno Mondadori, 2009), pp. 236–408.

II. Other primary sources

Achillini, A., *De orbibus libri quattuor* (Bologna, 1498).

–––, *Opera omnia in unum collecta: De intelligentia, De orbibus, etc.* (Venice, 1545).

Acta graduum academicorum Gymnasii Patavini. Vol. 1: Ab anno 1501 ad annum 1525, ed. E. Martelozzo Forin (Padua: Antenore, 1969).

Alberti, L. B., *De re aedificatoria* (Florence, 1485).

–––, *On the Art of Building in Ten Books*, tr. J. Rykwert, N. Leach, and R. Tavernor (Cambridge:The MIT Press, 1988).

Albert of Brudzewo, *Commentariolum super Theoricas novas planetarum Georgii Purbachii*, ed. L. A. Birkenmajer (Cracow, 1900).

Albert of Saxony, *Questiones subtilissimae in libros Aristotelis de caelo et mundo* (Venice, 1492).

Alcinous, *The Handbook of Platonism*, tr., introd., and comm. J. Dillon (Oxford: Clarendon Press, 2001).

Alfonsine Tables, in B. R. Goldstein and J. Chabás, *The Alfonsine Tables of Toledo*.

–––, "Extracts from the Alfonsine Tables and Rules for their Use," tr. V. Thoren and E. Grant, in E. Grant, ed., *A Source Book in Medieval Science* (Cambridge: Harvard University Press, 1974), pp. 465–87.

Amico, G. B., *De motibus corporum coelestium iuxta principia peripatetica sine eccentricis et epyciclis* (Venice, 1536).

–––, *De motibus corporum coelestium iuxta principia peripatetica*, in M. di Bono, *Le sfere omocentriche di Giovan Battista Amico nell'astronomia del Cinquecento* (Genoa: Università degli Studi Genova, 1990), pp. 131–200.

Anonymous, *The Theory of the Planets* [*Theorica planetarum*], tr. O. Pedersen, in E. Grant, ed., *A Source Book in Medieval Science*, pp. 451–65.

Apian, P., *Cosmographicus liber* (Antwerp, 1584).

Aristotle, *Aristotelis omnia quae extant opera* [...] *Averrois Cordubensis in ea opera omnes commentarij*, 13 vols. (Venice, 1562–74).

---, *The Complete Works of Aristotle*, 2 vols., ed. J. Barnes (Princeton: Princeton University Press, 1995).

---, *Physics*, tr. R. P. Hardie and R. K. Gaye, in J. Barnes, ed., *The Complete Works of Aristotle*, vol. 1, pp. 315–446.

---, *On the Heavens*, tr. J. L. Stocks, in J. Barnes, ed., *The Complete Works of Aristotle*, vol. 1, pp. 447–511.

---, *Meteorology*, tr. E. W. Webster, in J. Barnes, ed., *The Complete Works of Aristotle*, vol. 1, pp. 555–625.

---, *Physics: Book VIII*, tr. D. W. Graham (Oxford: Clarendon Press, 1999).

---, *Du ciel* [Greek-French], ed. and tr. P. Moraux (Paris: Les Belles Lettres, 1965).

---, *On the Heavens: Books I and II*, tr. S. Leggatt (Warminster: Aris & Phillips, 1995).

---, *Meteorologica* [Greek-English], tr. H. D. P. Lee (Cambridge: Harvard University Press, 1952).

Autolycus of Pitane, *La sphère en mouvement. Levers et couchers héliaques. Testimonia* [Greek-French], ed. and tr. G. Aujac (Paris: Les Belles Lettres, 1979).

Bacon, R., *Opus tertium*, in *Opera quaedam hactenus inedita*, vol. 1, ed. J. S. Brewer (London: Longman, Green, Longman, & Roberts, 1859), pp. 3–310.

---, [a fragment of] *Opus tertium*, in P. Duhem, *Un fragment inédit de l'Opus tertium de Roger Bâcon, précédé d'une étude sur ce fragment* (Quaracchi: Colegio S. Bonaventurae, 1909), pp. 73–193.

Bessarion, *In calumniatorem Platonis libri IV* [Greek-Latin], in L. Mohler, *Kardinal Bessarion als Theologe, Humanist and Staatsmann*, vol. 2 (Paderborn: Ferdinand Schöningh, 1927).

Bitrūjī, al- (Alpetragius), *De motibus celorum*, Latin translation of Michael Scot, ed. F. J. Carmody (Berkeley & Los Angeles: University of California Press, 1952).

---, *On the Principles of Astronomy* [Arabic-English], 2 vols., ed. and tr. B. Goldstein (New Haven & London: Yale University Press, 1971).

Buridan, J., *Questions on the Four Books on the Heavens and the World of Aristotle: Book II, Question 22*, in M. Clagett, *The Science of Mechanics in the Middle Ages*, pp. 594–8; reprod. in E. Grant, ed., *A Source Book in Medieval Science*, pp. 500–3.

---, *Ioannis Buridani Expositio et Quaestiones in Aristotelis De caelo*, ed. B. Patar (Louvain & Paris: Peeters, 1996).

Campanus of Novara, *Theorica planetarum*, in *Campanus of Novara and Medieval Planetary Theory: Theorica Planetarum* [Latin-English], ed., tr., and comm. F. S. Benjamin and G. J. Toomer (Madison: University of Wisconsin, 1971), pp. 127–363.

Capella, M., *The Marriage of Philology and Mercury*, tr. W. Stahl, R. Johnson, and

E. L. Burge, in W. Stahl, *Martianus Capella and the Seven Liberal Arts*, vol. 2 (New York: Columbia University Press, 1977).

---, *De Nuptiis Philologiae et Mercurii*, in M. Capella, *Le nozze di Filologia e Mercurio* [Latin–Italian], intr., tr., and comm. I. Ramelli (Milan: Bompiani, 2004).

Capuano, F. da Manfredonia, *Theorice nove planetarum Georgii Purbachii astronomi celebratissimi. Ac in eas [...] Domini Francisci Capuani de Manfredonia [...] sublimis expositio* (Paris, 1515).

Celtis, C., *Ars versificandi* (Leipzig, 1486).

Chalcidius, *In Platonis Timaeum Commentarius*, in Chalcidio, *Commentario al Timeo di Platone* [Latin-Italian], Latin text establ. J. H. Waszink, ed. and tr. C. Moreschini (Milan: Bompiani, 2003).

Chrysogonus, F., *Speculum astronomicum* [Latin-Croatian], ed. M. Girardi-Karšulin and O. Perić, tr. T. Ćepulić (Zagreb: Institut za filozofiju, 2007).

Cicero, M. T., *De natura Deorum & Academica* [Latin-English], tr. H. Rackham (Cambridge, MA & London: Harvard University Press & W. Heinemann, 1967).

---, *Scipio's Dream*, tr. W. Stahl, in *Macrobius's Commentary on the Dream of Scipio*, pp. 69–77.

Cleomedes, *The Heavens*, in A. C. Bowen and R. B. Todd, *Cleomedes' Lectures on Astronomy*, pp. 19–165.

Codro, A. U., *Orationes, Epistolae, Sylvae, Satyrae* (Venice, 1506).

Corpus Hermeticum [Greek-, Latin-, and Coptic-Italian], ed. of Greek and Latin texts A. D. Nock and A.-J. Festugière, ed. of Coptic texts I. Ramelli, tr. and ed. I. Ramelli (Milan: Bompiani, 2006).

D'Ailly, P., *14 quaestiones*, in J. Sacrobosco, *Sphera, cum commentis in hoc volumine contentis* (Venice, 1518).

---, *Tractatus de ymagine mundi* (Louvain, 1483); some snippets are translated in E. Grant, ed., *A Source Book in Medieval Science*, pp. 630–9.

De Sicilia, J., *Scriptum Johannis de Sicilia super canones Azarchelis de Tabulis Toletanis*, ed. F. S. Pedersen, in *Cahiers de l'Institut du Moyen Age Grec et Latin* 51–52 (1986), pp. 1–128 and 1–268.

Diogenes Laertius, *Lives of Eminent Philosophers* [Latin-English], 2 vols., ed. and tr. R. D. Hicks (Cambridge: Harvard University Press, 2000).

Euclid, *Opera omnia*, 9 vols., ed. I. L. Heiberg and H. Menge (Leipzig: Teubner, 1883–1916).

---, *Optica ex recensione Theonis* [Greek–Latin], ed. and tr. H. Menge, in *Opera omnia*, vol. 7, pp. 143–247.

---, *The Thirteen Books of the Elements*, 3 vols., tr. T. L. Heath (New York: Dover, 1956).

Eusebius of Caesarea, *Praeparatio evangelica* [Greek–Latin] (Cologne, 1687).

---, *Preparation for the Gospel*, 2 vols., tr. E. H. Gifford (Oxford: Clarendon Press, 1903).

Facciolati, J., *Fasti Gymnasii Patavini Jacobi Facciolati studio atque opera collecti*, 2 vols. (Padua, 1757).

Ficino, M., *Opera omnia*, 2 vols., with continuous pagination (Basel, 1576).

---, *Epistolae Libri XII* (Venice, 1495).

---, *Libellus de sole* (Leipzig, 1502).

---, *Platonic Theology* [Latin-English], 6 vols., ed. J. Hankins and W. Bowen, tr. M. J. B. Allen with J. Warden (Cambridge: Harvard University Press, 2001–2005).

---, *Commentaries on Plato. Vol. 1: Phaedrus and Ion* [Latin-English], ed. and tr. M. J. B. Allen (Cambridge: Harvard University Press, 2008).

---, *Commentaries on Plato. Vol. 2: Parmenides, Part 1* [Latin-English], ed. and tr. M. Vanhaelen (Cambridge: Harvard University Press, 2012).

---, *Commentaries on Plato. Vol. 2: Parmenides, Part 2* [Latin-English], ed. and tr. M. Vanhaelen (Cambridge: Harvard University Press, 2012).

---, *The Letters of Marsilio Ficino: Vol 1 (Liber I)*, tr. members of the Language Department of the School of Economic Science, London (London: Shepheard-Walwyn, 1975).

---, *The Letters of Marsilio Ficino: Vol 2 (Liber III)*, tr. members of the Language Department of the School of Economic Science, London (London: Shepheard-Walwyn, 1978).

---, *The Letters of Marsilio Ficino: Vol 3 (Liber IV)*, tr. members of the Language Department of the School of Economic Science, London (London: Shepheard-Walwyn, 1981).

---, *The Letters of Marsilio Ficino: Vol 4 (Liber V)*, tr. members of the Language Department of the School of Economic Science, London (London: Shepheard-Walwyn, 1988).

---, *The Letters of Marsilio Ficino: Vol 5 (Liber VI)*, tr. members of the Language Department of the School of Economic Science, London (London: Shepheard-Walwyn, 1994).

---, *The Letters of Marsilio Ficino: Vol 6 (Liber VII)*, tr. members of the Language Department of the School of Economic Science, London (London: Shepheard-Walwyn, 1999).

---, *The Letters of Marsilio Ficino: Vol 7 (Liber VIII)*, tr. members of the Language Department of the School of Economic Science, London (London: Shepheard-Walwyn, 2003).

---, *The Letters of Marsilio Ficino: Vol 8 (Liber IX)*, tr. members of the Language Department of the School of Economic Science, London (London: Shepheard-Walwyn, 2009).

---, *The Letters of Marsilio Ficino: Vol 9 (Liber X)*, tr. members of the Language

Department of the School of Economic Science, London (London: Shepheard-Walwyn, 2012).
---, *Disputa contro il giudizio degli astrologi*, in M. Ficino, *Scritti sull'astrologia*, ed. and tr. O. Pompeo Faracovi (Milan: BUR, 2008), pp. 49–174.
---, *When Philosophers Rule: Ficino on Plato's Republic, Laws, and Epinomis*, tr. A. Farndell (London: Shepheard-Walwyn, 2009).
---, *All Things Natural: Ficino on Plato's Timaeus*, tr. A. Farndell, notes and additional material P. Blumsom (London: Shepheard-Walwyn, 2010).
Fracastoro, G., *Homocentrica* (Venice, 1538).
---, *Opera Omnia, in unum proxime post illius mortem collecta, quorum nomina sequens pagina plenius indicat* (Venice, 1555).
---, *Navagero. Della Poetica* [Latin-Italian], ed. and tr. E. Peruzzi (Florence: Alinea, 2005).
Galileo, G., *Opere*, 20 vols., ed. A. Favaro (Florence: Giunti-Barbera, 1890–1909).
---, *Sidereus nuncius or the Sidereal Messenger*, tr. A. van Helden (Chicago & London: University of Chicago Press, 1989).
---, *Dialogo sopra i due massimi sistemi del mondo, tolemaico e copernicano*, ed. O. Besomi and M. Helbing (Padua: Antenore, 1998).
---, *Dialogue Concerning the Two Chief World Systems*, tr. S. Drake (Berkeley: University of California Press, 1967).
---, "Considerations on the Copernican Opinion," in M. A. Finocchiaro, *The Galileo Affair*, pp. 70–86.
---, *Letter to the Grand Duchess Christina*, in M. A. Finocchiaro, *The Galileo Affair*, pp. 87–118.
Gassendi, P., *Georgii Purbachii et Ioannis Mulleri Regiomontani vita*, in P. Gassendi, *Opera Omnia*, vol. 5 (Lyon, 1658), pp. 519–34.
Geminos, *Introduction to the Phenomena*, in J. Evans and J. L. Berggren, *Geminos's Introduction to the Phenomena*, pp. 113–230.
---, "Concise Exposition of the *Meteorology* of Poseidōnios" [the extant fragment], in J. Evans and J. L. Berggren, *Geminos's Introduction to the Phenomena*, pp. 252–5.
---, "Posidonius on Physics and Astronomy" [the extant fragment from *Concise Exposition of the Meteorology of Poseidōnios*], in A. C. Bowen and R. B. Todd, *Cleomedes' Lectures on Astronomy*, pp. 199–204.
---, *Introduction aux phénomènes* [Greek-French], ed. and tr. G. Aujac (Paris: Les Belles Lettres, 2002).
Gilbert, W., *De magnete magneticisque corporibus, et de Magno Magnete Telurre, Physiologia nova, plurimis argumentis et experimentis demonstrate* (London, 1600).
---, *De magnete*, tr. P. F. Mottelay (New York: Dover, 1958).

Horace, *Satires, Epistles & Ars Poetica* [Latin-English], ed. and tr. H. R. Fairclough (London & Cambridge, MA: Heinemann & Harvard University Press, 1942).
Ibn al-Haitham (Alhazen), *De configuratione mundi*, in J. M. Millàs Vallicrosa, *Las traductiones orientales en los manuscritos de la Bibliotheca Catedral de Toledo*, pp. 285–312.
---, *On the Configuration of the World* [Arabic-English] in Y. T. Langermann. *Ibn al-Haytham's On the Configuration of the World.*
---, *al-Shukūk 'alā Batlamiyūs (Dubitationes in Ptolemaeum)*, ed. A. I. Sabra and N. Shehaby (Cairo: Dār al-Kutub, 1971).
Ibn Rushd (Averroes), *Commentaria Magna in Aristotelem De caelo et mundo*, 2 vols., ed. J. Carmody (Louvain: Peeters, 2003).
---, *Metaphysics, Book Lām* in, C. Genequand, *Ibn Rushd's Metaphysics*, pp. 59–210.
Kepler, J., *Apologia pro Tychone contra Ursum* [Latin-English] in N. C. Jardine, *The Birth of History and Philosophy of Science*, pp. 85–133, and 134–207.
---, *Astronomia nova*, ed. M. Caspar, in *Gesammelte Werke*, vol. 3 (Munich: C. H. Beck, 1937).
---, *The New Astronomy*, tr. W. Donahue (Cambridge: Cambridge University Press, 1992).
Lactantius, *Opera quae extant omnia* (Cesena, 1646).
Leibniz, G. W. F., *Von dem Verhängnisse*, in *Selections*, ed. and tr. P. E. Wiener (New York: Charles Scribner's Sons, New York, 1951), pp. 370–76.
Leopold of Austria, *Compilatio Leupoldi ducatus Austrie filij de astrorum scientia Decem continens tractatus* (Venice, 1520).
---, *Li compilacions de le science des estoilles*, Books I–III., ed. F. Y. Carmody (Berkeley & Los Angeles: University of California Press, 1947).
Macrobius, *In somnium Scipionis* [Latin-Italian], in *Commento al Sogno di Scipione*, ed. and tr. M. Neri (Milan: Bompiani, 2007).
---, *Commentary on the Dream of Scipio*, tr. W. Stahl (New York: Columbia University Press, 1990).
Michael Scot [attributed], *Super auctore spere cum questionibus*, in L. Thorndike, *The Sphere of Sacrobosco and its Commentators*, pp. 147–342.
Nicolaus Cusanus, *De docta ignorantia, liber secundus* [Latin-German], ed. and tr. P. Wilpert, second edition ed. H. G. Senger (Hamburg: Felix Meiner Verlag, 1979).
---, *On The Learned Ignorance*, tr. J. Hopkins, in *Nicholas of Cusa on Learned Ignorance: A Translation and an Appraisal of De Docta Ignorantia* (Minneapolis: The Arthur J. Banning Press, 1990).
---, *De ludo globi (The Bowling-Game)*, tr. J. Hopkins, in Nicholas of Cusa, *Methaphysical Speculations*, vol. 2 (Minneapolis: The Arthur J. Banning Press, 2000).
Nifo, A., *In Aristotelis libros De coelo et mundo commentaria* (Venice, 1517).

---, *De Re Aulica*, in *La filosofia nella corte* [Latin-Italian], ed. and tr. E. De Bellis (Milan: Bompiani, 2000), pp. 141–638.

Oresme, Nicole, *Le Livre du ciel et du monde* [Old French-English], ed. A. D. Menut and A. J. Denomy, tr. A. D. Menut (Madison: University of Wisconsin Press, 1968).

Patrizi, F., *Nova de universis philosophia* (Ferrara, 1591).

Peurbach, G., *Theoricae novae planetarum Georgii Purbachii astronomi celebratissimi* (Nuremberg, 1472).

---, *Theoricae novae planetarum*, tr. and comm. E. J. Aiton, in "Peurbach's *Theoricae novae planetarum*: A Translation with Commentary," *Osiris* 3 (1987), pp. 9–43.

Pico della Mirandola, Giovanni, *Disputationes adversus Astrologiam Divinatricem* (Bologna, 1496); the key snippet tr. N. Swerdlow, in "Copernicus and Astrology," pp. 3–7.

---, *Disputationes adversus Astrologiam Divinatricem* [Latin-Italian], 2 vols., ed. and tr. E. Garin (Florence: Valechi, 1946–52).

Plato, *Diuini Platonis Opera omnia Marsilio Ficino interprete*, 2 vols. (Florence, ca. 1484).

---, *La République* (I–III) [Greek-French], ed. and tr. E. Chambry (Paris: Les Belles Lettres, 1970).

---, *La République* (IV–VII) [Greek-French], ed. and tr. E. Chambry (Paris: Les Belles Lettres, 1975).

---, *La République* (VIII–X) [Greek-French], ed. and tr. E. Chambry (Paris: Les Belles Lettres, 1973).

---, *Timée* and *Critias* [Greek-French], ed. and tr. A. Rivaud (Paris: Les Belles Lettres, 1970).

---, *Les Lois* (I–II) [Greek-French], ed. and tr. Ē. des Places (Paris: Les Belles Lettres, 1976).

---, *Les Lois* (III–VI) [Greek-French], ed. and tr. Ē. des Places (Paris: Les Belles Lettres, 1975).

---, *Les Lois* (VII–X) [Greek-French], ed. and tr. A. Dies (Paris: Les Belles Lettres, 1976).

---, *Les Lois* (XI–XII) and *Epinomis* [Greek-French], ed. and tr. A. Dies and Ē. des Places (Paris: Les Belles Lettres, 1956).

---, *Complete Works*, ed. J. M. Cooper (Indianapolis & Cambridge: Hackett Publishing Company, 1997).

---, *Phaedo*, tr. G. M. A. Grube, in Plato, *Complete Works*, pp. 49–100.

---, *Parmenides*, tr. M. L. Gill and P. Ryan, in Plato, *Complete Works*, pp. 359–97.

---, *Phaedrus*, tr. A. Nehams and P. Woodruff, in Plato, *Complete Works*, pp. 506–56.

---, *Georgias*, tr. D. J. Zeyl, in Plato, *Complete Works*, pp. 791–869.

---, *Republic*, tr. G. M. A. Grube, rev. C. D. C. Reeve, in Plato, *Complete Works*, pp. 971–1223.
---, *Timaeus*, tr. D. J. Zeyl, in Plato, *Complete Works*, pp. 1224–91.
---, *Laws*, tr. T. J. Saunders, in Plato, *Complete Works*, pp. 1318–1616.
---, *Epinomis*, tr. R. D. McKirahan, Jr., in Plato, *Complete Works*, pp. 1617–33.
---, *Timaeus*, in F. M. Cornford, *Plato's Cosmology. The Timaeus of Plato Translated with a Running Commentary* (London: Routledge & Kegan, 1956).
Pliny the Elder, *The Natural History*, 10 vols., vol. 1: *Praefatio, Libri I, II* [Latin-English], ed. and tr. H. Rackham (London & Cambridge, MA: W. Heinemann & Harvard University Press, 1949).
Plutarch, *Plutarci opuscula* LXXXXII (Venice, 1509).
---, *Concerning the Face Which Appears in the Orb of the Moon* [Greek-English], tr. H. Cherniss, in Plutarch, *Moralia*, vol. 12 (London & Cambridge, MA: W. Heinemann & Harvard University Press, 1957), pp. 34–223.
---, *Platonic Questions* [Greek-English], tr. H. Cherniss, in Plutarch, *Moralia*, vol. 13 (London & Cambridge, MA: W. Heinemann & Harvard University Press, 1976), pp. 1–129.
---, *On the Generation of the Soul in the Timaeus* [Greek-English], tr. H. Cherniss, in Plutarch, *Moralia*, vol. 13 (London & Cambridge, MA: W. Heinemann & Harvard University Press, 1957), pp. 131–345.
Proclus, *Hypotyposis astronomicarum positionum*, ed. C. Manitius (Leipzig: Teubner, 1909).
---, *A Commentary on the First Book of Euclid's Elements*, tr. G. L. Morrow (Princeton: Princeton University Press, 1970).
---, *Commentary on Plato's Timaeus. Vol. I: Book 1: Proclus on the Socratic State and Atlantis*, ed. and tr. H. Tarrant (Cambridge: Cambridge University Press, 2007).
---, *Commentary on Plato's Timaeus. Vol II: Book 2: Proclus on the Causes of the Cosmos and its Creation*, ed. and tr. D. T. Runia and M. Share (Cambridge: Cambridge University Press, 2008).
---, *Commentary on Plato's Timaeus. Vol III: Book 3, Part I: Proclus on the World's Body*, ed. and tr. D. Baltzly (Cambridge: Cambridge University Press, 2007).
---, *Commentary on Plato's Timaeus. Vol III: Book 3, Part II: Proclus on the World Soul*, ed. and tr. D. Baltzly (Cambridge: Cambridge University Press, 2009).
Ptolemy, *Almagesti Cl. Ptolemaei Pheludiensis Alexandrini* [...] *opus* [...] *omnes caelorum motus continens* (Venice, 1515).
---, *Almagestum seu magnae constructionis mathematicae opus* [...] *Latina donatum ab Georgio Trapezuntio* (Venice, 1528).
---, *Syntaxix mathematica*, ed. J. L. Heiberg (Leipzig: Teubner, 1898–1903).
---, *The Almagest*, tr. G. J. Toomer (Princeton: Princeton University Press, 1998).

---, *Hupotheseôn tôn planômenôn*, in *Opera astronomica minora*, ed. J. L. Heiberg and L. Nix, 2 vols. (Leipzig: Teubner, 1907), pp. 69–145.
---, *Optics*, in A. M. Smith, *Ptolemy's Theory of Visual Perception: An English Translation of the Optics, with Introduction and Commentary*, pp. 63–261.
---, *Tetrabiblos* [Greek-English], ed. and tr. F. E. Robins (Cambridge, MA & London: Harvard University Press & W. Heinemann, 1940).
Receptio Copernicana, ed. H. Nobis, M. Folkerts, S. Kirschner, and A. Kühne, *Nicolaus Copernicus Gesamtausgabe*, vol. 8/1 (Berlin: Akademie Verlag, 2002).
Regiomontanus, J., *Joannis Regiomontani opera collectanea*, ed. F. Schmeidler (Osnabrück: Otto Zeller, 1972).
---, *Defensio Theonis*, preliminary digital edition prep. R. Kremer and M. H. Shank, available at http://regio.dartmouth.edu.
Regiomontanus, J., [and G. Peurbach], *Epytoma in Almagestum Ptolemei* (Venice, 1596); a snippet tr. in N. Swerdlow, "Copernicus and Astrology," pp. 9–10.
Reinhold, E., *Theoricae G. Purbachii Germani ab Erasmo Reinholdo […] illustratae scholiis* (Paris, 1558).
Rheticus, G. J., *Ephemerides Novae seu Expositio Positus Diurni Siderum* (Leipzig, 1550).
---, *Narratio prima* [Latin-French], ed. tr. and comm. H. Hugonnard-Roche, J.-P. Verdet, with the collaboration of M.-P. Lerner, and A. Segonds (Wrocław & Warsaw: Polish Academy of Sciences, 1982).
---, *Narratio prima*, tr. E. Rosen, in E. Rosen, *Three Copernican Treatises*, pp. 107–96.
---, *Epistola cujusdam Anonymi de terrae motu* [Latin-English], in R. Hooykaas, *G. J. Rheticus' Treatise On Holy Scripture and the Motion of the Earth*, pp. 43–101; also in *Receptio Copernicana*, pp. 37–73.
Reisch, G., *Margarita philosophica nona* (Strasbourg, 1508).
Sacrobosco, J., *De sphaera* [Latin-English], in L. Thorndike, *The Sphere of Sacrobosco and Its Commentators*, pp. 76–117.
Schedel, H., *Liber Chronicarum* (Nuremberg, 1493).
Simplicius, *On Aristotle On the Heavens 2.10–14*, tr. I. Mueller (Ithaca: Cornell University Press, 2005).
---, *Commentaire au De caelo d'Aristote* [II, 12; Greek-French], ed. and tr. G. Aujac, in Avtolykus of Pitane, *La sphère en mouvement. Levers et couchers héliaques. Testimonia*, ed. and tr. G. Aujac (Paris: Les Belles Lettres, 1979). pp. 159–90.
Stobaeus, *Eclogarum libri duo* (Antwerp, 1575).
Thomas Aquinas, *In libros Aristotelis De caelo et mundo expositio*, in *Opera omnia*, 3 vols. (Rome, 1886).
Tolosani, G. M., *Opusculum quartum de coelo et elementis*, ed. E. Garin, *Studia Copernicana* 6 (1973), pp. 31–42; repr. in E. Garin, *Rinascite e rivoluzioni*, pp.

283–94; a new Latin edition and French translation in M.-P. Lerner, “Aux origines de la polemique anticopernicienne (I),” pp. 692–719.

Tūsī, Nasīr al-Dīn al-, *al-Tadkira fi´ilm al hay´a (Memoir on Astronomy)*, [Arabic-English], in F. J. Ragep, *Nasīr al-Dīn al-Tūsī's Memoir on Astronomy*, 2. vols. (New York & Berlin: Springer, 1993).

Tutti commenti a Marziano Capella [Latin-Italian], ed. and tr. I. Rameli (Milan: Bompiani, 2006).

Valla, G., *De expetendis et fugiendis rebus opus* (Venice, 1501).

Versoris, J., *Quaestiones super Metaphysicam Aristotelis* (Cologne, ca. 1486).

–––, *Quaestiones De celo et mundo, De generatione et corruptione, Metheororum, Parva naturalia* (Cologne, ca. 1488).

Vitruvius, *De Architectura* [Latin-French], 2 vols., ed. and tr. M. C.-L. Maufras (Paris: Panckoucke, 1847).

–––, *Ten Books on the Architecture*, ed. and tr. I. D. Rowland (New York: Cambridge University Press, 1999).

Witelo, *Opticae libri decem* (Basel, 1572).

III. Secondary sources

Africa, T. W., “Copernicus’ Relation to Aristarchus and Pythagoras,” *Isis* 52 (3/1961), pp. 403–409.

Aiton, E. J., “Celestial Spheres and Circles,” *History of Science* 19 (1981), pp. 75–114.

–––, “Peurbach’s *Theoricae novae planetarum*: A Translation with Commentary,” *Osiris* 3 (1987), pp. 4–43.

Allan, D. J., “Medieval Versions of Aristotle’s *De caelo*, and of the *Commentary* of Simplicius,” *Medieval and Renaissance Studies* 2 (1950), pp. 80–120.

Allen, M. J. B., and V. Rees, with M. Davies, eds., *Marsilio Ficino: His Theology, his Philosophy, his Legacy* (Leiden: Brill, 2002).

Andersen, H., P. Barker, and X. Chen, *The Cognitive Structure of Scientific Revolutions* (Cambridge: Cambridge University Press, 2006).

Anton, J. P., ed., *Science and the Sciences in Plato* (New York: Eidos, 1980).

Antonaci, A., “Aristotelismo e scienza medica a Padova nel primo Cinquecento. Il pensiero medico di Marcantonio Zimara,” in L. Olivieri, ed., *Aristotelismo veneto e scienza moderna*, pp. 415–34.

–––, “La posizione filosofica e scientifica di Marcantonio Zimara negli anni del suo primo insegnamento a Padova,” in A. Poppi, ed., *Scienza e filosofia all’Università di Padova nell Quattrocento*, pp. 315–28.

Applebaum, W., ed., *Encyclopedia of the Scientific Revolution from Copernicus to Newton* (New York & London: Garland, 2000).

Aujac, G., “Le géocentrisme en grèce ancienne?,” in *Avant, avec, après Copernic*, pp. 19–28.

–––, *Claude Ptolémée, astronome, astrologue, géographe* (Paris: Ed. du CTHS, 1993).

Avant, avec, après Copernic (Paris: Albert Blanchard, 1975).
Avezzù, G., "Le fonti greche di Copernico," in *Copernico a Padova*, pp. 123–47.
Babicz, J., "Die exakten Wissenschaften an der Universität zu Krakau und der Einfluss Regiomontans auf ihre Entwicklung," in G. Hamann, ed., *Regiomontanus-Studien*, pp. 301–4.
Baldwin, M. R., "Magnetism and the Anti-Copernican Polemic," *JHA* 16 (1985), pp. 155–74.
Barker, P., "Copernicus, the Orbs, and the Equant," *Synthese* 83 (1990), pp. 317–23.
---, "Copernicus and the Critics of Ptolemy," *JHA* 30 (1999), pp. 343–58.
---, "The Role of Religion in the Lutheran Response to Copernicus," in M. J. Osler, ed., *Rethinking the Scientific Revolution* (Cambridge: Cambridge University Press, 2000), pp. 59–88.
---, "Incommensurability and Conceptual Change during the Copernican Revolution," in P. Hoyningen-Huene and H. Sankey, eds., *Incommensurability and Related Matters* (Boston: Kluwer, 2001), pp. 241–73.
---, "Constructing Copernicus," *Perspectives on Science* 10 (2002), pp. 208–27.
---, "Astronomy, Providence and the Lutheran Contribution to Science," in A. J. L. Menuge, ed., *Reading God's World: The Scienctific Vocation* (St. Louis: Concordia Press, 2004), pp. 157–87.
---, "The Lutheran Contribution to the Astronomical Revolution: Science and Religion in the Sixteenth Century," in J. Brooke and E. İhsanoğlu, eds., *Religious Values and the Rise of Science in Europe* (Istanbul: Research Centre for Islamic History, Art and Culture, 2005), pp. 31–62.
---, "The Reality of Peurbach's Orbs: Cosmological Continuity in Fifteenth and Sixteenth Centuries Astronomy," in P. J. Boner, ed., *Change and Continuity in Early Modern Cosmology* (Dordrecht: Springer, 2011), pp. 7–32.
---, "Albert of Brudzewo's *Little Commentary* on George Peurbach's *Theoricae novae planetarum*," *JHA* 44 (2013), pp. 125–148.
Barker, P., and B. R. Goldstein, "Distance and Velocity in Kepler's Astronomy," *Annals of Science* 51 (1994), pp. 59–73.
---, "Realism and Instrumentalism in Sixteenth Century Astronomy: A Reappraisal," *Perspectives on Science* 6 (1998), pp. 232–58.
---, "Theological Foundations of Kepler's Astronomy," *Osiris* 16 (2001), pp. 88–113.
---, "Patronage and the Production of *De revolutionibus*," *JHA* 34 (2003), pp. 345–68.
Barker, P., and M. Vesel, "Essay Review of André Goddu, Copernicus and the Aristotelian Tradition: Education, Reading and Philosophy in Copernicus's Path to Heliocentrism," *Aestimatio* 9 (2012), pp. 304–36.
Barnes, J., ed., *The Cambridge Companion to Aristotle* (Cambridge: Cambridge University Press, 1995).
Baroncini, G., "Note sulla formazione del lessico della metafora *machina mundi*," *Nuncius* 4 (1989), pp. 3–30.

Barone, F., "La 'modernita' di Niccolò Copernico," in L. Pepe, ed., *Copernico e la questione copernicana in Italia*, pp. 1–37.

Bertozzi, M., ed., *Nello specchio del cielo. Giovanni Pico della Mirandola e le Disputationes contra l'astrologia divinatoria* (Florence: Olschki, 2008).

Biagioli, M., "The Social Status of Italian Mathematicians, 1450–1600," *History of Science* 27 (1989), pp. 41–95.

Bianchi, L., "Continuity and Change in the Aristotelian Tradition," in J. Hankins, ed., *The Cambridge Companion to the Renaissance Philosophy*, pp. 49–71.

Bianchi, L., and E. Randi, *Vérités dissonantes: Aristote à la fin du Moyen Age* (Paris & Freiberg: CERF & Editions Universitaires de Fribourg, 1993).

Biard, J., ed., *Itinéraires d'Albert de Saxe: Paris-Vienne au 14e siècle* (Paris: Vrin, 1991).

Bieńkowska, B., ed., *The Scientific World of Copernicus* (Dordrecht & Boston: Reidel, 1973).

Biliński, B., *Najstarszy życiorys Mikołaja Kopernika 1588 z roku pióra Bernardina Baldiego* (Wrocław & Warsaw: Polish Academy of Sciences, 1973).

----, *Il Pitagorismo di Niccolò Copernico* (Wrocław: Polish Academy of Sciences, 1977).

---, "Il periodo padovano di Niccolò Copernico (1501–3)," in A. Poppi, ed, *Scienza e filosofia all'Università di Padova nell Quattrocento*, pp. 223–85.

---, "Intorno ai problemi del Copernico padovano," in *Copernico a Padova*, pp. 77–122.

Birkenmajer, A., *Études d'histoire des sciences en Pologne* (Wrocław: Polish Academy of Sciences, 1972).

---, "Commentary [on Book I of *De revolutionibus*]," in N. Copernicus, *De revolutionibus libri sex*, ed. R. Gansiniec, pp. 351–91.

Birkenmajer, L. A., *Stromata Copernicana* (Cracow: Polish Academy of Arts, 1924).

Biskup, M., ed., *Regesta Copernicana (Calendar of Copernicus' Papers)*, tr. S. Puppel (Wrocław & Warsaw: Polish Academy of Sciences, 1973).

---, "Nouvelles recherches sur la biographie de Nicolas Copernic," *Revue d'histoire des sciences* 27 (4/1974), pp. 289–306.

---, "Biography and Social Background of Copernicus," *Studia Copernicana* 17 (1977), pp. 137–52.

---, "Polish Research on the Life and Wide-ranging Activity of Nicholas Copernicus," *Organon* 35 (2006), pp. 9–16.

Black, R., "The Philosopher and Renaissance Culture," in J. Hankins, ed., *The Cambridge Companion to the Renaissance Philosophy*, pp. 13–29.

Blair, A., "Organizations of Knowledge," in J. Hankins, ed., *The Cambridge Companion to the Renaissance Philosophy*, pp. 287–303.

---, "Natural Philosophy," in K. Park and L. Daston, eds., *The Cambridge History of Science. Vol. 3: Early Modern Science*, pp. 365–407.

Blumenberg, H., *Die Genesis der kopernikanischen Welt* (Berlin: Suhrkamp, 1996).

Bossier, F., "Traductions latines et influences du commentaire *In de caelo* en occident (XIIe–XIVe s.)," in I. Hadot, ed., *Simplicius, sa vie, son œuvre, sa survie* (Berlin & New York: Walter de Gruyter, 1987), pp. 289–325.
Bowen, A. C., "Simplicius and the Early History of Greek Planetary Theory," *Perspectives on Science* 10 (2/2002), pp. 155–67.
Bowen A. C., and R. B. Todd, *Cleomedes' Lectures on Astronomy: A Translation of The Heavens with an Introduction and Commentary* (Berkeley: University of California Press, 2004).
Bucciantini, M., and E. Torrini, eds., *La difusione del copernicanesimo in Italia 1543–1610* (Florence: Olschki, 1997).
Bucciantini, M., M. Camerota and S. Roux, eds., *Mechanics and Cosmology in the Medieval and Early Modern Period* (Florence: Olschki, 2007).
Bulmer-Thomas, I., "Plato's Astronomy," *Classical Quarterly* 34 (1984), pp. 107–12.
Byrne, J. S., "A Humanist History of Mathematics?: Regiomontanus's Padua Oration in Context," *Journal of the History of Ideas* 67 (1/2006), pp. 41–61.
Carmody, F. J., "The Planetary Theory of Ibn Rushd," *Osiris* 10 (1952), pp. 556–86.
Carone, G. R., *Plato's Cosmology and its Ethical Dimension* (Cambridge: Cambridge University Press, 2011).
Caroti, S., "Le fonti medievali delle *Disputationes adversus astrologiam divinatricem*," in M. Bertozzi, ed., *Nello specchio del cielo*, pp. 67–95.
Casanovas, J., "Copernicus and the Gregorian Calendar Reform," in L. Pepe, ed., *Copernico e la questione copernicana in Italia*, pp. 97–107.
Casini, P., "Copernicus, Philolaus and the Pythagoreans," *Memorie della Società Astronomica Italiana* 65 (1994), pp. 497–508.
–––, "The Pythagorean Myth: Copernicus to Newton," in L. Pepe, ed., *Copernico e la questione copernicana in Italia*, pp. 183–199.
Celenza, C., "Pythagoras in the Renaissance: The Case of Marsilio Ficino," *Renaissance Quarterly* 52 (3/1999), pp. 667–711.
Chabás, J., "The Diffusion of the Alfonsine Tables: The Case of *Tabulae resolutae*," *Perspectives on Science* 10 (2002), pp. 168–78.
Chalmers, A., "Planetary Distances in Copernican Theory," *The British Journal for the Philosophy of Science* 32 (4/1981), pp. 374–75.
Chen, X., and P. Barker, "Continuity through Revolutions: A Frame-based Account of Conceptual Change," *Philosophy of Science* 67 (2000), pp. 208–23.
Clagett, M., *The Science of Mechanics in the Middle Ages* (Madison: The University of Wisconsin Press, 1959).
Clavelin, M., *La philosophie naturelle de Galilée* (Paris: Albin Michel, 1996).
–––, "Le copernicianisme et la mutation de la philosophie naturelle," *Revue de Metaphysique et de Morale* (3/2004), pp. 353–70.
Clenza, C. S., "The Revival of Platonic Philosophy," in J. Hankins, ed., *The Cambridge Companion to the Renaissance Philosophy*, pp. 72–96.

Clutton-Brock, M., "Copernicus's Path to His Cosmology: An Attempted Reconstruction," *JHA* 36 (2/2005), pp. 197–216.
Cooper, J. M., "The Psychology of Justice in Plato," *American Philosophical Quarterly* 14 (2/1977), pp. 151–7.
Copenhaver, B. P., "Translation, Terminology, and Style in Philosophical Discourse," in C. B. Schmitt and Q. Skinner, eds., *The Cambridge History of Renaissance Philosophy*, pp. 75–110.
Copernico a Padova (Padua: Cleup, 1995).
Cornford, F. M., *Plato's Cosmolog: The Timaeus of Plato Translated with a Running Commentary* (London: Routledge & Kegan 1956).
Crowe, M. J., *Theories of the World from Antiquity to the Copernican Revolution* (New York: Dover, 2001).
Czartoryski, P., "The Library of Copernicus," *Studia Copernicana* 16 (1978), pp. 355–96.
D'Alverny, T., "Survivances du 'système d'Héraclide' au Moyen Age," in *Avant, avec, après Copernic*, pp. 39–50.
Dales, R. C., "The De-animation of the Heavens in the Middle Ages," *Journal of the History of Ideas* 41 (1980), pp. 531–50.
Dambska, I., "L'épistémologie de Ptolémée," in *Avant, avec, après Copernic*, pp. 31–7.
Danielson, D., "The Great Copernican Cliché," *American Journal of Physics* 69 (10/2001), pp. 1029–35.
---, *The First Copernican: Georg Joachim Rheticus and the Rise of the Copernican Revolution* (New York: Walker & Co., 2006).
Davidson, N., *Astronomy and the Imagination: A New Approach to Man's Experience of the Stars* (London: Routledge & Kegan, 1987).
De Bellis, D., "Niccolò Leonico Tomeo interprete di Aristototele naturalista," *Physis* 17 (1975), pp. 71–93.
---, "La vita e l'ambiente di Niccolò Leonico Tomeo," *Quaderni per la storia dell'Univeristà di Padova* 13 (1980), pp. 37–75.
De Bellis, E., "Monographia introduttiva," in A. Nifo, *La filosofia nella corte*, pp. 7–140.
De Pace, A., "Plutarco e la rivoluzione copernicana," in I. Gallo, ed., *L'eredità culturale di Plutarco dall'antichità al Rinascimento* (Naples: D'Auria, 1999), pp. 313–51.
---, "Copernicus against a Rhetorical Approach to the Beauty of the Universe: The Influence of the *Phaedo* on the *De revolutionibus*," in E. Kessler and J. Maclean, eds., *Res et Verba in der Renaissance* (Wiesbaden: Harrassowitz, 2002), pp. 77–115.
---, "Galileo, Ficino e la cosmologia. Ordine, moti ed elementi in due diverse interpretazioni platoniche," *Rivista di storia della filosofia* 61 (2006), pp. 469–507.
---, *Niccolò Copernico e la fondazione del cosmo eliocentrico con testo, traduzione*

e commentario del Libro I de Le rivoluzioni celesti (Milan: Bruno Mondadori, 2009).

Dear, P., *Discipline and Experience: The Mathematical Way in the Scientific Revolution* (Chicago: University of Chicago Press, 1995).

---, *Revolutionizing the Sciences: European Knowledge and its Ambitions, 1500–1700* (New York: Palgrave, 2001).

DeWitt, R., *Worldviews: An Introduction to the History and Philosophy of Science* (Oxford & Malden: Blackwell, 2004).

---, "The Beginnings of a Modern Copernican Revolution," *Filozofski vestnik* 25 (2/2004), pp. 9–24.

Di Bono, M., *Le sfere omocentriche di Giovan Battista Amico nell'astronomia del Cinquecento con il testo del "De motibus corporum coelestium"* (Genoa: Genoa University Press, 1990).

---, "Copernicus, Amico, Fracastoro and Tūsī's Device: Observations on the Use and Transmission of a Model," *JHA* 26 (1995), pp. 133–54.

---, "Teme e i fonti dell'*Homocentrica* di Fracastoro," in A. Pastore and E. Peruzzi, eds., *Girolamo Fracastroro*, pp. 117–39.

Di Napoli, G., "Il Cardinale Bessarione nella controversia tra platonici ed aristotelici," *Miscellanea francescana* 73 (1973), pp. 327–50.

Dillon, J., *The Middle Platonists: 80 B. C. to A. D. 220* (London: Duckworth, 1996).

Dobrzycki, J., ed., *The Reception of Copernicus' Heliocentric Theory* (Dordrecht & Boston: Reidel, 1972).

---, "The Aberdeen Copy of Copernicus's *Commentariolus*," *JHA* 4 (1973), pp. 124–7.

---, "The Centre of the Universe in the Copernican Astronomy," in *Avant, avec, après Copernic*, pp. 117–8.

---, "Notes on Copernicus's Early Heliocentrism," *JHA* 32 (2001), pp. 223–5.

Dobrzycki, J., and R. L. Kremer, "Peurbach and Marāgha Astronomy? The Ephemerides of Johannes Angelus and their Implications," *JHA* 27 (1996), pp. 187–237.

Dobrzycki, J., and L. Szczucki, "On the Transmission of Copernicus' *Commentariolus* in the Sixteenth Century," *JHA* 20 (1989), pp. 25–8.

Dod, B. G., "Aristoteles Latinus," in N. Kretzmann, A. Kenny and J. Pinborg, eds., *The Cambridge History of Later Medieval Philosophy*, pp. 45–79.

Domański, J., "La fortuna di Marsilio Ficino in Polonia nei secoli XV e XVI," in G. C. Garfagnini, ed., *Marsilio Ficino e il ritorno di Platone*, pp. 565–86.

Donahue, W. H., "The Solid Planetary Spheres in Post-Copernican Natural Philosophy," in R. S. Westman, ed., *The Copernican Achievement*, pp. 244–75.

---, *The Dissolution of the Celestial Spheres* (New York: Arno Press, 1981).

Dreyer, J. L. E., *History of Astronomy from Thales to Kepler* (New York: Dover, 1953).

Duhem, P., *Un fragment inédit de l'Opus tertium de Roger Bâcon, précédé d'une étude sur ce fragment* (Quaracchi: Colegio S. Bonaventurae, 1909).

---, *Le système du monde. Histoire des doctrines cosmologiques de Platon à Copernic*, 10 vols. (Paris: Hermann, 1913–1959).

---, *Medieval Cosmology: Theories of Infinity, Place, Time, Void, and the Plurality of Worlds*, tr. R. Ariew (Chicago: University of Chicago Press, 1985).

---, *SOZEIN TA PHAINOMENA: essai sur la notion de théorie physique de Platon à Galilée* (Paris: Vrin, 2003).

Duke, D., "Ptolemy's Treatment of the Outer Planets," *Archive for History of Exact Sciences* 59 (2005), pp. 169–87.

Easterling, H. J., "Homocentric Spheres in the *De caelo*," *Phronesis* 6 (1/1961), pp. 138–53.

Eastwood, B. S., "Kepler as Historian of Science: Precursors of Copernican Heliocentrism according to *De Revolutionibus*," *Proceedings of the American Philosophical Society* 126 (5/1982), pp. 367–94.

---, "Heraclides and Heliocentrism: Texts, Diagrams, and Interpretations," *JHA* 23 (1992), pp. 233–60.

---, "Plato and Circumsolar Planetary Motion in the Middle Ages," *Archives d'histoire doctrinale et littéraire du Moyen Age* 60 (1993), pp. 7–26.

---, *Ordering the Heavens: Roman Astronomy and Cosmology in the Carolingian Renaissance* (Leiden: Brill, 2007).

Elders, L., *Aristotle's Cosmology: A Commentary on the De caelo* (Assen: Van Gorcum, 1966).

Endress, G., "Averroes' *De Caelo*: Ibn Rushd's Cosmology in his Commentaries on Aristotle's *On the Heavens*," *Arabic Sciences and Philosophy* 5 (1/1995), pp. 9–49.

---, "Mathematics and Philosophy in Medieval Islam," in J. P. Hodendijk and A. I. Sabra, eds., *The Enterprise of Science in Islam: New Perspectives* (Cambridge:The MIT Press, 2003), pp. 121–76.

Evans, J., "The Division of the Martian Eccentricity from Hipparchus to Kepler: A History of the Approximations to Kepler Motion," *American Journal of Physics* 56 (1988), pp. 1009–24.

---, *The History and Practice of Ancient Astronomy* (New York: Oxford University Press 1998).

Evans, J., and J. L. Berggren, *Geminos's Introduction to the Phenomena: A Translation and Study of a Hellenistic Survey of Astronomy* (Princeton & Oxford: Princeton University Press, 2006).

Falcon, A., *Corpi e movimenti. Il De caelo di Aristotele é la sua fortuna nel mondo antico* (Naples: Bibliopolis, 2001).

Favaro, A., "Die Hochschule Padua zur Zeit des Coppernicus," tr. M. Curtze, *Mitteilungen des Copernicus-Vereins für Wissenschaft und Kunst zu Thorn* 3 (1881), pp. 18–46, and 50–60.

---, *Galileo e lo studio di Padova*, 2 vols. (Florence: Successori Le Monnier, 1883).

Federici-Vescovini, G., "Note sur circulation en Italie du commentaire d'Albert de Saxe sur le *De caelo*," in J. Biard, ed., *Itinéraires d'Albert de Saxe*, pp. 235–52.
---, "Michael Scot et la *Theorica Planetarum Gerardi*," *Early Science and Medicine* 1 (2/1996), pp. 272–82.
Finocchiaro, M. A., *The Galileo Affair: A Documentary History* (Berkeley: University of California Press, 1989).
Folkerts, M., "Regiomontanus's Role in the Transmission and Transformation of Greek Mathematics," in F. J. Ragep, S. Ragep, and S. Livesey, eds., *Tradition, Transmission, Transformation*, pp. 89–113.
Fraansen, M., "Did King Alfonso of Castile Really Want to Advise God against the Ptolemaic System? The Legend in History," *Studies in History and Philosophy of Science* 24 (3/1993), pp. 313–24.
Freud, S., *Introductory Lectures on Psychoanalysis*, tr. J. Strachey, ed. J. Strachey and A. Richards (Harmondsworth: Penguin Books, 1987).
Freudenthal, G., "'Instrumentalism' and 'Realism' as Categories in the History of Astronomy: Duhem vs. Popper, Maimonides vs. Gersonides," *Centaurus* 45 (2003), pp. 227–48.
Funkenstein, A., "Some Remarks on the Concept of Impetus and the Determination of Simple Motion," *Viator* 2 (1971), pp. 329–48.
---, "The Dialectical Preparation for Scientific Revolution," in R. S. Westman, ed., *The Copernican Achievement*, pp. 165–202.
---, *Théologie et imagination scientifique du moyen age au XVII*[e] *siècle*, tr. J.-P. Rothschild (Paris: Presses Universitaires de France, 1995).
Gaiser, K., "Plato's Enigmatic Lecture *On the Good*," *Phronesis* 25 (1980), pp. 5–37.
Gapaillard, J., *Et pourtant elle tourne! Le mouvement de la Terre* (Paris: Seuil, 1993).
Garfagnini, G. C., ed., *Marsilio Ficino e il ritorno di Platone*, 2 vols., continously paginated (Florence: Olschki, 1986).
---, ed., *Cieli e terre nei secoli XI-XII. Orizzonti, percezioni, rapporti* (Milan: Vita e Pensiero, 1998).
---,"Cielo e terra nella riflessione filosofica dei secoli XI e XII," in G. C. Garfagnini, ed., *Cieli e terre nei secoli XI-XII: Orizzonti, percezioni, rapporti*, pp. 51–73.
Garin, E., "Alle origini della polemica anticopernicana," *Studia Copernicana* 6 (1973), pp. 31–42.
---, "Copernico e Bessarione," *Rivista critica di storia della filosofia* 28 (1973), pp. 86–8.
---, "Marsilio Ficino e il ritorno di Platone," in G. C. Garfagnini, ed., *Marsilio Ficino e il ritorno di Platone*, pp. 3–13.
---, "Aristotelismo veneto e scienza moderna," in E. Garin, *Umanisti artisti scienziati. Studi sul Rinascimento italiano* (Rome: Editori Riuniti, 1989), pp. 205–28.
---, *Rinascite e rivoluzioni. Movimenti culturali dal XIV al XVIII secolo* (Rome: Laterza, 2007).

---, "Filippo Beroaldo il Vecchio: un universitario inquieto," in E. Garin, *Rinascite e rivoluzioni*, pp. 197–218.
---, "La rivoluzione copernicana e il mito solare," in E. Garin, *Rinascite e rivoluzioni*, pp. 255–81.
---, *L'umanesimo italiano* (Rome: Laterza, 2008).
Gatti, H., "Giordano Bruno's Copernican Diagrams," *Filozofski vestnik* 25 (2/2004), pp. 25–50.
Geanakoplos, D. J, "The Career of the Little-known Renaissance Greek Scholar Nicholas Leonicus Tomaeus and the Ascendancy of Greco-Byzantine Aristotelianism at Padua University (1497)," *Byzantina* 13 (1985), pp. 355–72.
Genequand, C., *Ibn Rushd's Metaphysics: A Translation with Introduction of Ibn Rushd's Commentary on Aristotle's Metaphysics, Book Lām* (Leiden: Brill, 1986).
Gentile, S., "Il ritorno delle culture classiche," in C. Vasoli, *Le filosofie del Rinascimento*, pp. 70–92.
---, "Il ritorno di Platone, dei platonici e del 'corpus' ermetico. Filosofia, teologia e astrologia nell'opera di Marsilio Ficino," in C. Vasoli, *Le filosofie del Rinascimento*, pp. 193–228.
Gerhard, M., "*Certe rationem ordinis non esse*? Zur Konstituierbarkeit einer vernünftigen Ordnung der Natur bei Copernicus, Bruno und Schelling," *Filozofski vestnik* 25 (2/2004), pp. 51–68.
Gierowski, J., ed., *The Cracow Circle of Nicholas Copernicus* (Warsaw & Cracow: Polish Science Publishers, 1973).
Gingerich, O., ed., *The Nature of Scientific Discovery* (Washington: Smithsonian Institution Press, 1975).
---, "'Crisis' versus Aesthetic in the Copernican Revolution," *Vistas in Astronomy* 17 (1975), pp. 85–94.
---, "Introductory Remarks on the Astronomy of Copernicus," in *Avant, avec, après Copernic*, pp. 101–4.
---, "A Tusi Couple from Schöner's *De revolutionibus*?" *JHA* 15 (1984), pp. 128–33.
---, "Sacrobosco as a Textbook," *JHA* 19 (4/1988), pp. 269–73.
---, *The Great Copernicus Chase and Other Adventures in Astronomical History* (Cambridge: Sky, 1992).
---, *The Eye of Heaven: Ptolemy, Copernicus, Kepler* (New York: American Institute of Physics, 1993).
---, *An Annotated Census of Copernicus' De revolutionibus* (Nuremberg, 1543 and Basel, 1566) (Leiden: Brill, 2002).
---, *The Book Nobody Read: Chasing the Revolutions of Nicolaus Copernicus* (New York: Penguin Books, 2004).
Goddu, A., "The Realism that Duhem Rejected in Copernicus," *Synthese* 83 (2/1990), pp. 301–15.

---, "Consequences and Conditional Propositions in John of Glogowia's and Michael of Biestrzykowa Commentaries on Peter of Spain and their Possible Influence on Nicholas Copernicus," *Archives d'histoire doctrinale et littéraire du Moyen Âge* 62 (1995), pp. 137–88.

---, "The Logic of Copernicus's Arguments and his Education in Logic at Cracow," *Early Science and Medicine* 1 (1996), pp. 26–68.

---, "Impetus," in W. Applebaum, ed., *Encyclopedia of the Scientific Revolution from Copernicus to Newton*, pp. 325–6.

---, "Hypotheses, Spheres, and Equants in Copernicus's *De revolutionibus*," in B. el Bouazzati, ed., *Les elements paradigmatiques thématiques et stylistiques dans la pensée scientifique* (Casablanca: Najah El Jadida, 2004), pp. 71–95.

---, "Copernicus's Annotations: Revisions of Czartoryski's 'Copernicana'," *Scriptorium* 58 (2004), pp. 202–26.

---, "Ptolemy, Copernicus, and Kepler on Linear Distances," *Organon* 35 (2006), pp. 45–59.

---, "Reflections on the Origin of Copernicus's Cosmology," *JHA* 37 (1/2006), pp. 37–53.

---, "Copernicus's Mereological Vision of the Universe," *Early Science and Medicine* 14 (2009), pp. 316–39.

---, *Copernicus and the Aristotelian Tradition: Education, Reading and Philosophy in Copernicus's Path to Heliocentrism* (Leiden: Brill, 2010).

Goldstein, B. R., "The Arabic Version of Ptolemey's Planetary Hypotheses," *Transactions of the American Philosophical Society* 57 (4/1967), pp. 3–55.

---, "The Status of Models in Ancient and Medieval Astronomy," *Centaurus* 24 (1980), pp. 132–47.

---, "Towards a Philosophy of Ptolemaic Planetary Astronomy," *Ancient Philosophy* 5 (1985), pp. 293–303.

---, "Historical Perspectives on Copernicus's Account of Precession," *JHA* 25 (1994), pp. 189–97.

---, "Saving the Phenomena: The Background to Ptolemy's Planetary Theory," *JHA* 28 (1997), pp. 1–12.

---, "Copernicus and the Origin of his Heliocentric System," *JHA* 33 (3/2002), pp. 219–35.

Goldstein, B. R., and A. C. Bowen, "A New View of Early Greek Astronomy," *Isis* 74 (1983), pp. 330–40.

Goldstein, B. R., and J. Chabás, *The Alfonsine Tables of Toledo* (Dordrecht: Kluwer, 2003).

---, "Ptolemy, Bianchini and Copernicus: Tables for Planetary Latitudes," *Archive for History of Exact Sciences* 58 (5/2004), pp. 453–73.

Goldstein, B. R, and G. Hon, "Kepler's Move from *Orbs* to *Orbits*: Documenting a Revolutionary Scientific Concept," *Perspectives on Science* 13 (2005), pp. 74–110.

Goldstein, T., "The Renaissance Concept of the Earth in its Influence upon Copernicus," *Terrae Incognitae* 4 (1972), pp. 19–51.

Grafton, A., "The New Science and the Traditions of Humanism," in J. Kraye, ed., *The Cambridge Companion to Renaissance Humanism* (Cambridge: Cambridge University Press, 2004), pp. 203–23.

Granada, M. A., "Giovanni Maria Tolosani e la prima reazione romana di fronte al *De revolutionibus*: la critica di Copernico nell'opuscolo *De caelo et elementis*," in M. Bucciantini and M. Torrini, eds., *La diffusione del copernicanesimo in Italia 1543–1610* (Florence: Olschki, 1997), pp. 11–35.

---, "Giovanni Pico e il mito della Concordia. La riflessione di Pico dopo il 1488 e la sua polemica antiastrologica," in C. Vasoli, *Le filosofie del Rinascimento*, pp. 229–46.

---, "Aristotle, Copernicus, Bruno: Centrality, the Principle of Movement and the Extension of the Universe," *Studies in History and Philosophy of Science* 35 (2004), pp. 91–114.

---, "New Visions of the Cosmos," in J. Hankins, ed., *The Cambridge Companion to the Renaissance Philosophy*, pp. 270–86.

Granada, M. A., and É. Mehl, eds., *Nouveau ciel, nouvelle terre: la révolution copernicienne dans l'Allemagne de la réforme (1530–1630)* (Paris: Les Belles Lettres, 2009).

Granada, M. A., and D. Tessicini, "Copernicus and Fracastoro: The Dedicatory Letters to Pope Paul III, the History of Astronomy, and the Quest for Patronage," *Studies in History and Philosophy of Science* 36 (2005), pp. 431–76.

Grant, E., "Late Medieval Thought: Copernicus, and the Scientific Revolution," *Journal of the History of Ideas* 23 (1962), pp. 197–220.

---, ed., *A Source Book in Medieval Science* (Cambridge: Harvard University Press, 1974).

---, "Cosmology," in D. C. Lindberg, ed., *Science in the Middle Ages* (Chicago: Chicago University Press, 1980), pp. 265–302.

---, *Much Ado about Nothing: Theories of Space and Vacuum from the Middle Ages to the Scientific Revolution* (Cambridge: Cambridge University Press, 1981).

---, "The Medieval Doctrine of Place: Some Fundamental Problems and Solutions," in A. Maierù and A. Paravicini Bagliani, eds., *Studi sul XIV secolo in memoria di Annalise Maier* (Rome: Storia e letteratura, 1981), pp. 59–64.

---, "In Defense of the Earth's Centrality and Immobility: Scholastic Reaction to Copernicanism in the Seventeenth Century," *Transactions of the American Philosophical Society* 74 (4/1984), pp. 1–69.

---, *Physical Science in the Middle Ages* (Cambridge: Cambridge University Press, 1986).

---, "Science and Theology in the Middle Ages," in D. C. Lindberg and R. R. Numbers, eds., *God and Nature*, pp. 49–75.

---, "Celestial Orbs in the Latin Middle Ages," *Isis* 78 (1987), pp. 153–94.

---, "Eccentrics and Epicycles in Medieval Cosmology," in E. Grant and J. E. Murdoch, eds., *Mathematics and its Applications to Science and Natural Philosophy in the Middle Ages* (Cambridge: Cambridge University Press, 1987), pp. 189–214.

---, "The Unusual Structure and Organisation of Albert of Saxony's *Questions on De caelo*," in J. Biard, ed, *Itinéraires d'Albert de Saxe*, pp. 205–18.

---, *Planets, Stars, and Orbs: The Medieval Cosmos, 1200–1687* (Cambridge MA: Cambridge University Press, 1996).

---, *The Foundations of Modern Science in The Middle Ages: Their Religious, Institutional and Intellectual Context* (Cambridge: Cambridge University Press, 1996).

---, "Celestial Motions in the Late Middle Ages," *Early Science and Medicine* 2 (2/1997), pp. 129–48.

Gregory, A., "Astronomy and Observation in Plato's *Republic*," *Studies in History and Philosophy of Science* 27 (4/1996), pp. 451–71.

---, *Plato's Philosophy of Science* (London: Duckworth, 2000).

---, *Ancient Greek Cosmogony* (London: Duckworth, 2007).

Grendler, P. F., *The Universities of the Italian Renaissance* (Baltimore: Johns Hopkins Press, 2002).

Griffiths, R. I., "Was There a Crisis Before the Copernican Revolution? A Reappraisal of Gingerich's Criticisms of Kuhn," *Proceedings of the Biennial Meeting of the Philosophy of Science Association* 1988, vol. 1 (1988), pp. 127–32.

Groh, D., *Göttliche Weltökonomie: Perspektiven der Wissenschaftlichen Revolution vom 15. bis 17. Jahrhundert* (Berlin: Suhrkamp, 2010).

Guerlac, H., "Copernicus and Aristotle's Cosmos," *Journal of the History of Ideas* 29 (1968), pp. 109–13.

Guessoum, N., "Copernicus and Ibn al-Shatir: Does the Copernican Revolution have Arabic Roots?," *The Observatory* 128 (6/2008), pp. 231–39.

Hallyn, F., "Copernic et le platonisme ficinien," in C.-G. Dubois, ed., *L'invention au XVIe siècle* (Bordeaux: Presses Univerisitaires de Bordeaux, 1987), pp. 137–51.

---, *The Poetic Structure of the World: Copernicus and Kepler*, tr. D. M. Leslie (New York: Zone Books, 1993).

Hamann, G., ed., *Regiomontanus-Studien* (Vienna: Verlag der Österreichischen Akademie der Wissenschaften, 1980).

Hankins, J., *Plato in the Italian Renaissance*, 2 vols. (Leiden: Brill, 1990).

---, "The Study of the *Timaeus* in Early Renaissance Italy," in A. Grafton and N. Sirasi, eds., *Natural Particulars: Nature and the Disciplines in Renaissance Europe* (Cambridge: The MIT Press, 1999), pp. 77–119.

---, "Galileo, Ficino and Renaissance Platonism," in J. Kraye and M. W. F. Stone,

eds., *Humanism and Early Modern Philosophy* (London & New York: Routledge, 2000), pp. 209–37.
---, ed., *The Cambridge Companion to the Renaissance Philosophy* (Cambridge: Cambridge University Press, 2007).
---, "Humanism, Scholasticism, and Renaissance Philosophy," in J. Hankins, ed., *The Cambridge Companion to the Renaissance Philosophy*, pp. 30–48.
---, *La riscoperta di Platone nel Rinascimento italiano* (Pisa: Edizioni della Normale, 2009).
Hanson, N. R., "The Copernican Disturbance and the Keplerian Revolution Source," *Journal of the History of Ideas* 22 (2/1961), pp. 169–84.
---, "Contra-Equivalence: A Defense of the Originality of Copernicus," *Isis* 55 (3/1964), pp. 308–25.
Hartner, W., "Copernicus, the Man, the Work, and Its History," *Proceedings of the American Philosophical Society* 117 (6/1973), pp. 413–22.
---, "Ptolemy's and Copernicus's Mercury Models: An Accuracy Test," *Archives internationales d'histoire des sciences* 24 (1974), pp. 367–9.
---, "Astronomy from Antiquity to Copernicus," in *Avant, avec, après Copernic*, pp. 11–7.
---, "The Islamic Astronomical Background of Nicholas Copernicus," *Studia Copernicana* 13 (1975), pp. 7–16.
---, "Ptolemäische Astronomie im Islam zur Zeit des Regiomontanus," in G. Hamann, ed., *Regiomontanus-Studien*, pp. 109–24.
Hasse, D. N., "Arabic Philosophy and Averroism," in J. Hankins, ed., *The Cambridge Companion to the Renaissance Philosophy*, pp. 113–36.
Hayli, A., "A propos du *Commentariolus*," in *Avant, avec, après Copernic*, pp. 107–10.
Heath, T. L., *Aristarhus of Samos, the Ancient Copernicus. A History of Greek Astronomy to Aristarchus together with Aristarchus' Treatise On the Sizes and Distances of the Sun and the Moon* (Oxford: Clarendon Press, 1913).
---, *Greek Astronomy* (New York: Dover, 1991).
Heidarzadeh, T., *A History of Physical Theories of Comets, From Aristotle to Whipple* (New York: Springer, 2008).
Hendrix, J., *Platonic Architectonics: Platonic Philosophies and the Visual Arts* (New York: Peter Lang, 2004).
Henninger, S. K., Jr., *Touches of Sweet Harmony: Pythagorean Cosmology and Renaissance Poetics* (San Marino: The Huntington Library, 1974).
Henry, J., "Animism and Empiricism: Copernican Physics and the Origins of William Gilbert's Experimental Method," *Journal of the History of Ideas* 62 (1/2001), pp. 99–119.
Henry, J., *The Scientific Revolution and the Origins of Modern Science* (New York: St. Martin Press, 1997).

---, *Moving Heaven and Earth: Copernicus and the Solar System* (Cambridge: Icon Books, 2001).

Hersey, G. L., *Pythagorean Palaces: Magic and Architecture in the Italian Renaissance* (Ithaca & London: Cornell University Press, 1976).

Hetherington, N. S., "Plato and Eudoxus: Instrumentalists, Realists, or Prisoners of Themata?," *Studies in History and Philosophy of Science* 27 (2/1996), pp. 271–89.

Hill, D. K., "The Projection Argument in Galileo and Copernicus: Rhetorical Strategy in the Defence of the New System," *Annals of Science* 41 (1984), pp. 109–33.

Hodendijk, J. P., and A. I. Sabra, eds., *The Enterprise of Science in Islam: New Perspectives* (Cambridge: The MIT Press, 2003).

Hon, G., and B. R. Goldstein, "Symmetry in Copernicus and Galileo," *JHA* 35 (3/2004), pp. 273–92.

Hooykaas, R., "Rheticus's Lost Treatise on Holy Scripture and the Motion of the Earth," *JHA* 15 (1984), pp. 77–80.

---, *G. J. Rheticus' Treatise On Holy Scripture and the Motion of the Earth* (Amsterdam: North-Holland, 1984).

---, "The Aristotelian Background to Copernicus's Cosmology," *JHA* 17 (1987), pp. 111–6.

Huffman, C. A., "The Pythagorean Tradition," in A. A. Long, ed., *The Cambridge Companion to Early Greek Philosophy* (Cambridge: Cambridge University Press, 1999), pp. 66–87.

Hugonnard-Roche, H., "Le problème de la rotation de la Terre au XIVe siècle," in *Avant, avec, après Copernic*, pp. 61–5.

---, "Remarques sur l'évolution doctrinale d'Averroès dans les commentaires au *De caelo*: le problème du mouvement de la terre," *Mélanges de la Casa de Velázquez* 13 (1977), pp. 103–17.

---, "L'Épitomé du De caelo d'Aristote par Averroès: questions de méthode et de doctrine," *Archives d'histoire doctrinale et littéraire du Moyen Age* 51 (1984), pp. 7–39.

---, "Problèmes méthodologiques dans l'astronomie au début du XIVe siècle," in G. Freudenthal, ed., *Studies on Gersonides: A Fourteenth-Century Jewish Philosopher-Scientist* (Leiden: Brill, 1992), pp. 55–70.

---, "The Influence of Arabic Astronomy in the Medieval West," in R. Rashed, ed., *Encyclopedia of the History of Arabic Science. Vol 1: Astronomy – Theoretical and Applied*, pp. 284–305.

---, "Contribution arabe à la cosmologie occidentale latine," in G. C. Garfagnini, ed., *Cieli e terre nei secoli XI-XII*, pp. 89–110.

Hugonnard-Roche, H., and J.-P. Verdet, *Introductions à l'astronomie de Copernic.*

Le Commentariolus de Copernic. La Narratio prima de Rheticus (Paris: A. Blanchard, 1975).

Hujer, K., "Sur l'histoire du manuscript copernicien: *De revolutionibus orbium coelestium*," *L'Astronomie* 90 (1976), pp. 181–90.

Hutchinson, K., "Copernicus, Apollo, and Heracles," in S. Gaukroger, ed., *The Uses of Antiquity: The Scientific Revolution and the Classical Tradition* (Dordrecht: Kluwer, 1990), pp. 1–24.

---, "Harmony and Authority: The Political Symbolism of Copernicus' Political Seal," in R. G. Mazzolini, ed., *Non-verbal Communication in Science Prior to 1900* (Florence: Olschki, 1993).

Jardine, N., "The Significance of the Copernican Orbs," *JHA* 13 (1982), pp. 168–94.

---, *The Birth of History and Philosophy of Science: Kepler's* A Defence of Tycho against Ursus, *with Essays on Its Provenance and Significance* (Cambridge: Cambridge University Press, 1984).

---, "Epistemology of Science," in C. B. Schmitt and Q. Skinner, eds., *The Cambridge History of Renaissance Philosophy*, pp. 685–711.

---, "The Places of Astronomy in Early-Modern Culture," *JHA* 29 (1998), pp. 49–62.

Jervis, J. L., *Cometary Theory in Fifteenth-Century Europe* (Dordrecht: Reidel, 1985).

Jones, A., "The Stoics and the Astronomical Sciences," in B. Inwood, ed., *The Cambridge Companion to the Stoics* (Cambridge: Cambridge University Press, 2003), pp. 328–44.

Joost-Gaugier, C. L, *Pythagoras and Renaissance Europe: Finding Heaven* (Cambridge: Cambridge University Press, 2009).

Judson, L., "Heavenly Motion and the Unmoved Mover," in M. L. Gill and J. G. Lennox, eds., *Self-Motion from Aristotle to Newton* (Princeton: Princeton University Press, 1994).

Kaske, C., V., "Ficino's Shifting Attitude towards Astrology in *De vita coelitus comparanda*, the Letter to Poliziano, and the *Apologia* to the Cardinals," in G. C. Garfagnini, ed., *Marsilio Ficino e il ritorno di Platone*, pp. 371–81.

Kennedy, E. S., and V. Roberts, "The Planetary Theory of Ibn Ash Shâtir," *Isis* 50 (1959), pp. 227–35.

---, "Late Medieval Planetary Theory," *Isis* 57 (1966), pp. 365–78.

Kerszberg, P., "La cosmologie de Copernic et les origines de la physique mathématique," *Revue d'histoire des sciences* 34 (1/1981), pp. 3–23.

Kessler, E., "Method in the Aristotelian Tradition: Taking a Second Look," in D. A. Di Liscia, E. Kessler and C. Methuen, eds., *Method and Order in Renaissance Philosophy of Nature: The Aristotle Commentary Tradition* (Aldershot: Ashgate, 1997), pp. 114–42.

Kleinert, A., "'Eine handgreifliche Geschichtslüge'. Wie Martin Luther zum

Gegner des copernicanischen Weltsystems gemacht wurde," *Berichte zur Wissenschaftsgeschichte* 26 (2003), pp. 101–11.

Knoll, P. W., "The Arts Faculty at the University of Cracow at the End of the Fifteenth Century," in R. Westman, ed., *The Copernican Achievement*, pp. 137–55.

Knorr, W., "Plato and Eudoxus on the Planetary Motions," *JHA* (1989), pp. 313–29.

---, "Notes on Rosen's Copernicus," *JHA* 21 (1990), pp. 203–11.

Knox, D., "Ficino, Copernicus and Bruno on the Motion of the Earth," *Bruniana & Campanelliana* 5 (1999), pp. 333–66.

---, "Ficino and Copernicus," in M. Allen and V. Rees, eds., *Marsilio Ficino: His Theology, his Philosophy, his Legacy*, pp. 399–418.

---, "Copernicus's Doctrine of Gravity and the Natural Circular Motion of the Elements," *Journal of the Warburg and Courtauld Institutes* 68 (2005), pp. 157–211.

Kobe, D. H., "Copernicus and Martin Luther: An Encounter between Science and Religion," *American Journal of Physics* 66 (3/1998), pp. 190–6.

Koestler, A., *The Sleepwalkers: A History of Man's Changing Vision of the Universe* (London: Penguin 1989).

Kokowski, M., *Copernicus's Originality: Towards Integration of Contemporary Copernican Studies* (Warsaw & Cracow: Polish Academy of Sciences Press, 2004).

---, "Nicholas Copernicus in Focus of Interdisciplinary Research: An Outline of Main Results," *Organon* 35 (2006), pp. 73–84.

Kosman, A., "Mechanisms in Ancient Philosophy of Science," *Perspectives on Science* 12 (3/2004), pp. 244–61.

Koyré, A., "Galileo and Plato," *Journal of the History of Ideas* 4 (1943), pp. 400–28.

---, *From the Closed World to the Infinite Universe* (Baltimore: The Johns Hopkins University Press, 1968).

---, *The Astronomical Revolution: Copernicus – Kepler – Borelli*, tr. R. E. W. Maddison (New York: Dover, 1992).

Krafft, F., "Copernicus Retroversus I," *Studia Copernicana* 13 (1975), pp. 113–24.

---, "Copernicus Retroversus II: Gravitation und Kohäsionstheorie," *Studia Copernicana* 14 (1975), pp. 65–78.

---, "Physikalische Realität oder mathematische Hypothese? Andreas Osiander und die physikalische Erneuerung der antiken Astronomie durch Nicolaus Copernicus," *Philosophia naturalis* 14 (1975), pp. 243–75.

Kraye, J., "The Philosophy of the Italian Renaissance," in G. H. R. Parkinson, ed., *Routledge History of Philosophy. Volume 4: The Renaissance and Seventeenth-century Rationalism* (London & New York: Routledge, 1993), pp. 15–64.

---, ed., *The Cambridge Companion to Renaissance Humanism* (Cambridge: Cambridge University Press, 2004).

---, "The Revival of Hellenistic Philosophies," in J. Hankins, ed., *The Cambridge Companion to the Renaissance Philosophy*, pp. 97–112.

Kren, C., "Homocentric Astronomy in the Latin West: The *De reprobatione ecentricorum et epiciclorum* of Henry of Hesse," *Isis* 59 (1968), pp. 269–81.

---, "A Medieval Objection to 'Ptolemy'," *British Journal for the History of Science* 4 (1969), pp. 378–93.

---, "Planetary Latitudes, the *Theorica Gerardi*, and Regiomontanus," *Isis* 68 (1977), pp. 194–205.

Kretzmann, N., A. Kenny and J. Pinborg, eds., *The Cambridge History of Later Medieval Philosophy* (Cambridge: Cambridge University Press, 1982).

Kristeller, P. O., "The First Printed Edition of Plato's Works and the Date of its Publication (1484)," *Studia Copernicana* 16 (1978), pp. 25–36.

---, "Marsilio Ficino and his Work after Five Hundred Years," in G. C. Garfagnini, ed., *Marsilio Ficino e il ritorno di Platone*, pp. 15–196.

Krókowski, J., "Laurentius Corvinus und seine Beziehungen zu Polen," in J. Irmscher, ed., *Renaissance und Humanismus in Mittel- und Osteuropa*, vol. 2 (Berlin: Deutsche Akademie der Wissenschaften, 1962), pp. 153–172.

Kuhn, T. S., *The Copernican Revolution: Planetary Astronomy in the Development of Western Thought* (Cambridge: Harvard University Press, 1957).

---, *The Structure of the Scientific Revolutions* (Chicago: University of Chicago Press, 1962).

Labowsky, L., "Bessarione," *Dizionario Biografico degli Italiani*, vol. 9 (1976), pp. 1–5.

Lakatos, I., and E. Zahar, "Why Did Copernicus' Research Program Supersede Ptolemy's," in R. Westman, ed., *The Copernican Achievement*, pp. 354–83.

Lang, H. S., *Aristotle's Physics and its Medieval Varieties* (Albany: SUNY Press, 1992).

Langermann, Y. T., *Ibn al-Haytham's On the Configuration of the World* (New York: Garland, 1990).

---, "Arabic Cosmology," *Early Science and Medicine* 2 (1997), pp. 185–213.

---, "A Compendium of Renaissance Science: *Ta'alumot Hokmah* by Moses Galeano," *Aleph* 7 (2007), pp. 285–318.

Laplanche, J., "The Unfinished Copernican Revolution," in J. Laplanche, *Essays on Otherness*, tr. L. Thurston (New York: Routledge, 2003), pp. 53–85.

Lattis, J. M., "Homocentrics, Eccentrics and Clavius' Refutation of Fracastoro," *Physis* 28 (1991), pp. 699–725.

---, *Between Copernicus and Galileo: Christoph Clavius and the Collaps of Ptolemaic Cosmology* (Chicago: The University of Chicago Press, 1994).

Lemay, R., "The Late Medieval Astrological School at Cracow and the Copernican System," *Studia Copernicana* 16 (1978), pp. 337–54.

Lerner, M.-P., *Tre saggi sulla cosmologia alla fine del Cinquecento* (Naples: Bibliopolis, 1992).

---, *Le monde des sphères*, 2 vols. (Paris: Les Belles Lettres, 1996–1997).
---, "Aux origines de la polemique anticopernicienne (I). L' *Opusculum quartum* de Giovanmaria Tolosani," *Revue des sciences philosophiques et théologiques* 86 (2002), pp. 681–721.
---, "La doctrine copernicienne et sa proscription (1616)," in F. Beretta, ed., *Galilée en procès, Galilée réhabilité?* (Saint Maurice: Éditions Saint-Augustin, 2005), pp. 13–49.
---, "The Origin and Meaning of 'World System'," *JHA* 36 (4/2005), pp. 407–41.
---, "The Heliocentric 'Heresy': From Suspicion to Condemnation," in E. McMulin, ed., *The Church and Galileo* (Notre Dame, IN: University of Notre Dame, 2005), pp. 11–27.
---, "Der Narr will die gantze kunst Astronomiae umkehren': sur un célèbre 'Propos de table' de Luther," in M. A. Granada and É. Mehl, eds., *Nouveau ciel, nouvelle terre*, pp. 41–65.
Lerner, M.-P., and A.-Ph. Segonds, "Éditer et traduire Copernic," *Bruniana & Campanelliana* 9 (2003), pp. 379–408.
Lindberg, D. C., ed., *Science in the Middle Ages* (Chicago: Chicago University Press, 1980).
Lindberg, D. C., "The Transmission of Greek and Arabic Learning to the West," in D. C. Lindberg, ed., *Science in the Middle Ages*, pp. 52–90.
---, *The Beginnings of Western Science: The European Scientific Tradition in Philosophical, Religious, and Institutional Context, 600 B. C. to A. D. 1450* (Chicago: The University of Chicago Press, 1992).
Lindberg, D. C., and R. R. Numbers, eds., *God and Nature: Historical Essays on the Encounter between Christianity and Science* (Berkeley & Los Angeles: University of California Press, 1986).
Lindberg, D. C., and R. S. Westman, eds., *Reappraisals of the Scientific Revolution* (Cambridge: Cambridge University Press, 1990).
Litt, T., *Les corps célestes dans l'univers de saint Thomas d'Aquin* (Louvain & Paris: Nauwelaerts, 1966).
Lloyd, G. E. R., *Greek Science after Aristotle* (London & New York: W. W. Norton, 1973).
---, "Saving the Apperarences," *Classical Quarterly* 28 (1978), pp. 202–22.
---, *Early Greek Science: Thales to Aristotle* (London: Chatto & Windus, 1982).
---, *Magic, Reason and Experience: Studies in the Origins and Development of Greek Science* (London: Duckworth, 1999).
Lohr, C. H., "Metaphysics," in C. B. Schmitt and Q. Skinner, eds., *The Cambridge History of Renaissance Philosophy*, pp. 537–638.
Lorch, R., "Some Remarks on the *Almagestum parvum*," in S. Demidov, ed., *Amphora: Festschrift für Hans Wussing zu seinem 65. Geburtstag* (Basel: Birckhäuser, 1992), pp. 407–38.

Luccheta, G., "Dinamica dell'impetus e aristotelismo veneto," in L. Olivieri, ed., *Aristotelismo veneto e scienza moderna*, pp. 701–15.

Maccagni, C., "L'astronomia a Padova e Niccolò Copernico," in *Copernico a Padova*, pp. 149–150.

Machamer, P. K., "Aristotle on Natural Place and Natural Motion," *Isis* 69 (1978), pp. 377–87.

Maeyama, Y., "Determination of the Sun's Orbit: Hipparchus, Ptolemy, al-Battānī, Copernicus, Tycho Brahe," *Archive for the History of Exact Sciences* 53 (1998), pp. 1–49.

Mahoney, E. P., "Marsilio Ficino's Influence on Nicolletto Vernia, Agostino Nifo and Marcantonio Zimara," in G. C. Garfagnini, ed., *Marsilio Ficino e il ritorno di Platone*, pp. 509–31.

Maier, A., *Zwei Grundprobleme der scholastichen Naturphilosophie* (Rome: Edizioni di Storia e Letteratura, 1951).

–––, *Ausgehendes Mittelalter. Gesammelte Aufsätze zur Geistgeschiechte des 14. Jahrhunderts* (Rome: Edizioni di Storia e Letteratura, 1964).

Markowski, M., *Burydanizm w Polsce w okresie przedkopernikańskim. Studium z historii filozofii i nauk ścisłych na uniwersytecie krakowskim w XV wieku* (Wrocław: Polish Academy of Sciences, 1971).

–––, "Marcin Biem of Olkusz," in J. Gierowski, ed., *The Cracow Circle of Nicholas Copernicus*, pp. 7–21.

–––, "The Earliest Unknown Excerpts from Nicholas Copernicus' *De revolutionibus*," *Studia Copernicana* 6 (1973), pp. 5–30.

Martijn, M., *Proclus on Nature: Philosophy of Nature and its Methods in Proclus' Commentary on Plato's Timaeus* (Leiden: Brill, 2010).

Millàs Vallicrosa, J. M., *Las traductiones orientales en los manuscritos de la Bibliotheca Catedral de Toledo* (Madrid, 1942).

Mittelstrass, J., *Die Rettung der Phänomene* (Berlin: Walter de Gruyter, 1962).

Modelska Strelska, B., "Bernard Wapowski," in J. Gierowski, ed., *The Cracow Circle of Nicholas Copernicus*, pp. 23–59.

Moesgaard, K. P., "Hunting the Immobile Centre of Copernicus' Planetary System," *Studia Copernicana* 14 (1975), pp. 93–100.

–––, "Thābit ibn Qurra between Ptolemy and Copernicus," in *Avant, avec, après Copernic*, pp. 67–70.

Mohler, L., *Kardinal Bessarion als Theologe, Humanist, und Staatsman: Funde und Forschungen*, 3 vols. (Paderborn: F. Schöningh, 1923–42).

Monfasani, J., *George of Trebizond: A Biography and a Study of his Rhetoric and Logic* (Leiden: Brill, 1976).

–––, "Bessarion Latinus", *Rinascimento* 21 (1981), pp. 165–209.

–––, "Still more on 'Bessarion Latinus'", *Rinascimento* 23 (1983), pp. 217–35.

---, ed., *Collectanea Trapezuntiana: Texts, Documents and Bibliographies of George of Trebizond* (Binghamton, NY: The Renaissance Society of America, 1984).
---, *Byzantine Scholars in Renaissance Italy: Cardinal Bessarion and Other Emigres* (Aldershot: Ashgate, 1995).
---, "Marsilio Ficino and the Plato-Aristotle Controversy," in M. J. B. Allen and V. Rees, with M. Davies, eds., *Marsilio Ficino: His Theology, his Philosophy, his Legacy*, pp. 179–202.
---, "A Tale of Two Books: Bessarion's I*n Calumniatorem Platonis* and George of Trebizond's *Comparatio Philosophorum Platonis et Aristotelis*," *Renaissance Studies* 22 (1/2008), pp. 1–15.
---, "Marsilio Ficino and Eusebius of Caesarea's *Praeparatio Evangelica*," *Rinascimento* 49 (2009), pp. 3–13.
Moraux, P., "Introduction," in Aristote, *Du ciel*, pp. VII–CXC.
---, "Copernic et Aristote," in *Platon et Aristote à la Renaissance* (Paris: Vrin, 1976), pp. 225–38.
Morelon, R., "General Survey of Arabic Astronomy," in R. Rashed, ed., *Encyclopedia of the History of Arabic Science. Vol. 1: Astronomy – Theoretical and Applied*, pp. 1–19.
---, "Eastern Arabic Astronomy between the Eighth and the Eleventh Centuries," in R. Rashed, ed., *Encyclopedia of the History of Arabic Science. Vol 1: Astronomy – Theoretical and Applied*, pp. 20–57.
Morison, B., *On Location: Aristotle's Concept of Place* (Oxford: Oxford University Press, 2002).
Morrison, R., "An Astronomical Treatise by Mūsā Jalīnūs alias Moses Galeano," *Aleph* 11 (2011), pp. 385–413.
Moss, J. D., *Novelties in the Heavens: Rhetoric and Science in the Copernican Controversy* (Chicago: Chicago University Press, 1993).
Mossakowski, S., "The Symbolic Meaning of Copernicus' Seal," *Journal of the History of Ideas* 34 (3/1973), pp. 451–60.
Mourelatos, A. P. D., "Plato's 'Real Astronomy': *Republic* VII. 527d–531d," in J. P. Anton, ed., *Science and the Sciences in Plato*, pp. 33–73.
---, "Astronomy and Kinematics in Plato's Project of Rationalist Explanation," *Studies in History and Philosophy of Science* 12 (1/1981), pp. 1–32.
Mueller, I., "Ascending to Problems: Astronomy and Harmonics in *Republic* VII," in J. P. Anton, ed., *Science and the Sciences in Plato*, pp. 103–22.
Murschel, A., "The Structure and Function of Ptolemy's Physical Hypotheses of Planetary Motion," *JHA* 26 (1995), pp. 33–61.
Nardi, B., "Appunti sull'averroista bolognese Alessandro Achillini," in *Saggi sull'aristotelismo padovano dal secolo XIV al XVI* (Florence: Sansoni, 1958), pp. 225–79.

———, *Saggi sulla cultura veneta del Quattrocento e Cinquecento*, ed. P. Mazzantini (Padua: Antenore, 1971).
———, "Copernico studente a Padova," in B. Nardi, *Saggi sulla cultura veneta del Quatrocento e Cinquecento*, pp. 99–111.
———, "I tre filosofi del Girgione," in B. Nardi, *Saggi sulla cultura veneta del Quatrocento e Cinquecento*, pp. 111–20.
Neugebauer, O., "On the Planetary Theory of Copernicus," *Vistas in Astronomy* 10 (1968), pp. 89–103.
———, *The Exact Sciences in Antiquity* (New York: Dover, 1969).
———, A *History of Ancient Mathematical Astronomy* (Berlin: Springer, 1975).
———, *Astronomy and History* (New York: Springer, 1983).
Norlind, W., "Copernicus and Luther: A Critical Study," *Isis* 44 (1953), pp. 273–6.
North, J., *God's Clockmaker: Richard Wallingford and the Invention of Time* (London: Humbeldon, 2006).
———, *Cosmos: An Illustrated History of Astronomy and Cosmology* (Chicago: The University of Chicago Press, 2008).
Nussbaum, M., "Saving Aristotle's Appearances," in M. Schofield and M. Nussbaum, eds., *Language and Logos* (Cambridge: Cambridge University Press, 2006), pp. 267–93.
O'Connell, D. J. K., "Copernicus and Calendar Reform," *Studia Copernicana* 13 (1975), pp. 189–200.
Olivieri, L., ed., *Aristotelismo veneto e scienza moderna* (Padua: Antenore, 1983).
Ongaro, G., "Girolamo Fracastoro e lo Studio di Padova," in A. Pastore and E. Peruzzi, eds., *Girolamo Fracastroro*, pp. 31–54.
Owen, G. E. L., "'Tithenai ta Phainomena'," in J. Barnes, M. Schofield and R. Sorabji, eds., *Articles on Aristotle. Vol. 1: Science* (London: Duckworth, 1975), pp. 113–26.
Palter, R., "An Approach to the History of Early Astronomy," *Studies in History and Philosophy of Science* 1 (1979), pp. 93–133.
Park, K., and L. Daston, eds., *The Cambridge History of Science. Vol. 3: Early Modern Science* (Cambridge: Cambridge University Press, 2008).
Parkinson, G.H.R., ed., *Routledge History of Philosophy. Vol. 4: The Renaissance and Seventeenth-century Rationalism* (London & New York: Routledge, 1993).
Pastore, A., and E. Peruzzi, eds., *Girolamo Fracastoro: fra medicina, filosofia e scienze della natura* (Florence: Olschki, 2006).
Pawlikowska Brożek, Z., "Wojciech of Brudzewo," in J. Gierowski, ed., *The Cracow Circle of Nicholas Copernicus*, pp. 61–75.
Pečjak, G., *Stoletna pratika dvajsetega stoletja, 1901–2000* (Ljubljana: Katoliško tiskovno društvo, 1901).
Pedersen, O., *A Survey of the Almagest* (Odense: Odense University Press, 1974).
———, "The *Corpus Astronomicum* and the Traditions of Mediaeval Latin Astronomy," *Studia Copernicana* 13 (1975), pp. 57–96.

---, "The Decline and Fall of the *Theorica Planetarum*: Renaissance Astronomy and the Art of Printing," *Studia Copernicana* 16 (1978), pp. 157–86.
---, "Astronomy," in D. C. Lindberg, ed., *Science in the Middle Ages*, pp. 303–37.
---, "The Origins of the *Theorica planetarum*," *JHA* 12 (1981), pp. 113–23.
---, "In Quest of Sacrobosco," *JHA* 16 (1985), pp. 175–220.
Pedersen, O., and M. Pihl, *Early Physics and Astronomy: A Historical Introduction* (New York: Science History Publications, 1974).
Pepe, L., ed., *Copernico e la questione copernicana in Italia dal XVI al XIX secolo* (Florence: Olschki, 1996).
Peruzzi, E., *La nave di Ermete. La cosmologia di Giordano Fracastoro* (Florence: Olschki, 1995).
---, "Introduzione," in G. Fracastoro, *Navagero: della Poetica*, critical text, tr., introd. and notes, E. Peruzzi (Florence: Alinea, 2005), pp. 7–43.
---, "La poetica del *Naugerius* tra platonismo e aristotelismo," in A. Pastore and E. Peruzzi, eds., *Girolamo Fracastoro*, pp. 217–28.
Polkowski, I., *Żywot Mikolaya Kopernika* (Gniezno: Drukiem J. B. Langiego, 1873).
Pompeo Faracovi, O., "In difesa dell'astrologia: risposte a Pico in Bellanti e Pontano," in M. Bertozzi, ed., *Nello specchio del cielo*, pp. 47–66.
Poppi, A., "L'aristotelismo negli anni del soggiorno padovano di Niccolò Copernico," in *Copernico a Padova*, pp. 49–76.
Poppi, A., ed, *Scienza e filosofia all'Università di Padova nell Quattrocento* (Padua: LINT, 1983).
---, "La filosofia naturale del primo Cinquecento nelle Università di Padova, Bologna e Ferrara," in L. Pepe, ed., *Copernico e la questione copernicana in Italia*, pp. 39–67.
Popplow, M., "Setting the World Machine in Motion: The Meaning of *Machina Mundi* in the Middle Ages and the Early Modern Period," in M. Bucciantini, M. Camerota and S. Roux, eds., *Mechanics and Cosmology in the Medieval and Early Modern Period*, pp. 45–70.
Poulle, E., "La théorie épicyclique selon Ptolémée au Moyen Age," in *Avant, avec, après Copernic*, pp. 51–60.
---, "L'astronomie du Moyen Âge et ses instruments," *Annali dell'Instituto e Museo di storia della scienza di Firenze* 6 (2/1981), pp. 3–16.
---, "The Alfonsine Tables and Alfonso X of Castille," *JHA* 19 (1989), pp. 97–113.
---, "L'astronomie latine au XIe siècle," in G. C. Garfagnini, ed., *Cieli e terre nei secoli XI-XII*, pp. 75–88.
Premuda, L., "Nicola Copernico studente di Medicina a Padova e medico," in *Copernico a Padova*, pp. 151–63.
Price, D. J., "Contra-Copernicus: A Critical Re-estimation of the Mathematical Planetary Theory of Ptolemy, Copernicus, and Kepler," in M. Clagett, ed.,

Critical Problems in the History of Sciences (Madison: University of Wisconsin Press, 1960), pp. 197–218.

Prowe, L., *Nicolaus Copernicus*, 2 vols. (Berlin: Weidmannsche Buchhandlung, 1883–1884).

Purnell, F., Jr., "The Theme of Philosophic Concord and the Sources of Ficino's Platonism," in G. C. Garfagnini, ed., *Marsilio Ficino e il ritorno di Platone*, pp. 397–415.

Ragep, F. J., "Duhem, the Arabs, and the History of Cosmology," *Synthese* 83 (1990), pp. 201–14.

---, *Nasīr al-Dīn al-Tūsī's Memoir on Astronomy*, 2. vols. (New York & Berlin: Springer, 1993).

---, "Freeing Astronomy from Philosophy: An Aspect of Islamic Influence on Science," *Osiris* 16 (2000), pp. 49–71.

---, "Tūsī and Copernicus: The Earth's Motion in Context," *Science in Context* 14 (1–2/2001), pp. 145–63.

---, "Copernicus and his Islamic Predecessors: Some Historical Remarks," *Filozofski vestnik* 25 (2/2004), pp. 125–42; revised and extended version published in *History of Science* 45 (2007), pp. 65–81.

---, "Alī Qushjī and Regiomontanus: Eccentric Transformations and Copernican Revolutions," *JHA* 36 (4/2005), pp. 359–71.

Ragep, F. J., S. Ragep, and S. Livesey, eds., *Tradition, Transmission, Transformation: Proceedings of Two Conferences on Pre-modern Science held at the University of Oklahoma* (Leiden: Brill, 1992).

Randall, J. H., *The School of Padua and the Emergence of Modern Science* (Padua: Antenore, 1961).

Randles, W. G. L., *De la Terre plate au globe terrestre. Une mutation épistémologique rapide, 1480–1520* (Paris: Colin, 1980)

Rashed, R., ed., *Encyclopedia of the History of Arabic Science. Vol. 1: Astronomy – Theoretical and Applied* (London & New York: Routledge, 1996).

Ravetz, J., *Astronomy and Cosmology in the Achievement of Nicolaus Copernicus* (Wrocław: Zakład Narodowy im. Ossolińskich, 1965).

Riddell, R. C., "Eudoxan Mathematics and the Eudoxan Spheres," *Archive for History of Exact Sciences* 20 (1979), pp. 1–19.

Rigo, A., "Bessarione, Giovanni Regiomontano e i loro studi su Tolomeo a Venezia e Roma (1462–1464)," *Studi veneziani* 21, pp. 49–110.

Robinson, R., *Plato's Earlier Dialectic* (Oxford: Oxford University Press, 1953).

Romer, A., "The Welcoming of Copernicus's *De revolutionibus*: The *Commentariolus* and its Reception," *Physics in Perspective* 1 (2/1999), pp. 157–83.

Rose, P. L., *The Italian Renaissance of Mathematics: Studies on Humanists and Mathematicians from Petrarch to Galileo* (Geneve: Droz, 1975).

Rosen, E., "The *Commentariolus* of Copernicus," *Osiris* 3 (1937), pp. 123–41.

---, "The Authentic Title of Copernicus's Major Work," *Journal of the History of Ideas* 4 (1943), pp. 457–74.
---, *Three Copernican Treatises* (New York: Dover, 1959).
---, "The Achievement of Copernicus," *Studia Copernicana* 13 (1975), pp. 125–36.
---, "Reply to N. Swerdlow," *Archives internationales d'histoire des sciences* 26 (1976), pp. 301–4.
---, *Copernicus and the Scientific Revolution* (Malabar: Krieger, 1984).
---, *Commentary*, in N. Copernicus, *On the Revolutions*, pp. 331–439.
---, *Copernicus and His Successors* (London & Rio Grande, OH: The Hambledon Press, 1995).
---, "Aristarchus of Samos and Copernicus," in E. Rosen, *Copernicus and His Successors*, pp. 1–9.
---, "Was Copernicus a Pythagorean?," in E. Rosen, *Copernicus and His Successors*, pp. 11–15.
---, "Copernicus and Al-Bitruji," in E. Rosen, *Copernicus and His Successors*, pp. 41–5.
---, "Copernicus' Earliest Astronomical Treatise," in E. Rosen, *Copernicus and His Successors*, pp. 71–9.
---, "Copernicus' Axioms," in E. Rosen, *Copernicus and His Successors*, pp. 99–104.
---, "Copernicus' Spheres and Epicycles," in E. Rosen, *Copernicus and His Successors*, pp. 117–26.
---, "Nicholas Copernicus and Giorgio Valla," in E. Rosen, *Copernicus and his Successors*, pp. 139–48.
Rosińska, G., "Nasīr al-Dīn al-Tūsī and Ibn al-Shātir in Cracow," *Isis* 65 (2/1974), pp. 239–43.
---, *Scientific Writings and Astronomical Tables in Cracow: A Census of Manuscript Sources: XIVth-XVIth Centuries* (Wrocław: Polish Academy of Sciences, 1984).
Russell, J. B., *Inventing the Flat Earth: Columbus and Modern Historians* (Westport & London: Praeger, 1997).
Russo, L., *La rivoluzione dimenticata. Il pensiero scientifico Greco e la scienza moderna* (Milan: Feltrinelli, 2003).
Sabra, A. I., "An Eleventh-Century Refutation of Ptolemy's Planetary Theory," *Studia Copernicana* 16 (1978), pp. 117–31.
---, "The Andalusian Revolt agains Ptolemaic Astronomy: Averroes and al-Bitruji," in E. Mendelsohn, ed., *Transformations and Traditions in Sciences: Essays in Honor of I. Bernard Cohen* (Cambridge: Cambridge University Press, 1984), pp. 133–53.
Saliba, G., "The Astronomical Tradition of Maragha: A Historical Survey and Prospects for Future Prospects," *Arabic Science and Philosophy* 1 (1/1991), pp. 67–99.

---, *A History of Arabic Astronomy: Planetary Theories during the Golden Age of Islam* (New York: New York University Press, 1994).
---, "Arabic Planetary Theories after Eleventh Century AD," in R. Rashed, ed., *Encyclopedia of the History of Arabic Science. Vol 1: Astronomy – Theoretical and Applied*, pp. 58–127.
---, *Islamic Science and the Making of the European Renaissance* (Cambridge: The MIT Press, 2007).
Sambursky, S., *The Physical World of the Greeks* (London: Routledge, 1963).
Sarnowsky, J., "Albert von Sachsen und die Astronomie des 14. Jahrhundert," in J. Biard, ed., *Itinéraires d'Albert de Saxe*, pp. 219–34.
---, "The Defence of the Ptolemaic System in Late Mediaeval Commentaries on John de Sacrobosco's *De Sphera*," in M. Bucciantini, M. Camerota and S. Roux, eds., *Mechanics and Cosmology in the Medieval and Early Modern Period*, pp. 29–44.
---, "Concepts of Impetus and the History of Mechanics," in W. R. Laird and S. Roux, eds., *Mechanics and Natural Philosophy before the Scientific Revolution* (Dordrecht: Springer, 2008), pp. 121–45.
Schmeidler, F., *Kommentar zu De revolutionibus* (Berlin: Akademie Verlag, 1998).
Schmitt, C. B., and Q. Skinner, eds., *The Cambridge History of Renaissance Philosophy* (Cambridge: Cambridge University Press, 1988).
Segel, H. B., *Renaissance Culture in Poland: The Rise of Humanism, 1470–1543* (Ithaca & London: Cornell University Press, 1989).
Shank, M. H., "The 'Notes on al-Bitruji' Attributed to Regiomontanus: Second Thoughts," *JHA* 23 (1992), pp. 15–30.
---, "The Classical Scientific Tradition in Vienna," in F. J. Ragep, S. Ragep and S. Livesey, eds., *Tradition, Transmission, Transformation*, pp. 115–36.
---, "Regiomontanus and Homocentric Astronomy," *JHA* 29 (2/1998), pp. 157–66.
---, "Regiomontanus on Ptolemy, Physical Orbs, and Astronomical Fictionalism: Goldsteinian Themes in the *Defense of Theon against George of Trebizond*," *Perspectives on Science* 10 (2/2002), pp. 179–207.
---, "Rings in a Fluid Heaven: The Equatorium-Driven Physical Astronomy of Guido de Marchia (fl. 1292–1310)," *Centaurus* 45 (2003), pp. 175–203.
---, "Mechanical Thinking in European Astronomy (13th–15th Centuries)," in M. Bucciantini, M. Camerota and S. Roux, eds., *Mechanics and Cosmology in the Medieval and Early Modern Period*, pp. 3–27.
---, "Regiomontanus as a Physical Astronomer: Samplings from the *Defence of Theon against George of Trebizond*," *JHA* 38 (2007), pp. 325–49.
---, "Setting up Copernicus? Astronomy and Natural Philosophy in Giambattista Capuano da Manfredonia's Expositio on the Sphere," *Early Science and Medicine* 14 (2009), pp. 290–315.

Shapere, D., "Copernicanism as a Scientific Revolution," *Vistas in Astronomy* 17 (1975), pp. 97–104.

Smith, A. M., *Ptolemy's Theory of Visual Perception: An English Translation of the Optics, with Introduction and Commentary, Transactions of the American Philosophical Society* 86 (2/1996).

Sorabji, R., ed., *Philoponus and the Rejection of Aristotelian Science* (London: Duckworth, 1987).

Souffrin, P., "Oresme, Buridan et le movement de rotation diurne de la terre ou des cieux," in B. Ribémot, ed., *Terres Médiévales* (Paris: Kliencksieck, 1993), pp. 105–40.

Spitz, L., *Conrad Celtis: The German Arch-Humanist* (Cambridge: Harvard University Press, 1957).

Stahl, W. H., *Martianus Capella and the Seven Liberal Arts. Vol. 1: The Quadrivium of Martianus Capella Latin Traditions in the Mathematical Sciences: 50 B.C.–A.D. 1250* (New York & London: Columbia University Press, 1971).

Steneck, N., *Science and Creation in the Middle Ages: Henry of Langenstein (d. 1397) on Genesis* (Notre Dame, IN: University of Notre Dame, 1976).

Stromer, W. von, "Hec opera fient in oppido Nuremberga Germanie ductu Ioannis de Monteregio: Regiomontan und Nuremberg, 1471–1475," in G. Hamann, ed, *Regiomontanus-Studien*, pp. 271–2.

Swerdlow, N. M., "Aristotelian Planetary Theory in the Renaissance: Giovanni Battista Amico's Homocentric Spheres," *JHA* 3 (1972), pp. 36–48.

---, "The Derivation and First Draft of Copernicus' Planetary Theory: A Translation of the *Commentariolus* with Commentary," *Proceedings of the American Philosophical Society* 67 (1973), pp. 423–512.

---, "The Holograph of *De revolutionibus* and the Chronology of its Composition," *JHA* 5 (1974), pp. 186–98.

---, "Copernicus's Four Models of Mercury," *Studia Copernicana* 13 (1975), pp. 141–55.

---, "Pseudoxia Copernicana: or, Inquiries into Very Many Received Tenets and Commonly Presumed Truths, Mostly Concerning Spheres," *Archives internationales d'histoire des sciences* 26 (1976), pp. 108–58.

---, "Translating Copernicus," *Isis* 72 (1981), pp. 73–82.

---, "On Establishing the Text of *De revolutionibus*," *JHA* 12 (1981), pp. 35–46.

---, "Regiomontanus on the Critical Problems of Astronomy," in T. H. Levere and W. R. Shea, eds., *Nature, Experiment, and the Sciences: Essays on Galileo and the History of Science in Honour of Stilman Drake* (Dordrecht: Kluwer, 1990), pp. 165–95.

---, "Amico's Homocentric Spheres," *JHA* 23 (1992), pp. 211–15.

---, "Science and Humanism in the Renaissance: Regiomontanus's Oration on the Dignity and Utility of the Mathematical Sciences," in P. Horwich, ed.,

World Changes: Thomas Kuhn and the Nature of Science (Cambridge: The MIT Press, 1993), pp. 131–68.

---, "Galilleo's Discoveries with the Telescope," in P. Machamer, ed., *The Cambridge Companion to Galileo* (Cambridge: Cambridge University Press, 1998), pp. 244–70.

---, "Regiomontanus's Concentric-Sphere Models for the Sun and Moon," *JHA* 30 (1/1999), pp. 1–23.

---, "Copernicus, Nicolaus (1473–1543)," in W. Applebaum, ed., *Encyclopedia of the Scientific Revolution from Copernicus to Newton*, pp. 154–68.

---, "The Empirical Foundations of Ptolemy's Planetary Theory," *JHA* 35 (2004), pp. 249–71.

---, "Copernicus and Astrology, with an Appendix of Translations of Primary Sources," *Perspectives on Science* 20 (3/2012), pp. 353–78.

Swerdlow, N. M., and O. Neugebauer, *Mathematical Astronomy in the Copernicus's De revolutionibus*, 2 vols. (New York: Springer, 1984).

Szczeciniarz, J.-J., *Copernic et la révolution copernicienne* (Paris: Flammarion, 1998).

---, *La Terre immobile* (Paris: Presses Universitaires de France, 2003).

Taylor, A. E., *Plato: The Man and his Work* (London: Methuen & Co, 1978).

Thom, J. C., *The Pythagorean Golden Verses* (Leiden: Brill, 1995).

Thorndike, L., *The Sphere of Sacrobosco and its Commentators* (Chicago: University of Chicago Press, 1949).

Tredwell, K. A., and P. Barker, "Copernicus' First Friends: Physical Copernicanism from 1543 to 1610," *Filozofski vestnik* 25 (2/2004), pp. 143–66.

Varanini, G. M., "Giralomo Fracastoro nel contesto. La famiglia e gli studi," in A. Pastore and E. Peruzzi, eds., *Girolamo Fracastoro*, pp. 7–24.

Vasoli, C., "Copernic et ses études italiennes," *Studia Copernicana* 14 (1975), pp. 161–71.

---, *Le filosofie del Rinascimento*, ed. C. Pissavino (Milan: Bruno Mondadori, 2002).

---, "Le tradizioni magiche ed esoteriche nel Quattrocento," in C. Vasoli, *Le filosofie del Rinascimento*, pp. 133–53.

---, "La polemica contro l'astrologia. Pomponazzi e il *De incantationibus*. Filosofia, medicina e profezia nella cultura del Cinquecento," in C. Vasoli, *Le filosofie del Rinascimento*, pp. 374–97.

---, "I tentative umanistici cinquecenteschi di nuovo 'ordine' del sapere," in C. Vasoli, *Le filosofie del Rinascimento*, pp. 398–415.

---, "La polemica antiastrologica di Giovanni Pico," in M. Bertozzi, ed., *Nello specchio del cielo*, pp. 1–17.

Verdet, J.-P., "Quelques remarques sur la *Narratio Prima* de J. Rheticus," in *Avant, avec, après Copernic*, pp. 111–15.

---, *Une histoire de l'astronomie* (Paris: Seuil, 1990).

Vermij, R. H., *The Calvinist Copernicans: The Reception of the New Astronomy in the Dutch Republic, 1575–1750* (Amsterdam: Koninklijke Nederlandse Akademie van Wetenschappen, 2002).

---, "Albertus Leoninus (1543–1614) and the Copernicus's 'third motion' of the Earth," *JHA* 37 (2006), pp. 101–9.

Vesel, M., "What is Revolutionary in Copernicus' *Revolutions*," *Filozofski vestnik* 25 (2/2004), pp. 167–86.

---, *Astronom-filozof: Nikolaj Kopernik, gibanje Zemlje in kopernikanska revolucija* (Ljubljana: Založba ZRC, 2007).

Veselovsky, I. N., "Copernicus and Naşīr al-Din al-Tūsī," *JHA* 4 (1973), pp. 128–30.

Vlastos, G., "The Role of Observation in Plato's Conception of Astronomy," in J. P. Anton, ed., *Science and the Sciences in Plato*, pp. 1–31.

---, *Socrates, Plato and their Tradition*, 2 vols. (Princeton: Princeton University Press, 1995).

Vollmann, W. T., *Uncentering the Earth: Copernicus and The Revolutions of the Heavenly Spheres* (New York: Atlas Books, 2006).

Von Erhardt, R., and E. von Erhardt-Siebold, "Archimedes' Sand-Reckoner: Aristarchos and Copernicus," *Isis* 33 (5/1942), pp. 578–602.

Walker, D. P., "Ficino and Astrology," in G. C. Garfagnini, ed., *Marsilio Ficino e il ritorno di Platone*, pp. 342–9.

Weinert, F., *Copernicus, Darwin and Freud: Revolutions in the History and Philosophy of Science* (Chichester: Wiley & Blackwell, 2009).

Weisheipl, J. A., "The Celestial Movers in Mediaeval Physics," *The Tomist* 24 (1961), pp. 286–326.

---, "The Classification of the Sciences in Medieval Thought," *Medieval Studies* 27 (1965), pp. 54–90.

Westman, R. S., ed., *The Copernican Achievement* (Berkeley & Los Angeles: University of California Press, 1975).

---, "Three Responses to the Copernican Theory: Johannes Praetorius, Tycho Brahe, and Michael Maestlin," in R. S. Westman, ed., *The Copernican Achivement*, pp. 285–345.

---, "The Melanchthon Circle, Rheticus, and the Wittenberg Interpretation of the Copernican Theory," *Isis* 66 (1975), pp. 165–93.

---, "The Astronomer's Role in the Sixteenth Century: A Preliminary Study," *History of Science* 18 (1980), pp. 105–47.

---, "The Copernicans and the Churches," in D. C. Lindberg and R. L. Numbers, eds., *God and Nature*, pp. 76–113.

---, "Proof, Poetics, and Patronage: Copernicus's Preface to *De revolutionibus*," in D. C. Lindberg and R. S. Westman, eds., *Reappraisals of the Scientific Revolution* (Cambridge: Cambridge University Press, 1990), pp. 167–206.

---, *The Copernican Question: Prognostication, Skepticism, and Celestial Order* (Berkeley: University of California Press, 2011).

Williams, C. J. F., "Aristotle and Copernican Revolutions," *Phronesis* 36 (3/1991), pp. 305–12.

Wingen-Trennhaus, A., "Regiomontanus als Frühdrucker in Nürnberg," *Mitteilungen des Vereins für Geschichte der Stadt Nürnberg* 78 (1991), pp. 17–87.

Wolff, M., *Geschichte der Impetus Theorie: Untersuchungen zum Ursprung der Klassischen Mechanik* (Frankfurt/M: Suhrkamp, 1980).

---, "Impetus Mechanics as a Physical Argument for Copernicanism: Copernicus, Benedetti, Galileo," *Science in Context* 1 (2/1987), pp. 215–56.

---, "Philoponus and the Rise of Preclassical Dynamics," in R. Sorabji, ed., *Philoponus and the Rejection of Aristotelian Science*, pp. 84–120.

Wright, L., "The Astronomy of Eudoxus: Geometry or Physics?" *Studies in History and Philosophy of Science* 4 (1973), pp. 165–72.

Wright, M. R., *Cosmology in Antiquity* (London & New York: Routledge, 1995).

Wrightsman, B., "Andreas Osiander's Contribution to the Copernican Achievement," in R. S. Westman, ed., *The Copernican Achievement*, pp. 213–43.

Yavetz, I., "On the Homocentric Spheres of Eudoxus," *Archive for History of Exact Sciences* 51 (1998), pp. 221–78.

Zathey, J., "Marcin Bylica of Ołkusz," in J. Gierowski, ed, *The Cracow Circle of Nicholas Copernicus*, pp. 77–94.

Zhmud, L., *The Origin of the History of Science in Classical Antiquity*, tr. A. Chernoglazov (Berlin & New York: Walter de Gruyter, 2006).

Zilsel, E., "Copernicus and Mechanics," *Journal of the History of Ideas* 1 (1940), pp. 11–8.

Zimmermann, F., "Philoponus' Impetus Theory in the Arabic Tradition," in R. Sorabji, ed., *Philoponus and the Rejection of Aristotelian Science*, pp. 121–39.

Zimmermann, G., "Die Gottesvorstellung des Nicolaus Copernicus," *Studia Leibnitiana* 20 (1/1988), pp. 63–79.

Zinner, E., *Leben und Wirken des Joh. Müller von Königsberg, genannt Regiomontanus* (Osnabrück: O. Zeller Verlag, 1968).

Zwiercan, M., "Jan of Glogów," in J. Gierowski, ed., *The Cracow Circle of Nicholas Copernicus*, pp. 95–110.

INDEX OF NAMES

Pre-1800

Post-1800

INDEX OF PLACES

INDEX OF SUBJECTS

R

S

T

U

www.ingramcontent.com/pod-product-compliance
Ingram Content Group UK Ltd.
Pitfield, Milton Keynes, MK11 3LW, UK
UKHW041822031026
15587UKWH00002B/508